Foundations of Robotics and Automation

THEORY & PRACTICE

Foundations of Robotics and Automation

THEORY & PRACTICE

Prof. A. B. Bhattacharya

M. Sc., Ph. D. (Cal Univ.), Pot Doc. (MIT, USA), FIETE
Former: Professor & Head, Department of Physics, Kalyani University;
Dean of Research Techno India University;
Pro Vice Chancellor, JIS University;
Director of Brainware University

&

Dr. Debasish Roy, IPS

BE, MCA, M Tech, Ph. D., MB, IPM, PPM
Director General & Inspector General of Police Railways, West Bengal

KHANNA BOOK PUBLISHING CO. (P) LTD.

Publisher of Engineering and Computer Books

4C/4344, Ansari Road, Darya Ganj, New Delhi-110002
Phone : 011-23244447-48 **Mobile:** +91-9910909320
E-mail : contact@khannabooks.com
Website : www.khannabooks.com

Dear Readers,

To prevent the piracy, this book is secured with HIGH SECURITY HOLOGRAM on the front title cover. In case you don't find the hologram on the front cover title, please write us to at contact@khannabooks.com or whatsapp us at +91-99109 09320 and avail special gift voucher for yourself.

Specimen of Hologram on front Cover title:

Moreover, there is a SPECIAL DISCOUNT COUPON for you with EVERY HOLOGRAM.

How to avail this SPECIAL DISCOUNT:

Step 1: Scratch the hologram
Step 2: Under the scratch area, your "coupon code" is available
Step 3: Logon to www.khannabooks.com
Step 4: Use your "coupon code" in the shopping cart and get your copy at a special discount
Step 5: Enjoy your reading!

Thanks,
Khanna Publishing House

Price: ₹ 575.00

ISBN: 978-93-55388-00-1

Foundations of Robotics and Automation (Theory & Practice)
by Prof. A. B. Bhattacharya,
Dr. Debasish Roy

First Edition: 2025

To view complete list of books, please scan the QR Code:

Published by:
Khanna Book Publishing Co. (P) Ltd.
CIN: U22110DL1998PTC095547

Visit us at: www.khannabooks.com
Write us at: contact@khannabooks.com

Printed in Bharat.

Preface

Robotics and Automation, an ever-evolving field that blends engineering, computer science, and creativity to redefine the way we interact with technology. This textbook serves as your comprehensive guide into the realm of Robotics and Automation, offering a thorough exploration of the principles, technologies, and applications that drive this dynamic discipline. The 21st century has ushered in a technological renaissance, with robotics and automation at its forefront. From manufacturing floors to surgical theaters, from exploration of distant planets to our very own homes, robots have transformed the way we live, work, and perceive the boundaries of human capability. In this rapidly evolving era, where technology is advancing at an unprecedented pace, robotics and automation have emerged as the driving forces behind innovation, transforming industries and reshaping the way we interact with the world around us. This textbook serves as your comprehensive guide to understanding the fundamental principles, advanced concepts, and real-world applications of robotics and automation.

This book aims to provide an in-depth understanding of the underlying concepts, enabling you to unravel the complexities of robotics and automation and contribute to the ongoing revolution. This textbook is structured to take you on a journey from the fundamental building blocks of robotics and automation to their advanced applications in various domains. The content is organized into coherent sections that gradually build upon one another, allowing for a logical and intuitive progression of knowledge.

The primary objective of this textbook is to provide students, researchers, engineers, and enthusiasts with a solid foundation in the field of robotics and automation. It covers a wide range of topics, from the basics of robotics and automation to cutting-edge advancements in artificial intelligence, machine learning, and human-robot interaction. Whether you are a novice seeking an introduction to the field or a seasoned professional aiming to deepen your expertise, this book offers valuable insights and knowledge to cater to your needs. This textbook is designed to progressively guide you through the intricacies of robotics and automation.

As you embark on this educational journey through the pages of this textbook, you will gain a profound understanding of the principles, methodologies, and innovations driving the field of robotics and automation. The knowledge you acquire here will not only empower you to contribute to advancements in technology but also enable you to navigate the challenges and opportunities presented by our increasingly automated world. Let this book be your guide to unlocking the limitless possibilities that wait in the realm of robotics and automation.

Prof. A. B. Bhattacharya

Dr. Debasish Roy

Preface

Robotics and Automation, an ever-evolving field that blends engineering, computer science, and creativity to redefine the way we interact with technology. This textbook serves as your comprehensive guide into the realm of Robotics and Automation, offering a thorough exploration of the principles, technologies, and applications that drive this dynamic discipline. The 21st century has ushered in a technological renaissance, with robotics and automation at its forefront. From manufacturing floors to surgical theaters, from exploration of distant planets to our very own homes, robots have transformed the way we live, work, and perceive the boundaries of human capability. In this rapidly evolving era, where technology is advancing at an unprecedented pace, robotics and automation have emerged as the driving forces behind innovation, transforming industries and reshaping the way we interact with the world around us. This textbook serves as your comprehensive guide to understanding the fundamental principles, advanced concepts, and real-world applications of robotics and automation.

This book aims to provide an in-depth understanding of the underlying concepts, enabling you to unravel the complexities of robotics and automation and contribute to the ongoing revolution. This textbook is structured to take you on a journey from the fundamental building blocks of robotics and automation to their advanced applications in various domains. The content is organized into different sections that gradually build upon one another, allowing for a logical and intuitive progression of knowledge.

The primary objective of this textbook is to provide students, researchers, engineers, and enthusiasts with a solid foundation in the field of robotics and automation. It covers a wide range of topics, from the basics of robotics and automation to cutting-edge advancements in artificial intelligence, machine learning, and human-robot interaction. Whether you are a novice seeking an entry point into the field or a seasoned professional aiming to deepen your expertise, this book offers valuable insights and knowledge to cater to your needs. The textbook is designed to progressively guide you through the intricacies of robotics and automation.

As you embark on this educational journey through the pages of this textbook, you will gain a profound understanding of the principles, methodologies, and innovations driving the field of robotics and automation. The knowledge you acquire here will not only empower you to contribute to advancements in technology but also enable you to navigate the challenges and opportunities presented by an increasingly automated world. Let this book be your guide to unlocking the limitless possibilities that await in the realm of robotics and automation.

Prof. A. B. Bhattacharya

Dr. Debasish Roy

Acknowledgments

Writing a book on robotics and automation has been an incredible journey and we are deeply grateful to many individuals who have contributed to its creation. Their expertise, support and encouragement have been invaluable. We are profoundly grateful for the support and contributions of those individuals and entities that have made this endeavor possible. We would like to thank the dedicated researchers and engineers in the field of robotics and automation, whose groundbreaking work serves as the foundation of this book. We extend our sincere gratitude to the authors of many standard texts whose guidance and mentorship have shaped the direction of the present book.

We would like to express our heartfelt gratitude to Google for their unwavering commitment to advancing technology and fostering innovation. The resources, research and tools provided by them have been instrumental in shaping the content of this book. Their dedication to pushing the boundaries of technology has undoubtedly enriched the field of robotics and automation. We also want to extend our appreciation to innovative creation of Open AI whose capabilities in understanding and generating knowledge are invaluable for insightful interactions and suggestions to the clarity and coherence of the content. The participants of various conferences, workshops, and seminars, where ideas were shared and debated shaping the direction of this book are acknowledged. We are also thankful to the open-source community and contributors, whose freely shared knowledge and tools are driving advancements in robotics and automation.

We are indebted to the reviewers and colleagues who provided constructive feedback and insightful suggestions, enhancing the quality of this book. Their expertise and perspectives have been vital in shaping the final product. Furthermore, we want to express our thanks to our families and friends for their unwavering support throughout this endeavor. Their encouragement has been a constant source of motivation.

Lastly, our heartfelt appreciation goes to the readers of this book. It is our hope that this work contributes to your understanding of robotics and automation and inspires new ideas and possibilities in this ever-evolving field.

Prof. A. B. Bhattacharya

Dr. Debasish Roy

About the Authors

Prof. A. B. Bhattacharya

Prof. A. B. Bhattacharya did his M Sc and Ph D degrees from the University of Calcutta in Physics and Post-doc from the Massachusetts Institute of Technology (MIT), Cambridge, USA. He worked in close collaboration with the leading laboratories like Lincoln Laboratory; Milestone Hill Observatory; Earth, Atmospheric and Planetary Sciences division of MIT and many other well recognized laboratories in India and abroad. He served as Professor & Head of the Department of Physics, University of Kalyani; Dean of Research and Head of the ECE Department, Techno India University; Pro Vice Chancellor of JIS University and Director of Brainware University. He has published 32 text books for Science and Engineering students, two popular books for general readers and over 400 research articles in reputed International Journals. Under his supervision 24 scholars received their doctoral degree. He is a fellow of the Institute of Electronics and Telecommunication Engineers and member of many scientific organizations. India International Friendship Society awarded him the prestigious "Shiksha Ratna Award" and Certificate of Excellence for meritorious services, outstanding performance and significant contribution to education. The International Institute of Success Awareness honored him with their globally recognized "Glory of India Gold Medal" for his outstanding contribution to the national stature of India. A few words from the certificate issued by Prof. Williams of MIT may be quoted "Our collaborative effort was directed at the wavelength dependence of radar echoes observations with the Millstone Hill UHF radar in Westford. He made use of the Charles River Data Systems computer in the Weather Radar Laboratory for these calculations. His results help to make a convincing case that the volume reflectivity associated with lightning is a diminishing function of radar wavelength, contrary to contemporary thinking".

Dr. Debasish Roy, IPS

Dr. Debasish Roy is a highly accomplished and motivated professional with a robust academic background in Information Technology, including an MCA, MBA, MTech, and a PhD. He was currently the Director General (Railways) at a top-level position in the Indian Police Service (IPS). He received the Indian Police Medal and President's Police Medal in 2000 and 2010.He is well known for pioneering contributions in land mobile radio systems. He is highly recognized for spearheading strategic planning and execution of security measures for the railway network, implementing innovative technologies to enhance railway security and counteract emerging threats, overseeing a team of professionals, and managing day-to-day operations efficiently. He has provided expert guidance on implementing land mobile radio systems for public safety agencies and conducted groundbreaking research in cryptanalysis, leading to the development of advanced techniques for enhancing cyber security. He successfully led a team in designing and implementing secure communication protocols for land mobile radio systems, ensuring robust and encrypted communication. He has published numerous research papers in reputed journals and presented many works at international conferences. He mentored and guided students at the postgraduate and doctoral levels. He has a proven track record of effective leadership and team management, besides all his strategic planning and execution in high-stakes environments. He received many honors and awards for his outstanding contributions to the field and was recognized for exceptional leadership and innovation in technology-driven security initiatives. He completed an Executive Program in Quantum Computing and Machine Learning from IIT Delhi in 2023. He participated in the Workshop on Machine Learning for Cryptology (ML4Crypto 2023) at ISI, Kolkata, from November 8-10, 2023, organized by SAG, DRDO. His recent paper on "Iterative Quantum Algorithm for Linear System Solving" has been accepted at the International Conference SAI, Computing, 2024 London, organized by Microsoft Corporation.

A Special Note from the Authors

Need for the Robotics and Automation Book

Books on robotics and automation are desperately needed for a number of reasons, including the dynamic nature of these subjects and their influence across multiple industries. The following are some of the main causes of the increased need for robotics and automation publications:

1. **Rapid Technological Advancements:** Rapid technical developments in hardware, software, and artificial intelligence are transforming robotics and automation. The book offers a means of recording and elucidating these developments.
2. **Interdisciplinary Nature:** Mechanical engineering, electrical engineering, computer science, and artificial intelligence are all combined in robotics and automation. Books give professionals and scholars a thorough grasp by bridging multidisciplinary gaps.
3. **Educational Resources:** Books are vital teaching tools for instructors and students taking robotics and automation courses. They provide exercises, examples, and organized content to help with learning and comprehension.
4. **Industry Applications:** The need for literature addressing real-world applications, implementation techniques, and sector-specific case studies is expanding as industry use robotics and automation technology more and more.
5. **Research Contributions:** Scholars can disseminate in-depth analyses, methodology, and theoretical frameworks through books. They add to the scholarly conversation by introducing fresh concepts and research results.
6. **Addressing Challenges:** Automation and robotics must overcome a number of obstacles, including integration problems, safety worries, and ethical dilemmas. Books can examine these issues and provide researchers and practitioners with ideas and answers.
7. **Global Impact:** Publications that address worldwide viewpoints, standards, and best practices are necessary given the global influence that robotics and automation are having on economies, industries, and society.
8. **Continuous Learning:** Experts in the domain must remain current with the newest advancements and industry best practices. Books are an invaluable resource for lifelong learning and career advancement.
9. **Policy and Ethical Considerations:** There is a growing demand for literature that discusses the ethical issues, legal frameworks, and policy consequences of robots and automation as these technologies become more and more integrated into daily life.
10. **Knowledge Dissemination:** Books play a crucial role in disseminating knowledge to a broad audience. They help transfer expertise from experienced professionals and researchers to those entering or advancing in the field.
11. **Innovation and Entrepreneurship:** By showcasing success stories, entrepreneurial tactics, and new business models, books can encourage creativity and entrepreneurship in the robotics and automation industry.

In conclusion, the demand for in-depth information distribution, industry applications, educational requirements, the interdisciplinary nature of robotics and automation, and the ongoing growth of technology all contribute to the necessity for book releases in these domains. The authors and publishers that add to this corpus of work have a significant influence on how automation and robots develop in the future.

Particular Categories of People Who may find the Book Beneficial

Because robotics and automation are multidisciplinary and have a significant impact, "Robotics and Automation: Theory & Practice" might be helpful to a wide range of people in different categories. The following particular groups of persons might benefit from the book:

1. **Students:** The book can be used as a thorough textbook or reference by undergraduate and graduate students studying robotics, automation, mechanical engineering, electrical engineering, computer science, or related fields.
2. **Researchers and Academics:** The book is a great resource for academic and industrial researchers working on cutting edge robotics and automation innovations to stay current on theoretical frameworks, techniques, and new discoveries.
3. **Engineers and Technologists:** The book can help professionals in the robotics and automation fields—such as engineers, technicians, and practitioners—better understand the newest innovations, practical applications, and industry best practices.
4. **Entrepreneurs and Innovators:** When looking into business opportunities, start-ups, or innovative solutions in the robotics and automation space, entrepreneurs and innovators can benefit from gaining insights into emerging trends and potential areas for disruption.
5. **Educators and Instructors:** Teachers, educators, and curriculum designers who are teaching robotics and automation courses will find the book to be a helpful resource. Lessons and additional resources for pupils could be made with it.
6. **Policy Makers and Regulators:** Those concerned in drafting laws, norms, and standards relevant to robots and automation will find the book's consideration of ethical dilemmas, safety requirements, and governance implications to be beneficial.
7. **Industry Professionals:** Professionals using robotics and automation technologies in a variety of industries, such as manufacturing, logistics, healthcare, and more, can gain knowledge from case studies, implementation strategies, and practical applications.
8. **Technological Enthusiasts:** Tech enthusiasts who are generally interested in robotics and automation can learn more about the fundamental concepts, challenges, and potential future developments in these fields by reading this book.
9. **Investors and Decision Makers:** The book will assist investors and decision-makers in technology-driven sectors in making well-informed decisions about strategic planning, investments, and alliances related to automation and robotics.
10. **Policy Advocates and Ethicists:** For individuals who support ethical considerations, responsible innovation, and the societal impact of robots and automation, the book is beneficial in comprehending the ethical dimensions and implications of these technologies.
11. **Lifelong Learners:** The book can assist individuals engaged in lifelong learning and professional development programs in staying up to date on new developments in the field of robotics and automation, as well as in broadening their knowledge and exploring uncharted territory.

By appealing to these diverse audience groups, the book can be a beneficial resource for both inexperienced professionals and seasoned experts, encouraging a deeper understanding of the theoretical and practical aspects of robotics and automation.

Possible Stakeholders Who could Benefit from the Book

The book "Robotics and Automation: Theory & Practice" may be helpful to a wide range of stakeholders due to the multidisciplinary nature of robotics and automation and its considerable impact. The following individuals could read the book:

1. **Academic Institutions:** The book can be used as a textbook, source material, or resource for robotics and automation study by academics, researchers, and students in universities and research facilities.
2. **Engineering and Technology Companies:** The book will assist engineers, developers, and professionals employed by robotics and automation technology companies in staying current with the most recent innovations, methods, and best practices in the field.
3. **Startups and Entrepreneurs:** Startup founders and other entrepreneurs in the robotics and automation space can learn about new trends, business models, and approaches to creative problem-solving.
4. **Government Agencies and Policymakers:** The book can be used by government agencies, regulatory organizations, and policymakers to better comprehend the technological, moral, and societal ramifications of robots and automation.
5. **Investors and Venture Capitalists:** The book can be used by investors and venture capitalists to evaluate technology trends, market opportunities, and potential dangers in order to fund projects and companies in the robotics and automation area.
6. **Manufacturing and Industrial Sectors:** Professionals and decision-makers implementing robots and automation technologies in the manufacturing and industrial sectors will find the book's practical applications, implementation methodologies, and case studies useful.
7. **Logistics and Supply Chain Professionals:** The book will help professionals in supply chain management and logistics comprehend how automation and robotics are changing these fields, increasing productivity, and resolving logistical issues.
8. **Healthcare Professionals:** Healthcare practitioners and professionals can learn about the uses, advantages, and particular difficulties of this industry by investigating the integration of robots and automation in healthcare settings.
9. **Technology Enthusiasts and Hobbyists:** Individuals with a general interest in technology, robotics, and automation, including hobbyists and enthusiasts, can use the book to deepen their understanding of fundamental principles and explore potential DIY projects.
10. **Ethics and Policy Advocates:** The book will be helpful to those who support ethical issues, responsible innovation, and the social effects of automation and robots. It explains the ethical aspects and ramifications of these technologies.
11. **International Organizations:** The book can be used by international organizations that priorities technical development, standardization, and cooperation to evaluate international trends, standards, and best practices in automation and robotics.
12. **Educational Organizations and Training Centers:** For all-encompassing learning, educational institutions and training facilities that provide professional development robots and automation courses might include the book in their curricula.

The book can be an invaluable tool for individuals and organizations looking to navigate and take advantage of robotics and automation breakthroughs, as it addresses the demands and interests of these many stakeholders.

Importance of Professional Societies or Organizations for Readership

The readership and influence of a book on robotics and automation are greatly enhanced by professional groups and organizations. Their significance is multifaceted and contributes to the book's prominence, credibility, and accessibility in the pertinent professional community. For readers of a book on robotics and automation, professional associations and organizations are crucial for the following reasons:

1. **Networking Opportunities:** Professional societies offer a forum for academics, researchers, and practitioners in the subject to network. By partnering with these groups, the writer and the book can become more visible in the neighborhood, which will attract more readers.
2. **Promotion and Endorsement:** Professional societies frequently provide forums for publicizing notable publications, newsletters, and communication channels. Such organizations' endorsements or promotions can greatly increase the book's visibility and credibility within its intended readership.
3. **Audience Engagement:** Professional societies use workshops, seminars, and conferences to help members interact with one another. Through the alignment of the book with these organizations' goals and interests, the author establishes a direct line of communication with potential readers.
4. **Conference Presentations and Book Launches:** Authors can present their work at conferences and other events that are arranged by numerous professional associations. This offers a chance for book releases, talks, and presentations that may catch the interest of researchers and professionals in attendance.
5. **Access to Resources:** Professional societies frequently have resource centers, libraries, or online platforms that act as repositories for important publications. By being included in these resources, the book will become more visible and easily available to a larger audience.
6. **Collaborative Opportunities:** Authors and professional associations working together on projects can create combined efforts like webinars, workshops, or cooperative research. These partnerships increase the field's body of knowledge in addition to increasing the book's visibility.
7. **Publication Awards and Recognition:** A few professional associations recognize or prize exceptional articles. Gaining recognition from an esteemed institution and possibly winning an award will greatly increase the book's credibility and draw in readers.
8. **Discussion Forums and Community Interaction:** Professional societies frequently provide social media groups, online communities, or discussion platforms for its members to discuss the most recent advancements in their profession. By taking part in these forums, writers can interact directly with the intended readership and spread the word about their book.
9. **Continuing Education and Certification:** Professional societies participate in certification procedures and ongoing education initiatives. In these kinds of programs, the book may be assigned reading or made mandatory, which will increase its uptake among professionals looking to advance their expertise.
10. **Global Reach:** International exposure for the book is made possible by the global reach of numerous professional societies. This is especially crucial for an area like automation and robotics, which affects numerous businesses worldwide.

To sum up, professional associations and societies play a crucial role in creating a readership community around a book about automation and robotics. Their assistance increases the book's exposure, legitimacy, and usability, fostering a mutually beneficial partnership that benefits the writer and the academic community.

How our Publication Can Stand Out

In order to make our robotics and automation publication stand out, we have included a strategic approach that prioritizes interaction, distinctiveness, and relevancy. The following are crucial tactics to help set our publication apart:

1. **Identify Unique Contributions:** Clearly we have focusssed how our publication differs from previously published studies. Emphasize any original viewpoints, approaches, case studies, or applications that aren't thoroughly discussed in previous publications.
2. **Interdisciplinary Approach:** Accept the fact that automation and robotics are multidisciplinary fields. Think about the ways in which our work can close gaps between other academic fields, offering a comprehensive analysis of the topic and reaching a wider readership.
3. **Practical Relevance and Applications:** We have made sure the theoretical ideas and real-world applications in our publication in a balanced manner. Case studies, real-world examples, and actionable insights that highlight the content's practical applicability are frequently emphasized for out readers.
4. **Incorporate Emerging Technologies:** We keep up with the most recent advancements and new breakthroughs in automation and robotics. Our publication will draw readers who are looking for forward-thinking information if it offers insights into cutting-edge technologies or examines emerging trends.
5. **Engage with Ethical and Societal Implications:** We talk about the societal effects and ethical issues around automation and robotics. A careful examination of these subjects can set our publication apart by offering a thorough comprehension of the wider consequences of the technology.
6. **Visual and Multimedia Elements:** We incorporate visual components like pictures, and diagrams to make our publication easier to read. Visual aids can improve the effectiveness of explaining difficult ideas and visually interest readers.
7. **User-Friendly Structure:** We make our publication easy to read and navigate. An enjoyable reading experience is enhanced by our content's logical flow, distinct chapter formats, and well-defined parts.
8. **Diverse Perspectives and Contributors:** Our joint venture with different experiences, backgrounds, and viewpoints has helped to contribute significantly. A publication that presents multiple points of view can offer a more inclusive and thorough understanding of automation and robotics.
9. **Interactive Components:** We have considered about adding interactive components like links to more content, online tools, or hands-on activities. This can make reading more engaging for readers and make the learning process more dynamic.
10. **Global Relevance:** We examine how automation and robotics are affecting the world by taking into account global viewpoints, applications, and norms. A book of this type that is relevant to the entire world can draw in more readers.
11. **Collaborate with Professional Societies:** We engage in cooperative efforts with pertinent robotics and automation professional societies and organizations. By collaborating with these organizations, we can increase our publication's legitimacy and visibility in the professional community..

12. **Promotional and Outreach Strategies:** Gathering knowledge from conferences, social media, academic networks, and other venues we have sincerely attempted to implement those to spread the word about and spark interest in our publication.

In the competitive field of robotics and automation we can position our publication to stand out by carefully weighing these and highlighting what makes it special, timely, and interesting.

The International Appeal of our Book

In order to augment the global attractiveness of our Robotics and Automation book, we emphasize about putting the following parameters into practice:

1. **Global Perspectives:** Priorities are given so that the book's material takes a global perspective into account. To make the material globally applicable, we talk about applications, case studies, and examples from various geographical areas, business sectors, and cultural situations..
2. **International Collaboration:** We have consulted work together with writers, scholars, and specialists from other nations. An international audience may be reached and a comprehensive understanding of robotics and automation can be impacted by incorporating comments from a varied range of professionals.
3. **Multilingual Considerations:** We have to think of translating the book into several languages, if at all possible. This has the potential to greatly increase its reach and open it up to a larger international audience.
4. **Addressing Global Challenges:** We talk about the worldwide issues surrounding robotics and automation, including international standards, regulatory variances, and ethical concerns. By addressing these problems, the content can become more appealing and applicable worldwide.
5. **International Case Studies:** We provide case studies that illustrate how automation and robotics technologies have been successfully used in different nations and sectors of the economy. This aids readers in comprehending the usefulness and advantages in various international contexts.
6. **Cross-Cultural Sensitivity:** To make the information more approachable to an international audience, we make sure that the examples, anecdotes, and pictures are culturally appropriate.
7. **Global Industry Trends:** We talk about and evaluate current robotics and automation industry trends worldwide. We examine the ways in which various locations are implementing and modifying these technologies and investigate the global ramifications for enterprises and communities.
8. **Collaboration with International Publishers:** We think of working with foreign publishers to distribute and market the book in different areas. This can improve visibility across marketplaces and make use of existing networks.
9. **International Conferences and Events:** We prefer to deliver the book at worldwide robotics and automation conferences and events. Engaging in international forums offers prospects for networking, cooperation, and visibility to a varied viewership.
10. **Engage with International Professional Societies:** We like to work together with global organizations and professional groups in the sector. Interacting with these organizations may present chances for advertising, critiques, and recommendations that strengthen the book's reputation abroad.

By implementing all these, we have made our book more appealing to readers and professionals throughout the world who work in the robotics and automation fields, regardless of their nationality or industry.

Contents

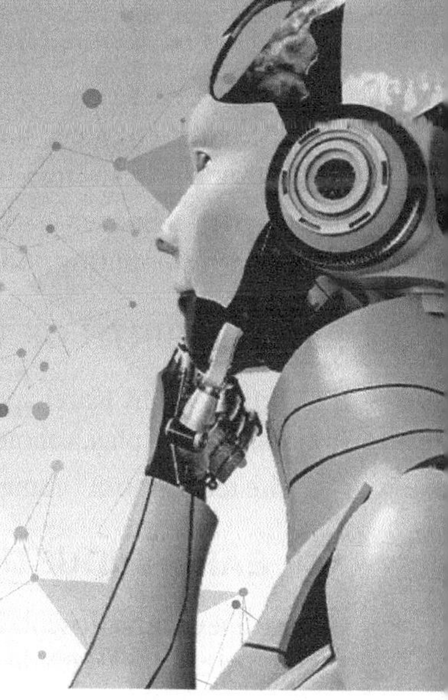

Evolution of Robotics and Automation

1.1 INTRODUCTION

The concept of automata, self-operating machines, and mechanical devices first appeared in antiquity, which is where robotics got its start. Here, we have provided a thorough summary of the development of robotics.

1.2 ANCIENT AUTOMATA (CIRCA 300 BCE – 500 CE)

The first accounts of automata may be found in the cultures of ancient China, Egypt, and Greece. These early automata were frequently created as decorative objects, amusements, or religious artefacts.

In China, the renowned mechanical engineer Yan Shi is credited with creating a human-like automaton that could perform a variety of tasks, as described in texts from the third century BCE.

In ancient Greece, inventors like Ctesibius and Hero of Alexandria created hydraulic and pneumatic devices that demonstrated simple automated movements.

1.3 MEDIEVAL AUTOMATA (500 CE – 1500 CE)

In Europe and the Islamic world, developments in mechanical engineering and clockmaking during the Middle Ages produced more complex automata.

Abbas Ibn Firnas, a polymath from Islamic Spain, is credited with creating a flying machine and several mechanised technologies in the ninth century.

The well-known Muslim inventor Al-Jazari (1136–1206) developed programmable humanoid automata in his important work "The Book of Knowledge of Ingenious Mechanical Devices."

1.4 INDUSTRIAL REVOLUTION (18th – 19th CENTURIES)

The development of automation saw a considerable change throughout the Industrial Revolution. More and more duties that were previously completed by hand were being carried out by machines and mechanisms.

A French engineer named Jacques de Vaucanson created automata like the "Flute Player" and "Digesting Duck" in the 18th century, exhibiting increasingly sophisticated and lifelike movements.

Charles Babbage's work on the "Analytical Engine" in the early 19th century created the groundwork for modern computing and the idea of programmable computers.

1.5 THE BIRTH OF THE TERM "ROBOT" (1920S)

Czech playwright Karel apek first used the word "robot" in his 1920 play "R.U.R. (Rossum's Universal Robots)." In the play, robots were artificial, biologically produced organisms that resembled humans.

The term "robot" comes from the Czech word "robota," which denotes slavery or forced labour.

1.6 EARLY INDUSTRIAL ROBOTS (1950S – 1960S)

George Devol and Joseph Engelberger, who developed the first industrial robot, Unimate, in the late 1950s, are credited with developing modern industrial robotics.

Unimate was initially used in die-casting equipment to handle hot metal. It signalled the start of industrial manufacturing's robotic automation.

1.7 ADVANCEMENTS IN ROBOTICS (1970S – 2000S)

Significant breakthroughs in robotics were made throughout the ensuing decades as a result of developments in electronics, computation, and materials.

Researchers like Marc Raibert, Joseph F. Engelberger, and Hiroshi Makino helped develop a variety of robotic technologies, such as robotic arms, moving robots, and walking robots.

1.8 MODERN ROBOTICS (21ST CENTURY)

In the twenty-first century, robots proliferated in a variety of sectors, including industry, healthcare, space exploration, and more.

The prevalence of collaborative robots (cobots), which can safely operate alongside people, increased, enabling automation for smaller firms.

1.9 PRESENT DEVELOPMENT

With current research concentrating on artificial intelligence, machine learning, and cutting-edge sensor technologies, robotics is still evolving quickly.

Robots are being developed today to carry out intricate duties in a variety of fields, from autonomous automobiles and surgical robots to service robots in homes and public places.

The development of robotics throughout history is evidence of human creativity and the desire to build tools that can duplicate or improve human talents. The development of robotics, from the primitive automata to the sophisticated robots of today, shows our persistent pursuit of advancement and automation in all facets of life.

1.10 HOW THE FIELD OF ROBOT HAS EVOLVED OVER TIME

As technology, computation, and artificial intelligence have advanced throughout time, the discipline of robotics has significantly evolved. Here is a thorough examination of the development of robotics:

Early Automata (Ancient Times – 18th Century)

As was previously noted, the idea of automata and autonomous devices first appeared in ancient civilizations such ancient Greece, Egypt, and China.

These early automata were straightforward mechanical creations made for amusement, religious rituals, and technological display.

Industrial Revolution (18th – 19th Centuries)

Robotics underwent a paradigm shift during the Industrial Revolution as numerous industrial operations were automated using machines and mechanisms.

Manufacturing processes were revolutionised and productivity rose thanks to industrial automation, which was first powered by steam and then later electricity.

Early Robotics (1940s – 1960s)

Isaac Asimov first used the word "robotics" in his science fiction writing in the 1940s, and it rose to popularity in the 1960s.

The first industrial robot, Unimate, was created in the 1950s and revolutionised production by automating processes like material handling and welding.

1.10.1 Advancements in Robotics

Significant advancements in robot control, sensing, and manipulation abilities were made in the 1970s.

Robots have become smarter and more adaptable as a result of the development of microprocessors and programmable logic controllers (PLCs).

Robotics in Space (1970s – 1980s)

Robots have been used by space agencies like NASA since the Viking and Voyager missions.

The Mars Rover Sojourner was the first mobile robot to operate on another planet in 1989.

Robotics Research and AI (1990s – Early 2000s)

The 1990s saw a surge in robotics research, with a focus on machine learning and artificial intelligence to enhance robot autonomy and adaptability.

Research into swarm robotics and multi-agent systems emerged, enabling the coordination of multiple robots for complex tasks.

Service and Domestic Robotics (Late 2000s – 2010s)

In the latter part of the 2000s, service robots began to appear, including humanoid robots like Pepper and self-driving hoover cleaners like Roomba.

These robots were created to help out in homes, workplaces, and public areas.

Collaborative Robotics (Cobots) (2010s)

Collaborative robots, or cobots, became common in sectors including manufacturing and healthcare because they were made to securely work alongside humans.

Cobots have cutting-edge sensors and obedient systems that make interactions with people safe.

Advancements in AI and Machine Learning (2010s – Present)

Robotics has advanced significantly as a result of the combination of artificial intelligence and machine learning.

Robots are now able to learn from data and adapt to varied settings, increasing their flexibility and usefulness in a variety of applications.

Robotics in Healthcare (Present)

The development of surgical robots for minimally invasive treatments and robotic prosthetics for improved mobility has revolutionized healthcare.

Emerging Technologies (Present – Future)

Soft robotics, bio-inspired robotics, and autonomous vehicles are some of the developing technologies in the fast evolving field of robotics.

Ethical issues, human-robot interaction, and social acceptance of robots are all gaining relevance in research.

Overall, robotics has advanced significantly from the primitive automata to the complex, intelligent machines of the present. The applications of robotics in numerous industries and daily life are essentially endless as technology advances, so it will be exciting to watch how this dynamic field develops in the future.

1.11 COMPONENTS OF ROBOTIC SYSTEMS

Robotic systems are made up of many parts that cooperate to let the robot successfully carry out its tasks. Typically, a robotic system's primary parts are:

1. **Sensors:** Without sensors, a robot cannot sense or communicate with its surroundings. Vision sensors (cameras) for object detection and perception are frequently employed in robotics.
   Infrared sensors for obstacle avoidance and proximity detection.
   3D mapping and navigation using LIDAR (light detection and ranging).
   Force feedback and touch-sensitive tactile sensors.
   Proximity sensors for spotting obstructions or other things in the way.
2. **Actuators:** The robot's actual physical motions are controlled by actuators. They translate mechanical actions from electrical impulses coming from the control system. Pneumatic actuators for supple and compliant motions and electric motors (such as DC motors and servo motors) for precise and controlled movements are common actuators in robotics.
3. **Control System:** There are different types of control systems, such as: (i) Closed-loop control systems, which use feedback to modify the robot's behavior; and (ii) Open-loop control systems, which carry out predefined sequences of actions without feedback. The control system is the brain of the robotic system and processes sensory information, generates commands for the actuators, and ensures the robot follows the desired trajectory.
4. **End Effectors:** The component of the robot that works directly with the environment to complete the task is called the end effector. It might be a gripper, a tool, or any other specialised device made to perform a particular function. Examples include welding guns for use in industrial welding applications and grippers for holding things of various sizes and forms.

5. **Mechanical Structure**: The robot's body is supported by its mechanical structure, which also controls its dexterity and mobility. Robotic arms are multi-jointed devices that enable precise movement and reach.

   Mobile bases: Bases that let robots move about on different terrains.

   Legged structures: These are used in walking robots to help them traverse difficult terrain.
6. **Power Supply**: Robots need a steady source of power to run their actuators and electronics. Power sources can be batteries, power cables, or a combination of both, depending on the application.
7. **Communication Interface**: The robot can converse with people or other systems thanks to its communication interface. It could consist of communication protocols for robot-to-robot or robot-to-computer communication, or it might be a graphical user interface (GUI) for controlling and programming the robot.
8. **Software and Programming**: An essential component of the robotic system is the software. It contains higher-level logic for task planning and decision-making as well as the robot's control and perception systems.
9. **Safety Systems**: Safety is a crucial factor in robotic systems, especially when people and robots collaborate. Emergency stop buttons, collision-detection motion sensors, and software safety features are some examples of safety systems.
10. **Data Storage and Processing**: Robots frequently require large amounts of processing power to evaluate sensor data and make quick choices. Depending on the capabilities of the robot, either onboard processors or cloud-based computing may be used for this.

The robot can perceive its environment, plan actions, and carry out tasks with accuracy and efficiency and thereby for a coordinated movement of these parts. The capabilities and performance of the robotic system in different applications are determined by the interaction of these components.

1.12 AUTOMATION CONCEPTS

Automation is the process of using machinery and technology to complete activities with little or no human involvement. It strives to decrease the need for manual labour while enhancing efficiency, accuracy, and production. The notions of automation are vital to many different businesses and have a significant impact on contemporary society. The following are some essential automation-related ideas:

1. **Automation Process**: It takes a methodical strategy to identify jobs that are eligible for automation, develop the automated system, and integrate it into current workflows. The automation process involves replacing manual stages with automated systems, generally employing machines, robots, or software.
2. **Types of Automation**: There are three main categories for automation:

   Fixed automation: Often referred to as hard automation, is created for a particular purpose and is not easily reprogrammable or adaptable for other tasks.

   Programmable Automation: Systems that are more adaptable than fixed automation because they can be reprogrammed or customised to carry out various tasks.

   Flexible Automation: Systems with a high degree of adaptability that can be swiftly reconfigured and adjusted to handle a range of jobs with little human involvement.

3. **Benefits of Automation**: Among the many benefits of automation are:
    1. Increased productivity and efficiency through the reduction of manual labour and cycle times.
    2. By reducing human error and variance, the quality and uniformity of the final product were improved.
    3. Increased safety by automating risky or boring operations.
    4. Long-term cost reductions brought on by decreased labour costs and enhanced process management.
    5. Improved ability to compete and satisfy rising production demands.
4. **Challenges of Automation**: While automation has numerous advantages, there are some drawbacks as well:
    1. High initial technology and training costs; • Potential job displacement; and • Potential negative effects on the workforce and society.
    2. The complexity of integration and compatibility difficulties when automating existing operations; the requirement for trained technicians and engineers to design, deploy, and maintain automated systems.
5. **Industry 4.0 and Smart Manufacturing**: The integration of automation, data sharing, and digital technology in manufacturing is known as "industry 4.0." In order to optimise manufacturing processes and enable predictive maintenance, it wants to build "smart factories" that make use of cutting-edge sensors, Internet of Things (IoT) gadgets, and artificial intelligence.
6. **Robotic Process Automation (RPA)**: RPA refers to the use of computer software "bots" to automate routine, rule-based tasks in corporate processes. It enables businesses to improve administrative processes and lessen human error.
7. **Artificial Intelligence (AI) and Machine Learning in Automation**: Automation is greatly aided by AI and machine learning because they allow systems to analyse data, draw conclusions, and adjust to changing circumstances. More autonomous and smarter automation is possible as a result.
8. **Ethics and Social Implications of Automation**: The increased use of automation creates moral questions and social repercussions. It entails dealing with challenges like employment loss, data privacy, AI bias, and making sure that the advantages of automation are dispersed fairly.

    In general, automation ideas are a driving force in contemporary industries, ranging from production to services, and they are always changing as technology progresses. As automation spreads, it's critical to find a balance between its advantages and potential drawbacks to make sure it helps both society and the workers.

1.13 ROLE OF AUTOMATION IN MODERN INDUSTRIES

Modern industries are fundamentally changed by automation, which has a profound impact on many elements of production, services, and overall effectiveness. In today's sectors, automation plays a number of important functions, including:

1. **Increased Productivity**: Automation speeds up production cycles, lowers human error, and streamlines procedures. Employees can concentrate on higher-value jobs when repetitive and time-consuming tasks are handled by automated systems, which increases total productivity.

2. **Improved Quality and Consistency**: Automated systems carry out activities precisely and consistently, reducing variations and mistakes that could happen with manual labour. Higher product quality and dependability result from this.
3. **Reduced Labor Costs:** The demand for human labour and associated costs can be decreased by automating labor-intensive processes. Automation technology may require an initial investment, but it frequently results in long-term cost reductions.
4. **Enhanced Safety**: Automation eliminates physically taxing, repetitive, and dangerous tasks, lowering the possibility of workplace accidents and injuries. This encourages making workplaces safer for workers.
5. **Increased Production Efficiency**: By reducing downtime, increasing equipment utilization, and enabling continuous production runs, automation optimises production processes and results in a more effective use of resources.
6. **Flexibility and Scalability**: Modern automation systems are created to be adaptable and scalable, enabling industries to readily accommodate changes in demand and production volume.
7. **Real-time Data and Analytics**: Massive amounts of data are produced by automation systems, and these data can be utilized to monitor, analyze, and improve production processes in real time. Businesses may identify areas for improvement and make educated decisions thanks to data-driven insights.
8. **Industry 4.0 and Smart Manufacturing**: The fourth industrial revolution, known as Industry 4.0, is greatly facilitated by automation. Advanced technologies like IoT, AI, and data analytics are used into smart manufacturing to build linked, data-driven production environments that improve productivity and adaptability.
9. **Robotics in Manufacturing**: Robots are frequently used in manufacturing to automate processes including welding, assembly, packaging, and material handling. Modern production lines value the precision, speed, and reproducibility that industrial robots provide.
10. **Robotics in Logistics and Warehousing**: Automated guided vehicles (AGVs), autonomous mobile robots (AMRs), and robotic arms efficiently handle material moving, sorting, and order fulfillment. Automation plays a vital role in logistics and warehousing.
11. **Customer Service and Support**: Chatbots, virtual assistants, and automated customer service responses all work together to offer clients speedy and effective support.
12. **Data Analysis and Decision-making**: Real-time data analysis and automated decision-making are made possible by automation, allowing firms to react quickly to shifting market conditions and client preferences.
13. **Robotic Process Automation (RPA)**: RPA automates repetitive and rule-based processes in administrative chores, minimizing manual labor and human error.

Although automation has many advantages, it also creates significant issues, including the effect on the labor, data protection, and ethical issues. To fully utilise automation in contemporary businesses and ensure beneficial societal impact, a balance must be struck between human labour and automated technology.

1.14 MATHEMATICAL PRELIMINARIES ON AUTOMATION

Automation is the process of performing work without the direct involvement of humans utilising machines, computers, or other technical systems. The development and analysis of automation systems heavily rely on a variety of mathematical ideas and tools. A summary of some mathematical foundations that are pertinent to automation are given here:

1. **Algebra**: Due to its focus on mathematical expressions and symbols, algebra is a prerequisite for automation. In automated systems, relationships between various factors and components are modeled using variables and equations.
2. **Calculus**: Calculus is necessary to comprehend how dynamic systems behave and to govern them. Derivatives and integrals are concepts that aid in trajectory planning, optimisation, and rate of change analysis.
3. **Linear Algebra**: Automation, particularly in the area of control systems, depends heavily on linear algebra. It deals with linear transformations, matrices, and vectors, which are essential for expressing and resolving issues with control and feedback systems.
4. **Probability and Statistics**: Automation frequently entails handling ambiguous events and data processing. The modelling of randomness is aided by probability theory, while decision-making based on observations and inferences from data are aided by statistics.
5. **Differential Equations**: Differential equations are used to simulate the behavior of dynamic systems over time. They are frequently used in control theory to explain the connection between the state, inputs, and outputs of the system.
6. **Optimization**: To effectively accomplish specified objectives, automation systems frequently need to have certain parameters optimized. The goal of optimization techniques is to maximize or minimize a function while satisfying constraints in order to identify the optimal solution from a group of possibilities.
7. **Graph Theory**: The structure of automation systems is represented and examined using graph theory, particularly when components are coupled and have dependencies.
8. **Numerical Methods**: Solving difficult mathematical equations or non-analytical systems is a common task in automation difficulties. Iterative strategies are utilised to estimate these solutions using numerical methods.
9. **Discrete Mathematics**: Designing and analysing digital systems, algorithms, and logic circuits need the use of discrete mathematics.

Let's see an illustration of a mathematical process used to automate a basic control system:

Example: Automated Temperature Control System

Consider the automation challenge of creating a straightforward room temperature control system. The objective is to maintain the room's temperature within a certain range.

1.14.1 Mathematical Steps on Automation Preliminaries

1. **Modeling the System**: We begin by creating a mathematical model of the dynamics of the room's temperature. Let's assume that the difference between the setpoint temperature (T_setpoint) and the present room temperature (T_room) determines the pace of change of the room's temperature. This can be stated as a differential equation.:

   $$dT\_room/dt = k * (T\_setpoint - T\_room)$$

   Where 'k' is a constant,representing the system's dynamics.
2. **Control Algorithm**: We require a control algorithm that computes the proper action to change the room temperature in order to automate the system. PID controllers, which calculate the control action based on the error (the difference between the setpoint and the current

temperature), integral of the error over time, and derivative of the error over time, are a prevalent type of control method.

Control output = Kp * error + Ki * integral(error) + Kd * derivative(error)

Here, Kp, Ki, and Kd are the tuning parameters of the PID controller.

3. **Sensor Data**: Temperature sensors are utilized to measure the current room temperature (T_ room).
4. **Feedback Loop**: The control algorithm generates the control output using the sensor data as input, then modifies the heating or cooling system as necessary.
5. **Optimization**: To get the desired control performance, it might be necessary to optimise the PID controller's Kp, Ki, and Kd parameters. If we have a mathematical model of the system and want to optimise a certain objective function, we can do this using approaches like trial and error or more sophisticated ones like gradient descent.
6. **Safety Constraints**: Safety in automation is crucial. The definition of safety constraints, such as limiting the temperature to a given range, can be aided by mathematical analysis.
7. **Iterative Testing and Refinement**: Iterative testing and improvement are common during the development of automation systems. This could entail changing the system model, the control parameters, or adding new features.
8. **Real-time Processing**: Automation systems must process data and make decisions in real-time for use in practical applications. Hardware considerations and effective algorithms are needed for this.

The automation of a control system involves a number of mathematical operations. Obviously, the level of sophistication required in the control algorithms and system models, as well as the specific automation task, will have a significant impact on the complexity of the mathematics used.

NUMERICAL PROBLEMS WITH SOLUTIONS

Algebra

Problem 1. Solve the equation: $3x + 7 = 22$.

Solution: Subtract 7 from both sides and divide by 3 when we get $x = (22 - 7)/3 = 5$.

Problem 2. Simplify the expression: $(2x + 3)(x - 5)$.

Solution: Use the distributive property: $2x^2 - 7x - 15$

Calculus

Problem 3. Find the derivative of $f(x) = 4x^3 - 2x^2 + 7x - 1$

Solution: $f'(x) = 12x^2 - 4x + 7$

Problem 4. Calculate the integral of $g(x) = 3x^2 + 4x + 2$.

Solution: $\int g(x)dx = x^3 + 2x^2 + 2x + C$ (where C is the constant of integration)

Linear Algebra

Problem 5. Perform matrix multiplication: $A = \begin{bmatrix} 2 & 3 \\ 4 & 1 \end{bmatrix}, B = \begin{bmatrix} 1 & 0 \\ 2 & 5 \end{bmatrix}$.

Solution:

$$A \times B = \begin{bmatrix} 8 & 15 \\ 6 & 5 \end{bmatrix}.$$

Problem 6. Find the determinant of the matrix: $M = \begin{bmatrix} 3 & 1 \\ 2 & 5 \end{bmatrix}$.

Solution: $\det(M) = 3 \times 5 - 1 \times 2 = 15 - 2 = 13$.

Probability and Statistics

Problem 7. Tossing a fair coin, what is the probability of getting heads?

Solution: The probability of getting heads is 0.5 (or 50%).

Problem 8. A factory produces light bulbs, and 5% of them are defective. If you randomly select two bulbs, what's the probability that both are defective?

Solution: Probability = 0.05 (defective) * 0.05 (defective) = 0.0025 or 0.25%

Differential Equations

Problem 9. Solve the differential equation: $dy/dx = 2x$

Solution: Integrate both sides with respect to x when we get $y = x^2 + C$ (where C is the constant of integration).

Problem 10. Find the particular solution to the differential equation: $dy/dx = 3x$, $y(0) = 2$

Solution: Integrate with respect to x when we get $y = (3/2)x^2 + C$.

Apply the initial condition: $2 = (3/2)(0)^2 + C$, i.e., $C = 2$

The particular solution is $y = (3/2)x^2 + 2$.

Optimization

Problem 11. Find the minimum value of the function $f(x) = x^2 - 6x + 8$

Solution: To find the minimum, take the derivative: $f'(x) = 2x - 6$

Set $f'(x) = 0$ to find critical points: $2x - 6 = 0$, $x = 3$

Check the second derivative: $f''(x) = 2$, which is positive, confirming the minimum.

Evaluate f(3): $f(3) = 3^2 - 6*3 + 8 = -1$

The minimum value is –1.

Problem 12. A rectangular fence is to be built with 100 meters of fencing material. Find the dimensions that maximize the area enclosed by the fence.

Solution: Let the sides of the rectangle be x and y. The perimeter is $2x + 2y = 100$, so $y = 50 - x$. The area $A = xy = x(50 - x) = 50x - x^2$.

To find the maximum, take the derivative: $dA/dx = 50 - 2x$

Set $dA/dx = 0$ to find critical points: $50 - 2x = 0$, $x = 25$.

Check the second derivative: $d^2A/dx^2 = -2$, which is negative, confirming the maximum.

Evaluate A(25): $A(25) = 25(50 - 25) = 625$ square meters. The dimensions for maximum area are $x = 25$ meters and $y = 50 - x = 25$ meters.

Graph Theory

Problem 13. Draw the graph representing a network of five interconnected devices (nodes).

Solution: The graph has five nodes connected in some pattern, represented as a visual graph.

Problem 14. Find the shortest path between two nodes in a given graph.

Solution: Use graph algorithms like Dijkstra's algorithm or the Breadth-First Search (BFS) to find the shortest path.

Numerical Methods

Problem 15. Solve the equation $x^3 - 4x^2 + 5x - 2 = 0$ using the Newton-Raphson method.

Solution: Start with an initial guess x_0, then apply the iteration formula: x_(n+1) = x_n – f(x_n)/f'(x_n) where $f(x) = x^3 - 4x^2 + 5x - 2$ and $f'(x) = 3x^2 - 8x + 5$ Iterate until you reach a solution within the desired tolerance.

Problem 16. Use the bisection method to find a root of the equation $e^x - 3x^2 = 0$ in the interval [0, 1].

Solution: Check the signs of f(0) and f(1): $f(0) = e^0 - 3 \times 0^2 = 1 > 0$ and $f(1) = e^1 - 3 \times 1^2 = e - 3 < 0$. Since f(0) and f(1) have opposite signs, there must be at least one root in the interval [0, 1]. Apply the bisection method to narrow down the interval until you find a root within the desired tolerance.

Discrete Mathematics

Problem 17. How many ways can you arrange the letters in the word "AUTOMATE"?

Solution: There are 8 letters in the word "AUTOMATE," so the number of arrangements is 8!.

Problem 18. Given a Boolean expression: (A AND B) OR (NOT A AND C), construct the truth table.

Solution: The truth table will have $2^3 = 8$ rows representing all possible combinations of truth values for variables A, B, and C.

A	B	C	(A AND B) OR (NOT A AND C)
0	0	0	1
0	0	1	1
0	1	0	0
0	1	1	0
1	0	0	0
1	0	1	1
1	1	0	0
1	1	1	1

Real-time Processing

Problem 19. A sensor generates temperature data every 0.5 seconds. If you want to achieve real-time control with a response time of 1 second, how frequently should you update the control system?

Solution: Update the control system every 1 second to achieve real-time processing.

Problem 20. Calculate the processing time required to perform a matrix multiplication of two 100 × 100 matrices on a computer that takes 10 microseconds to perform one multiplication.

Solution: The total number of multiplications is $100^3 = 10,000,000$. Given each multiplication takes 10 microseconds, the total processing time is 10,000,000 * 10 microseconds = 100 seconds.

Safety Constraints

Problem 21. A robotic arm operates at a speed of 2 meters per second. Implement a safety constraint to prevent the arm from colliding with any obstacle in its path.

Solution: Define a safety boundary around the robotic arm's path, and if any part of the arm breaches this boundary, trigger an immediate stop or reroute the arm.

Problem 22. An autonomous vehicle must maintain a minimum distance of 5 meters from any obstacle. Implement a safety constraint to control the vehicle's speed based on the distance from obstacles.

Solution: Use sensors to measure the distance to the nearest obstacle continuously. If the distance is less than 5 meters, reduce the vehicle's speed to maintain the safety distance.

Iterative Testing and Refinement

Problem 23. The algorithm for a drone's flight control is being improved. Its flight path is noted after each iteration. How can the effectiveness of the algorithm be evaluated by data analysis?

Solution: To see the trajectories of the observed flight paths, plot them on a graph. Examine the trajectories for smoothness, divergence from the intended course, and any oscillation patterns.

Problem 24. To minimize production flaws, an automated assembly line is being improved. The quantity of defects is counted throughout numerous production cycles after improvements are implemented. How can the data be analyzed to assess the improvements?

Solution: Calculate the defect rate at each stage of production and contrast it with the rates prior to the adjustments. To establish whether there has been a considerable improvement, use statistical techniques.

Mathematical Models in Automation

Problem 25. A ball is thrown vertically upward with an initial velocity of 20 m/s. Develop a mathematical model to describe the ball's height as a function of time considering gravity ($g = 9.8$ m/s^2).

Solution: The ball's height (h) can be modeled by the equation: $h(t) = 20t - 0.5gt^2$

Problem 26. A population of bacteria doubles every hour. Create a mathematical model to represent the population size (P) at any time (t).

Solution: The bacterial population size can be modeled by the equation: $P(t) = P_0 * 2^t$, where P_0 is the initial population size.

Control Algorithms

Problem 27. Design a PID controller to maintain a water tank's liquid level at a setpoint. Determine the control output based on the error, integral of the error, and derivative of the error, using appropriate tuning parameters.

Solution: The control output can be calculated as: Control output = Kp * error + Ki * integral(error) + Kd * derivative(error), where Kp, Ki, and Kd are the tuning parameters.

Problem 28. An autonomous vehicle needs to navigate to a destination while avoiding obstacles. Propose a control algorithm that takes sensor data to adjust the vehicle's steering and speed.

Solution: Develop an algorithm that uses sensor data (e.g., LiDAR, cameras) to detect obstacles. Based on this data, control the steering and speed of the vehicle to follow a safe path to the destination.

Feedback Loop

Problem 29. Explain the concept of a feedback loop in the context of automation.

Solution: In automation, a feedback loop involves using the output of a system to adjust its input. The system continuously measures its output, compares it to a desired setpoint, and makes necessary corrections to achieve the desired outcome.

Problem 30. Describe a feedback loop used in a temperature control system to maintain a stable room temperature.

Solution: The feedback loop measures the current room temperature using temperature sensors, compares it to the setpoint temperature, calculates the error (difference), and adjusts the heating or cooling system accordingly to minimize the error and maintain a stable room temperature.

Control Systems

Problem 31. A control system has a transfer function $G(s) = (s + 3)/(s^2 + 2s + 5)$. Find the poles and zeros of the system.

Solution: The poles are the values of 's' that make the denominator zero, i.e., $s^2 + 2s + 5 = 0$. Solving this quadratic equation, we get the poles as $-1 + 2i$ and $-1 - 2i$. There are no zeros in this transfer function.

Problem 32. Given a block diagram of a control system, determine the overall transfer function.

Solution: To find the overall transfer function, multiply all the individual transfer functions in the block diagram.

Signal Processing

Problem 33. A sound wave is represented by the equation $y(t) = 5\sin(2\pi ft)$, where f is the frequency in hertz and t is the time in seconds. Calculate the frequency of the sound wave.

Solution: The frequency (*f*) is the number of cycles per second, which is given by the coefficient of 't'. In this case, $f = 2\pi f$, so the frequency is 2π Hz.

Problem 34. A digital signal is sampled at a rate of 1000 samples per second. What is the Nyquist frequency, and what is the maximum frequency that can be accurately represented?

Solution: The Nyquist frequency is half of the sampling rate, so the Nyquist frequency is 1000/2 = 500 Hz. The maximum frequency that can be accurately represented is also 500 Hz.

Convolution

Problem 35. Compute the convolution of the two signals: $x(t) = u(t)$ and $h(t) = e^{(-2t)u(t)}$, where u(t) is the unit step function.

Solution: The convolution integral is given by $(x * h)(t) = \int[x(\tau)h(t - \tau)]d\tau$. Evaluate the integral to find the result.

Laplace Transform

Problem 36. Find the Laplace transform of the function $f(t) = 3t^2 + 2te^{(-t)}$.

Solution: Apply the definition of the Laplace transform to find F(s).

Z-Transform

Problem 37. Determine the z-transform of the discrete-time signal $x[n] = \{1, 2, 3, 4, 5\}$.

Solution: Apply the definition of the z-transform to find X(z).

Markov Chains

Problem 38. A website visitor can be in one of three states: A, B, or C. The transition probabilities are $P(A \to A) = 0.4$, $P(A \to B) = 0.3$, $P(A \to C) = 0.3$, $P(B \to A) = 0.2$, $P(B \to B) = 0.6$, $P(B \to C) = 0.2$, $P(C \to A) = 0.1$, $P(C \to B) = 0.4$, and $P(C \to C) = 0.5$. If the visitor is currently in state A, what is the probability of being in state C after two transitions?

Solution: Use the transition matrix to find the probability of being in state C after two transitions.

Game Theory

Problem 39. Two players play a zero-sum game. Player 1 wins 10 points if they choose strategy A and player 2 chooses strategy X; player 1 wins 5 points if they choose A and player 2 chooses Y; player 1 wins 3 points if they choose B and player 2 chooses X; player 1 wins 8 points if they choose B and player 2 chooses Y. Determine the optimal strategies for both players.

Solution: Use the concept of the minimax strategy to find the optimal strategies for both players.

Queuing Theory

Problem 40. An hourly average of 12 clients is served at a supermarket checkout counter, with an average hourly arrival rate of 10 consumers. Determine the typical number of consumers waiting in queue.

Solution: Calculate the average number of people in the wait using formulas from queueing theory.

Boolean Algebra

Problem 41. Simplify the Boolean expression: (A + B)(A + C)(B + C).

Solution: Use Boolean algebra rules to simplify the expression.

Logic Gates

Problem 42. Implement the XOR gate using only AND, OR, and NOT gates.

Solution: Create a circuit that performs the XOR operation using the given logic gates.

Combinatorial Logic

Problem 43. Convert a 4-bit binary number to its 2's complement by creating a combinational circuit.

Solution: Create a circuit that converts a 4-bit binary number to its 2's complement.

State Machines

Problem 44. Construct a state diagram and a state table for a simple traffic light control system with three states: red, green, and yellow.

Solution: Draw a state diagram and create a state table for the traffic light control system.

Fourier Transform

Problem 45. Find the Fourier transform of the signal $x(t) = e^{(-2\pi t)u(t)}$.

Solution: Apply the definition of the Fourier transform to find $X(\omega)$.

Wavelet Transform

Problem 46. Use the Haar wavelet transform to decompose a given discrete signal into its wavelet coefficients.

Solution: Apply the Haar wavelet transform to find the wavelet coefficients.

Principal Component Analysis (PCA)

Problem 47. Given a dataset with two variables, apply PCA to find the principal components and their corresponding eigenvalues.

Solution: Perform PCA on the dataset to find the principal components and eigenvalues.

Regression Analysis

Problem 48. Given a set of data points (x, y), fit a linear regression model $y = mx + b$ and find the values of m and b that minimize the sum of squared errors.

Solution: Apply the least squares method to find the values of m and b that minimize the sum of squared errors.

Probability Distributions

Problem 49. A random variable X follows a normal distribution with a mean of 50 and a standard deviation of 10. Calculate the probability $P(X > 70)$.

Solution: Use the properties of the normal distribution to find the probability.

Hypothesis Testing

Problem 50. The average lifespan of a manufacturer's light bulbs is stated to be 5000 hours, with a standard deviation of 200 hours. Test this assertion using a sample of 50 light bulbs with an average lifespan of 4950 hours and a significance threshold of 0.05.

Solution: Test your hypothesis to see if the sample data supports the manufacturer's assertion.

Basic Problems on Automation with Hints

Algebra

Problem 1. Solve the equation: $2x - 5 = 15$.

Hint: Add 5 to both sides and then divide by 2 to isolate x.

Problem 2. Simplify the expression: $3(x + 2) - 2(3x - 1)$.

Hint: Apply the distributive property and then combine like terms.

Calculus

Problem 3. Find the derivative of $f(x) = 4x^2 + 3x - 2$.

Hint: Use the power rule for differentiation.

Problem 4. Calculate the integral of $g(x) = 2x + 5$.

Hint: Use the power rule for integration.

Linear Algebra

Problem 5. Perform matrix addition: $A = \begin{bmatrix} 2 & 3 \\ 5 & 6 \end{bmatrix}, B = \begin{bmatrix} 1 & 4 \\ 2 & 1 \end{bmatrix}$.

Hint: Add corresponding elements in each matrix.

Problem 6. Calculate the determinant of the matrix: $M = \begin{bmatrix} 3 & 2 \\ 1 & 4 \end{bmatrix}$.

Hint: The determinant of a 2×2 matrix is given by $\det(M) = (3 \times 4) - (2 \times 1) = 10$.

Probability and Statistics

Problem 7. A fair six-sided die is rolled. What is the probability of rolling an odd number?

Hint: Count the number of favorable outcomes and divide by the total number of outcomes.

Problem 8. A bag contains 5 red balls and 3 blue balls. If a ball is randomly drawn, what's the probability of getting a red ball?

Hint: Count the number of red balls and divide by the total number of balls.

Differential Equations

Problem 9. Solve the differential equation: $dy/dx = 2e^x$.

Hint: Integrate both sides with respect to x.

Problem 10. Find the particular solution to the differential equation: $dy/dx = 2x$, $y(0) = 3$.

Hint: Integrate with respect to x and apply the initial condition.

Optimization

Problem 11. Find the maximum value of the function $f(x) = -x^2 + 4x + 3$.

Hint: To find the maximum, take the derivative and set it equal to zero.

Problem 12. A rectangular box is to be constructed with a fixed surface area of 100 square meters. Find the dimensions that maximize the volume of the box.

Hint: Express the volume of the box in terms of one variable and use optimization techniques.

Graph Theory

Problem 13. Draw the graph representing a network of four interconnected devices (nodes).

Hint: Connect the nodes with edges to represent the connections.

Problem 14. Find the shortest path between two nodes in a given graph.

Hint: Use graph algorithms like Dijkstra's algorithm or the Breadth-First Search (BFS).

Numerical Methods

Problem 15. Solve the equation $2x - 5 = 0$ using the bisection method within an interval [1, 3].

Hint: Check the signs of f(1) and f(3) to ensure a root lies within the interval.

Problem 16. Use the Newton-Raphson method to find a root of the equation $e^x - 3x = 0$, starting with an initial guess of $x = 1$.

Hint: Derive the iterative formula and perform iterations until convergence.

Discrete Mathematics

Problem 17. How many ways can you arrange the letters in the word "AUTOMATE"?

Hint: Use the permutation formula for arrangements of distinct objects.

Problem 18. Given a Boolean expression: (A AND B) OR (NOT A AND C), construct the truth table.

Hint: List all possible combinations of truth values for variables A, B, and C.

Real-time Processing

Problem 19. A sensor generates temperature data every 0.2 seconds. If you want to achieve real-time control with a response time of 0.5 seconds, how frequently should you update the control system?

Hint: Determine the update frequency based on the desired response time.

Problem 20. Calculate the processing time required to perform a matrix multiplication of two 50 × 50 matrices on a computer that takes 5 milliseconds to perform one multiplication.

Hint: Calculate the total number of multiplications and then the total processing time.

Safety Constraints

Problem 21. A robotic arm operates at a speed of 1 meter per second. Implement a safety constraint to prevent the arm from colliding with any obstacle in its path.

Hint: Define a safety boundary around the robotic arm's path and check for potential collisions.

Problem 22. An autonomous vehicle must maintain a minimum distance of 10 meters from any obstacle. Implement a safety constraint to control the vehicle's speed based on the distance from obstacles.

Hint: Use sensors to measure the distance to the nearest obstacle continuously and adjust the vehicle's speed accordingly.

Iterative Testing and Refinement

Problem 23. A drone's flight control algorithm is being refined. After each iteration, its flight path is recorded. How can you analyze the data to assess the algorithm's performance?

Hint: Plot the recorded flight paths and analyze for smoothness and accuracy.

Problem 24. An automated assembly line is being optimized to reduce manufacturing defects. After implementing changes, the number of defects is recorded over several production cycles. How can you analyze the data to evaluate the improvements?

Hint: Calculate the defect rate for each cycle and compare it to the rate before changes.

Mathematical Models in Automation

Problem 25. A ball is thrown vertically upward with an initial velocity of 15 m/s. Develop a mathematical model to describe the ball's height as a function of time considering gravity (g = 9.8 m/s²).

Hint: Apply the kinematic equation for vertical motion.

Problem 26. A population of bacteria doubles every 30 minutes. Create a mathematical model to represent the population size (P) at any time (t).

Hint: The population size can be modeled as an exponential growth function.

Control Algorithms

Problem 27. Design a PID controller to maintain a water tank's liquid level at a setpoint. Determine the control output based on the error, integral of the error, and derivative of the error, using appropriate tuning parameters.

Hint: Use the PID control algorithm and adjust the tuning parameters to achieve the desired response.

Problem 28. An autonomous vehicle needs to navigate to a destination while avoiding obstacles. Propose a control algorithm that takes sensor data to adjust the vehicle's steering and speed.

Hint: Develop an algorithm that uses sensor data to detect obstacles and navigate the vehicle safely.

Feedback Loop

Problem 29. Explain the concept of a feedback loop in the context of automation.

Hint: Describe how a system uses feedback to adjust its behavior based on the output.

Problem 30. Describe a feedback loop used in a temperature control system to maintain a stable room temperature.

Hint: The feedback loop continuously measures the room temperature, compares it to the setpoint, and adjusts the heating or cooling system to maintain the desired temperature.

Control Systems

Problem 31. A control system has a transfer function G(s) = (s + 2) / (s^2 + 3s + 5). Find the poles and zeros of the system.

Hint: Poles are the roots of the denominator polynomial, and zeros are the roots of the numerator polynomial.

Problem 32. Given a block diagram of a control system, determine the overall transfer function.

Hint: Combine the individual transfer functions using block diagram algebra.

Signal Processing

Problem 33. A sound wave is represented by the equation y(t) = 3sin(4πft), where f is the frequency in hertz and t is the time in seconds. Calculate the frequency of the sound wave.

Hint: The frequency (*f*) is the coefficient of the argument of the sine function.

Problem 34. A digital signal is sampled at a rate of 2000 samples per second. What is the Nyquist frequency, and what is the maximum frequency that can be accurately represented?

Hint: The Nyquist frequency is half of the sampling rate, and the maximum frequency is also half of the sampling rate.

Convolution

Problem 35. Compute the convolution of the two signals: x(t) = $e^{(-t)u(t)}$ and h(t) = $2e^{(-2t)u(t)}$, where u(t) is the unit step function.

Hint: Apply the convolution integral formula for continuous-time signals.

Laplace Transform

Problem 36. Find the Laplace transform of the function f(t) = $2e^{(3t)}$.

Hint: Apply the definition of the Laplace transform.

Z-Transform

Problem 37. Determine the z-transform of the discrete-time signal x[n] = {1, 1, 2, 3}.

Hint: Apply the definition of the z-transform.

Markov Chains

Problem 38. A website visitor can be in one of two states: A or B. The transition probabilities are P(A → A) = 0.8 and P(A → B) = 0.2. If the visitor is currently in state A, what is the probability of being in state B after three transitions?

Hint: Use the transition matrix to find the probability of being in state B after three transitions.

Game Theory

Problem 39. Two players play a zero-sum game. Player 1 wins 10 points if they choose strategy A and player 2 chooses strategy X; player 1 wins 5 points if they choose A and player 2 chooses Y; player 1 wins 3 points if they choose B and player 2 chooses X; player 1 wins 8 points if they choose B and player 2 chooses Y. Determine the optimal strategies for both players.

Hint: Use the concept of the minimax strategy to find the optimal strategies.

Queuing Theory

Problem 40. A supermarket checkout counter serves customers with an average arrival rate of 20 customers per hour and an average service rate of 25 customers per hour. Calculate the average number of customers in the queue.

Hint: Use queuing theory formulas to find the average number of customers in the queue.

Boolean Algebra

Problem 41. Simplify the Boolean expression: (A + B)(A + C)(B + C).

Hint: Use Boolean algebra rules to simplify the expression.

Logic Gates

Problem 42. Implement the XNOR gate using only AND, OR, and NOT gates.

Hint: Create a circuit that performs the XNOR operation using the given logic gates.

Short Questions and Answers

Q.1. When did the concept of automata first emerge?

Ans. The concept of automata first emerged in ancient civilizations, around 300 BCE.

Q.2. Who is credited with coining the term "robot"?

Ans. The term "robot" was coined by Czech playwright Karel Čapek in his 1920 play "R.U.R. (Rossum's Universal Robots)."

Q.3. What was the first industrial robot, and when was it developed?

Ans. The first industrial robot was Unimate, developed in the late 1950s by George Devol and Joseph Engelberger.

Q.4. What did Unimate initially do in industrial settings?

Ans. Unimate was used for tasks like handling hot metal in die-casting machines.

Q.5. When did the term "robotics" become popular?

Ans. The term "robotics" became popular in the 1960s.

Q.6. What is Industry 4.0?

Ans. Industry 4.0 refers to the integration of automation, data exchange, and digital technologies in manufacturing to create "smart factories."

Q.7. What are the main types of automation?

Ans. The main types of automation are fixed, programmable, and flexible automation.

Q.8. How does automation enhance safety in industries?

Ans. Automation can handle hazardous tasks, reducing the risk of workplace accidents and injuries.

Q.9. What is the purpose of robotic process automation (RPA)?

Ans. RPA automates repetitive and rule-based tasks in business processes, reducing manual workload and errors.

Q.10. What is kinematics in robotics?

Ans. Kinematics deals with the study of motion in robots without considering the forces causing it.

Q.11. What is the significance of computer vision in robotics?

Ans. Computer vision enables robots to process and interpret visual data from cameras, aiding in tasks like object recognition.

Q.12. What is inverse kinematics used for in robotics?

Ans. Inverse kinematics is used to determine the joint angles required to achieve a desired end-effector position in robot arms.

Q.13. What is the role of control theory in robotics?

Ans. Control theory is used to design control systems that govern the behavior of robots and ensure precise movements.

Q.14. When did the first walking robot, the Mars Rover Sojourner, operate on another planet?

Ans. The Mars Rover Sojourner operated on another planet in 1997.

Q.15. What are collaborative robots, or cobots?

Ans. Collaborative robots are designed to work safely alongside humans, making them suitable for human-robot collaboration.

Q.16. What is the historical significance of Al-Jazari in robotics?

Ans. Al-Jazari, a Muslim inventor from the 12th century, created programmable humanoid automata in his book "The Book of Knowledge of Ingenious Mechanical Devices."

Q.17. How did the Industrial Revolution impact the development of automation?

Ans. The Industrial Revolution marked a turning point for automation by using machines to replace manual labor in various industries.

Q.18. What is the main advantage of using automation in manufacturing?

Ans. Automation in manufacturing increases productivity and efficiency by reducing human labor and cycle times.

Q.19. What are the challenges of automation?

Ans. Challenges of automation include job displacement, high initial investment, and complex integration.

Q.20. What role do robotics and automation play in space exploration?

Ans. Robotics are crucial in space exploration, enabling tasks like satellite maintenance, planetary exploration, and assembly.

Q.21. What is the significance of the term "robota" in robotics?

Ans. The term "robota" means forced labor or servitude and is the origin of the word "robot."

Q.22. How do autonomous vehicles utilize automation technology?

Ans. Autonomous vehicles use automation to navigate, sense their environment, and make decisions without human intervention.

Q.23. What is the difference between fixed and flexible automation?

Ans. Fixed automation is designed for specific tasks and cannot be easily reprogrammed, while flexible automation can adapt to different tasks.

Q.24. What is the role of machine learning in modern robotics?

Ans. Machine learning enables robots to learn from data, adapt to different environments, and improve their performance.

Q.25. How does automation impact job markets?

Ans. Automation can lead to job displacement in certain industries, but it can also create new job opportunities in other sectors.

Q.26. What is the historical significance of the ancient Greek engineer Hero of Alexandria?

Ans. Hero of Alexandria designed hydraulic and pneumatic devices, showcasing some of the earliest automata in ancient times.

Q.27. How do robots utilize sensors in their operations?

Ans. Robots use sensors to perceive their environment, detect obstacles, and gather data for decision-making.

Q.28. What is the difference between open-loop and closed-loop control systems in robotics?

Ans. Open-loop control systems execute predefined actions without feedback, while closed-loop systems use feedback to adjust their behavior.

Q.29. How do robots use grippers in their tasks?

Ans. Grippers are end effectors used by robots to grasp and manipulate objects in various applications.

Q.30. What is the purpose of robotics in the healthcare sector?

Ans. Robotics in healthcare are used for surgical procedures, rehabilitation, telemedicine, and assisting healthcare professionals.

Q.31. How does automation contribute to environmental sustainability?

Ans. Automation can optimize resource usage, reduce waste, and improve energy efficiency, contributing to environmental sustainability.

Q.32. What is the potential impact of robotics in agriculture?

Ans. Robotics in agriculture can enhance efficiency in planting, harvesting, and monitoring crops, leading to increased food production.

Q.33. How do mobile robots utilize automation in various industries?

Ans. Mobile robots are used in industries like logistics and warehousing for efficient material movement and order fulfillment.

Q.34. What role did Charles Babbage play in the evolution of automation?

Ans. Charles Babbage's work on the "Analytical Engine" laid the foundations for modern computing and programmable machines.

Q.35. How does automation influence the development of smart cities?

Ans. Automation technologies can optimize infrastructure, transportation, and utilities, contributing to the development of smart cities.

Q.36. How do robots use machine vision in industrial settings?

Ans. Machine vision is used in industrial robots for quality inspection, object tracking, and process control.

Q.37. What role do humanoid robots play in human-robot interaction?

Ans. Humanoid robots are designed to resemble human characteristics, making them suitable for more natural human-robot interactions.

Q.38. How are robots integrated into educational curricula?

Ans. Robots are integrated into educational curricula to teach programming, problem-solving, and teamwork to students.

Q.39. How do collaborative robots enhance worker safety in manufacturing?

Ans. Collaborative robots have built-in safety features that allow them to work alongside humans safely without protective barriers.

Q.40. What role did Isaac Asimov play in shaping the concept of robotics?

Ans. Isaac Asimov's science fiction stories introduced the Three Laws of Robotics, which are influential in discussions about robotic ethics.

Q.41. How do robotic process automation (RPA) bots automate repetitive tasks in business processes?

Ans. RPA bots interact with software applications and perform tasks, such as data entry and document processing, just like human users.

Q.42. What is the role of automation in the entertainment industry?

Ans. Automation is used in the entertainment industry for special effects, animatronics, and theme park attractions.

Q.43. How do robots contribute to disaster response and recovery efforts?

Ans. Robots can be used to search for survivors in hazardous environments and perform tasks in disaster-stricken areas.

Q.44. How do robots facilitate advancements in space exploration?

Ans. Robots are used in space exploration to conduct experiments, collect samples, and explore distant celestial bodies.

Q.45. How do industrial robots improve manufacturing processes?

Ans. Industrial robots automate repetitive tasks, increase production speed, and improve product quality in manufacturing.

Q.46. What are some potential ethical considerations of robotics and automation?

Ans. Ethical considerations include job displacement, data privacy, AI bias, and ensuring robotic systems act responsibly and safely.

Q.47. How do robots contribute to medical surgeries and procedures?

Ans. Surgical robots provide enhanced precision, dexterity, and minimally invasive capabilities in medical procedures.

Q.48. What are the social implications of widespread automation adoption?

Ans. Widespread automation adoption may impact employment patterns, necessitating upskilling and reskilling of the workforce.

Q.49. How does artificial intelligence enhance the capabilities of robots?

Ans. Artificial intelligence enables robots to learn from data, recognize patterns, and make decisions in complex environments.

Q.50. What is the future outlook for robotics and automation?

Ans. The future of robotics and automation is expected to bring advancements in AI, human-robot collaboration, and autonomous systems, transforming various industries and aspects of daily life.

Multiple Choice Questions and Answers

Q.1. Who coined the term "robot"?

(*a*) Isaac Asimov (*b*) Karel Čapek
(*c*) George Devol (*d*) Joseph Engelberger

Q.2. The first industrial robot, Unimate, was developed in which decade?

(*a*) 1930s (*b*) 1950s (*c*) 1970s (*d*) 1990s

Q.3. What was the initial use of Unimate in industrial settings?

(*a*) Assembly line tasks (*b*) Material handling (*c*) Welding (*d*) Painting

Q.4. Which ancient civilization had inventors like Ctesibius and Hero of Alexandria who created automata?

(*a*) Roman Empire (*b*) Ancient Egypt
(*c*) Ancient Greece (*d*) Han Dynasty (China)

Q.5. What does the term "robota" mean, from which "robot" is derived?

(*a*) Workforce (*b*) Automation
(*c*) Mechanical device (*d*) Artificial intelligence

Q.6. In which play was the term "robot" first used?

(*a*) Macbeth (*b*) Frankenstein
(*c*) R.U.R. (Rossum's Universal Robots) (*d*) Hamlet

Q.7. The Mars Rover Sojourner became the first mobile robot to operate on another planet in which year?

(*a*) 1984 (*b*) 1997 (*c*) 2003 (*d*) 2012

Q.8. Which type of automation is designed for a specific task and cannot be easily reprogrammed?

(*a*) Fixed automation (*b*) Programmable automation
(*c*) Flexible automation (*d*) Autonomous automation

Q.9. What is the primary role of collaborative robots (cobots)?

(*a*) Working alongside humans (*b*) High-speed production
(*c*) Hazardous tasks (*d*) Long-distance transportation

Q.10. Which aspect of robotics focuses on the study of motion in robots without considering forces causing it?

(*a*) Kinematics (*b*) Dynamics
(*c*) Control theory (*d*) Computer vision

Q.11. Who is considered the "father of robotics"?

(*a*) Nikola Tesla (*b*) George Devol
(*c*) Joseph Engelberger (*d*) Leonardo da Vinci

Q.12. What type of control system uses feedback to adjust a robot's behavior?

(*a*) Open-loop control system (*b*) Closed-loop control system
(*c*) Programmable logic controller (PLC) (*d*) Industrial computer system

Q.13. What is the term used for robots that are designed to resemble human characteristics?

(*a*) Androids (*b*) Cyborgs (*c*) Nanobots (*d*) Drones

Q.14. What is the fourth industrial revolution, incorporating automation and data exchange in manufacturing?

(*a*) Industrial Evolution 4.0 (*b*) Industry 4.0
(*c*) Robotics Revolution (*d*) Digital Transformation

Q.15. What is the study of forces and torques that cause motion in robots?

(*a*) Kinematics (*b*) Dynamics
(*c*) Control theory (*d*) Computer vision

Q.16. Which country is credited with developing the first humanoid robot named WABOT–1?

(*a*) Japan (*b*) United States (*c*) Germany (*d*) South Korea

Q.17. The Three Laws of Robotics were introduced by:

(*a*) Isaac Asimov (*b*) Arthur C. Clarke
(*c*) Ray Bradbury (*d*) Philip K. Dick

Q.18. In which century did Leonardo da Vinci design mechanical knights and robotic devices?

(*a*) 12th century (*b*) 14th century (*c*) 16th century (*d*) 18th century

Q.19. What is the primary purpose of robotic process automation (RPA) in business?

(*a*) Quality control
(*b*) Enhancing creativity
(*c*) Automating repetitive tasks
(*d*) Complex decision-making

Q.20. How did the Industrial Revolution impact the development of automation?

(*a*) It slowed down automation progress.
(*b*) It had no impact on automation.
(*c*) It accelerated the development of automation.
(*d*) It shifted focus away from automation.

Q.21. The concept of automata and self-operating machines first emerged in which ancient civilization?

(*a*) Ancient Egypt
(*b*) Ancient Greece
(*c*) Roman Empire
(*d*) Han Dynasty (China)

Q.22. What is the primary advantage of using automation in manufacturing?

(*a*) Increased labor costs
(*b*) Reduced productivity
(*c*) Improved product quality
(*d*) Less reliance on technology

Q.23. What type of automation is reconfigurable and can handle different tasks with minimal human intervention?

(*a*) Fixed automation
(*b*) Programmable automation
(*c*) Flexible automation
(*d*) Autonomous automation

Q.24. What was the purpose of Charles Babbage's "Analytical Engine"?

(*a*) Industrial automation
(*b*) Weather prediction
(*c*) Modern computing
(*d*) Robotic arms

Q.25. What role do robots play in space exploration?

(*a*) Analyzing data from Earth
(*b*) Delivering mail in space
(*c*) Collecting samples and conducting experiments
(*d*) Piloting spacecraft

Q.26. How does robotics impact job markets?

(*a*) Robots take over all human jobs.
(*b*) It creates new job opportunities in different fields.
(*c*) Automation has no impact on job markets.
(*d*) All jobs become remote.

Q.27. What aspect of robotics focuses on forces and torques causing motion in robots?

(*a*) Kinematics
(*b*) Dynamics
(*c*) Control theory
(*d*) Computer vision

Q.28. Which type of automation is designed for specific tasks and cannot be reprogrammed easily?

(*a*) Fixed automation (*b*) Programmable automation
(*c*) Flexible automation (*d*) Autonomous automation

Q.29. What term is used for robots designed to work safely alongside humans?

(*a*) Cobots (*b*) Androids (*c*) Cyborgs (*d*) Drones

Q.30. Which ancient civilization had inventors like Ctesibius and Hero of Alexandria who created automata?

(*a*) Roman Empire (*b*) Ancient Egypt
(*c*) Ancient Greece (*d*) Han Dynasty (China)

Q.31. What does the term "robota" mean, from which "robot" is derived?

(*a*) Workforce (*b*) Automation
(*c*) Mechanical device (*d*) Artificial intelligence

Q.32. Who coined the term "robot"?

(*a*) Isaac Asimov (*b*) Karel Čapek
(*c*) George Devol (*d*) Joseph Engelberger

Q.33. In which play was the term "robot" first used?

(*a*) Macbeth (*b*) Frankenstein
(*c*) R.U.R. (Rossum's Universal Robots) (*d*) Hamlet

Q.34. The Mars Rover Sojourner became the first mobile robot to operate on another planet in which year?

(*a*) 1984 (*b*) 1997 (*c*) 2003 (*d*) 2012

Q.35. What was the initial use of Unimate in industrial settings?

(*a*) Assembly line tasks (*b*) Material handling
(*c*) Welding (*d*) Painting

Q.36. The first industrial robot, Unimate, was developed in which decade?

(*a*) 1930s (*b*) 1950s (*c*) 1970s (*d*) 1990s

Q.37. Who is considered the "father of robotics"?

(*a*) Nikola Tesla (*b*) George Devol
(*c*) Joseph Engelberger (*d*) Leonardo da Vinci

Q.38. What is the primary role of collaborative robots (cobots)?

(*a*) Working alongside humans (*b*) High-speed production
(*c*) Hazardous tasks (*d*) Long-distance transportation

Q.39. What is the fourth industrial revolution, incorporating automation and data exchange in manufacturing?

(*a*) Industrial Evolution 4.0 (*b*) Industry 4.0
(*c*) Robotics Revolution (*d*) Digital Transformation

Q.40. What is the primary advantage of using automation in manufacturing?

(*a*) Increased labor costs
(*b*) Reduced productivity
(*c*) Improved product quality
(*d*) Less reliance on technology

Q.41. What type of automation is reconfigurable and can handle different tasks with minimal human intervention?

(*a*) Fixed automation
(*b*) Programmable automation
(*c*) Flexible automation
(*d*) Autonomous automation

Q.42. What was the purpose of Charles Babbage's "Analytical Engine"?

(*a*) Industrial automation
(*b*) Weather prediction
(*c*) Modern computing
(*d*) Robotic arms

Q.43. What role do robots play in space exploration?

(*a*) Analyzing data from Earth
(*b*) Delivering mail in space
(*c*) Collecting samples and conducting experiments
(*d*) Piloting spacecraft

Q.44. How does robotics impact job markets?

(*a*) Robots take over all human jobs.
(*b*) It creates new job opportunities in different fields.
(*c*) Automation has no impact on job markets.
(*d*) All jobs become remote.

Q.45. What aspect of robotics focuses on forces and torques causing motion in robots?

(*a*) Kinematics
(*b*) Dynamics
(*c*) Control theory
(*d*) Computer vision

Q.46. Which type of automation is designed for specific tasks and cannot be reprogrammed easily?

(*a*) Fixed automation
(*b*) Programmable automation
(*c*) Flexible automation
(*d*) Autonomous automation

Q.47. What term is used for robots designed to work safely alongside humans?

(*a*) Cobots
(*b*) Androids
(*c*) Cyborgs
(*d*) Drones

Q.48. Which ancient civilization had inventors like Ctesibius and Hero of Alexandria who created automata?

(*a*) Roman Empire
(*b*) Ancient Egypt
(*c*) Ancient Greece
(*d*) Han Dynasty (China)

Q.49. What does the term "robota" mean, from which "robot" is derived?

(*a*) Workforce
(*b*) Automation
(*c*) Mechanical device
(*d*) Artificial intelligence

Q.50. Who coined the term "robot"?

(*a*) Isaac Asimov
(*b*) Karel Čapek
(*c*) George Devol
(*d*) Joseph Engelberger

Key to Multiple Choice Questions

1. (b)	2. (b)	3. (c)	4. (c)	5. (a)	6. (c)	7. (b)	8. (a)	9. (a)
10. (a)	11. (c)	12. (b)	13. (a)	14. (b)	15. (b)	16. (a)	17. (a)	18. (c)
19. (c)	20. (c)	21. (b)	22. (c)	23. (c)	24. (c)	25. (c)	26. (b)	27. (b)
28. (a)	29. (a)	30. (c)	31. (a)	32. (b)	33. (c)	34. (b)	35. (c)	36. (b)
37. (c)	38. (a)	39. (b)	40. (c)	41. (c)	42. (c)	43. (c)	44. (b)	45. (b)
46. (a)	47. (a)	48. (c)	49. (a)	50. (b)				

Project Works

1. **Historical Timeline of Robotics**: Create a comprehensive timeline of significant milestones in robotics and automation, starting from ancient automata to modern-day robotics achievements. Include key events, inventions, and influential figures.
2. **Robotics Pioneers and Their Contributions**: Research and present in-depth profiles of pioneers in the field of robotics and automation. Highlight their contributions, inventions, and the impact of their work on the development of robotics.
3. **Comparative Analysis of Early Automata**: Choose multiple examples of early automata from different civilizations, such as ancient Greece, China, and the Islamic world. Compare and contrast their designs, functionalities, and societal implications.
4. **Evolution of Industrial Robotics**: Investigate the evolution of industrial robots in manufacturing processes. Study how robotics transformed industries over time, leading to increased efficiency and productivity.
5. **The Role of Robotics in Space Exploration**: Explore the history of robotics in space exploration and the contributions of robotic missions to scientific discoveries and understanding our solar system.
6. **Impact of Robotics on Labor Markets**: Analyze the historical impact of robotics and automation on labor markets. Investigate how automation affected job opportunities, workforce skills, and the changing nature of work.
7. **Robotics in Popular Culture**: Investigate the portrayal of robots in literature, movies, and other forms of media throughout history. Examine how these depictions influenced public perception and shaped the development of real-world robotics.
8. **Robotic Arms Through the Ages**: Study the evolution of robotic arms, from their early designs to modern-day sophisticated robotic manipulators. Explore the applications and industries where robotic arms are commonly used.
9. **Robotics and Ethics**: Examine the ethical considerations surrounding robotics and automation throughout history. Discuss ethical dilemmas posed by emerging technologies and the evolution of ethical guidelines in robotics.
10. **Robotics in Education**: Explore the history of robotics in educational settings. Discuss how robots have been used as educational tools and how they have influenced STEM (Science, Technology, Engineering, and Mathematics) education.

02

Linear Algebra for solving Complex Problems of Robotics and Automation

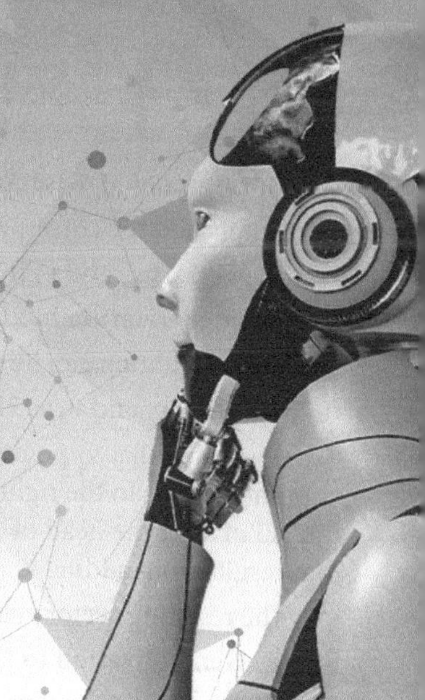

2.1 INTRODUCTION

Vector spaces, linear transformations, and linear equation systems are all covered in the area of mathematics known as linear algebra. It is an essential tool used in many branches of science and engineering, such as physics, computer graphics, robotics, and machine learning. Let's walk through a discussion of linear algebra step by step. Here, we've provided a succinct introduction to linear algebra, which offers a potent toolkit for tackling challenging issues across a range of robotics and automation-related domains.

2.2 SCALARS AND VECTORS

A single value, such as a real number, is known as a scalar. It has size but lacks direction.

An ordered group of scalars is referred to as a vector and is typically represented as a column or row matrix. Vectors are represented by bold lowercase letters or arrows and have both magnitude and direction.

2.3 VECTOR OPERATIONS

Addition: A new vector can be created by competently adding two vectors of the same dimension. In linear algebra, adding vectors is a fundamental operation. It is crucial to comprehend how to add vectors while working with them. To execute computations accurately, it's imperative to understand the laws that govern vector addition. Let's walk through vector addition step-by-step with examples:

Step 1: Understanding Vector Representation:

Either row matrices or column matrices are used to represent vectors.

Let's use two vectors in two dimensions as an illustration: $v_1 = [3, 5]$ (Represented in columns) $v_2 = [1, -2]$ (Represented in columns)

Step 2: Ensure Both Vectors Have the Same Dimension:

Two vectors must have the same dimension, or the same number of elements, in order to be added. You cannot add vectors if they don't have the same dimension.

Step 3: Component-wise Addition:

Simply add the corresponding components (elements) of the two vectors to make a new vector.

The vectors v_1 and v_2 from Step 1 are as follows: $v_1 + v_2 = [3 + 1, 5 + (-2)] = [4, 3]$

Step 4: Interpretation in 2D Space:

Vector addition can be seen geometrically in a 2-dimensional Cartesian coordinate system.

Consider v_1 and v_2 as 2D space displacement vectors.

Assuming that v_1 points 1 unit to the right along the x-axis and 2 units down along the y-axis, and v_2 points 3 units to the right along the x-axis and 5 units up along the y-axis, starting at the origin. Place the tail of v_2 at the head of v_1 and draw a line from the origin to the head of v_2 to get the displacement that results from adding v_1 and v_2.

The resulting vector $v_1 + v_2$ points 4 units to the right along the x-axis and 3 units up along the y-axis.

Step 5: Generalization to Higher Dimensions:

Higher dimensional vector addition, like that in 3D or nD, also operates similarly.

The 3D row or column matrices are used to represent vectors in 3D space.

In 3D, you may simply add a vector's x, y, and z components.

A 3D illustration is $v_1 = [1, 2, 3]$, $v_2 = [4, 5, 6]$

$$v_1 + v_2 = [1 + 4, 2 + 5, 3 + 6] = [5, 7, 9]$$

Keep in mind that since vector addition is commutative, the order in which the elements are added is irrelevant. In other words, $v_1 + v_2$ yields the same outcome as $v_2 + v_1$.

You can improve your comprehension of linear algebra by practising working with vectors and vector addition. This knowledge is essential when solving many mathematical and engineering problems.

Scalar Multiplication: A vector's magnitude can be increased without affecting its direction by multiplying it by a scalar. Another essential mathematical operation in linear algebra is scalar multiplication. A scaled version of the original vector is produced by multiplying a vector by a scalar (a single value). Let's walk through the vector scalar multiplication method step by step using examples:

Step 1: Understanding Vector Representation:

Vectors are represented using either column matrices or row matrices.

For example, let's consider a 3-dimensional vector: $v = [2, -3, 5]$ (Column representation)

Step 2: Choose a Scalar Value:

A scalar is a single value (e.g., a real number) that will be used to scale the vector.

For example, let's choose a scalar value of 3:

Step 3: Multiply the Scalar with Each Component of the Vector:

To perform vector scalar multiplication, simply multiply the scalar with each component (element) of the vector.

For the vector v and scalar 3 from Step 1 and Step 2: $3v = 3 * [2, -3, 5] = [3 * 2, 3 * (-3), 3 * 5] = [6, -9, 15]$

Step 4: Interpretation in 3D Space:

In a 3-dimensional Cartesian coordinate system, vector scalar multiplication can be visualized geometrically.

Let's consider v as a displacement vector in 3D space.

The original vector v points from the origin to the point (2, –3, 5).

When we do scalar multiplication with a value of 3 the resulting vector 3v points from the origin to the point (6, –9, 15).

The vector 3v is three times longer than the original vector v and points in the same direction.

Step 5: Generalization to Higher Dimensions:

Vector scalar multiplication works similarly in higher dimensions (e.g., 2D, 4D, or nD).

In nD space, vectors are represented using n-dimensional column matrices or row matrices.

To perform scalar multiplication in nD, simply multiply the scalar with each component of the vector.

Example in 4D: v = [1, –2, 3, 0] Scalar = –2

$$-2v = -2 * [1, -2, 3, 0] = [-2, 4, -6, 0]$$

Keep in mind that scalar multiplication only scales a vector's magnitude, not its direction. Additionally, for any scalar c and any pair of vectors u and v, c (u + v) = cu + cv, demonstrating that scalar multiplication is distributive over vector addition.

Scalar multiplication is a fundamental operation in linear algebra, and practising it with various scalar values and vectors can help you comprehend it better.

2.3.1 Vector Spaces

A set of vectors that can perform addition and scalar multiplication as well as satisfy certain qualities, such as closure, associativity, and commutativity as well as the presence of a zero vector and additive inverses, is referred to as a vector space.

2.4 MATRICES AND MATRIX OPERATIONS

A matrix is a rectangular array of numbers (scalars) arranged in rows and columns. It is represented using uppercase letters (e.g., A).

Matrix Addition: A second matrix of the same size can be created by adding elements from matrices of the same dimensions.

A fundamental operation in linear algebra is the addition of two matrices, which is done element-by-element. The final matrix will be the same size as the starting matrices after addition. The matrices' rows and columns must match in order for there to be a matrix addition to be possible.

The sum C (also m × n) of two identically sized matrices A and B is created by adding the corresponding elements from both matrices. It can be modelled mathematically as follows:

If $A = [a_{ij}]$ and $B = [b_{ij}]$ are two matrices of dimensions m × n, then the sum C = A + B is a matrix of dimensions m × n where:

$C = [c_{ij}]$, where $c_{ij} = a_{ij} + b_{ij}$ for all i = 1, 2, ..., m and j = 1, 2, ..., n.

For example, consider the following two matrices:

$$A = \begin{bmatrix} 2 & 4 \\ 1 & 3 \end{bmatrix} \quad \text{and} \quad B = \begin{bmatrix} 5 & 1 \\ 2 & 7 \end{bmatrix}$$

Their sum C = A + B will be:

$$C = \begin{bmatrix} 2+5 & 4+1 \\ 1+2 & 3+7 \end{bmatrix} = \begin{bmatrix} 7 & 5 \\ 3 & 10 \end{bmatrix}$$

Note that matrix addition is commutative, which means that A + B = B + A. Additionally, it follows the associative property, so (A + B) + C = A + (B + C).

In many disciplines where matrices are used to describe data and connections between variables, such as computer graphics, physics, engineering, and many other domains, matrix addition is a fundamental function.

Matrix Multiplied by a Scalar: When a scalar is multiplied by a matrix, each matrix element is multiplied by the scalar.

Another essential operation in linear algebra is matrix multiplication by a scalar. It entails multiplying each matrices element by a constant, also referred to as a scalar value. The original matrix's dimensions will be carried over into the final matrix.

When a scalar value k and a matrix A with dimensions m × n are combined, the result is shown as follows:

If $A = [a_{ij}]$ is a matrix of dimensions m × n, and k is a scalar value, then the scalar multiplication kA is a matrix of dimensions m × n where:

$$kA = [ka_{ij}]$$

where $ka_{ij} = ka_{ij}$ for all i = 1, 2, ..., m and j = 1, 2, ..., n.

In other words, each element of the original matrix A is multiplied by the scalar value k to obtain the corresponding element in the resulting matrix kA.

For example, consider the following matrix A:

$$A = \begin{bmatrix} 2 & 4 \\ 1 & 3 \end{bmatrix}$$

Now, let's perform scalar multiplication with k = 3:

$$3A = \begin{bmatrix} 3\times2 & 3\times4 \\ 3\times1 & 3\times3 \end{bmatrix} = \begin{bmatrix} 6 & 12 \\ 3 & 9 \end{bmatrix}$$

As shown in the example, each element in matrix A was multiplied by 3 to obtain the result.

Properties of Scalar Multiplication

1. **Commutative:** kA = Ak. The order of scalar multiplication and the matrix does not matter.
2. **Associative:** (kl)A = k(lA). Scalar multiplication is associative.
3. **Distributive over matrix addition:** k(A + B) = kA + kB. Scalar multiplication distributes over matrix addition.
4. **Distributive over scalar addition:** (k + l)A = kA + lA. Scalar multiplication distributes over scalar addition.

Scaling changes in computer graphics, rescaling data in statistics, and many other situations where we need to uniformly change the magnitude of matrix elements call for the usage of matrix multiplication by a scalar.

Matrix Multiplication: When the number of rows in the second matrix equals the number of columns in the first matrix, the product of the two matrices is known. The final matrix has the dimensions (m x p), where m is the sum of the number of rows in the first matrix and the number of columns in the second matrix.

One of the most important operations in linear algebra is matrix multiplication, which is distinct from matrix addition and scalar multiplication. It entails using particular principles to combine two matrices to create a new matrix. The sequence of multiplication matters because matrix multiplication is not commutative.

Given two matrices A and B, where A is of dimensions m × n and B is of dimensions n x p, their multiplication C (also known as the product) is a new matrix of dimensions m x p. The element at the intersection of the i-th row and j-th column of matrix C is calculated using the dot product of the i-th row of A and the j-th column of B.

Here is how matrix multiplication is performed step by step:

1. **Ensure compatibility:** To multiply matrices A and B, the number of columns in A must be equal to the number of rows in B. In other words, for A (m × n) and B (n x p), the "n" value must be the same.
2. **Calculate the dot product:** For each element C_{ij} of the resulting matrix C, perform the dot product of the ith row of A and the jth column of B.

   Let's say C_{ij} is the element at row i and column j in matrix C. Then the dot product is calculated as follows:

   $$C_{ij} = a_{i_1} \times b_{1j} + a_{i_2} \times b_{2j} + \dots + a_{in} \times b_{nj}$$

   Where a_{ik} is the element in row i and column k of matrix A, and b_{kj} is the element in row k and column j of matrix B.
3. **Repeat for all elements:** Repeat step 2 for all elements in matrix C to get the complete product matrix.

   Here's an example of matrix multiplication:

   Matrix A (2 × 3):

   $$A = \begin{pmatrix} 2 & 1 & 3 \\ 4 & 0 & 2 \end{pmatrix}$$

   Matrix B (3 × 2):

   $$B = \begin{pmatrix} 1 & 2 \\ 3 & 5 \\ 0 & 6 \end{pmatrix}$$

   To calculate the product C (2 × 2):

   $$C_{11} = (2 \times 1) + (1 \times 3) + (3 \times 0) = 2 + 3 + 0 = 5$$
   $$C_{12} = (2 \times 2) + (1 \times 5) + (3 \times 6) = 4 + 5 + 18 = 27$$
   $$C_{21} = (4 \times 1) + (0 \times 3) + (2 \times 0) = 4 + 0 + 0 = 4$$
   $$C_{22} = (4 \times 2) + (0 \times 5) + (2 \times 6) = 8 + 0 + 12 = 20$$

   So, the resulting matrix C is:

   $$C = \begin{pmatrix} 5 & 27 \\ 4 & 20 \end{pmatrix}$$

   Solving systems of linear equations, transforming computer images, conducting data analysis, and many other tasks requiring multiple linear relationships between variables all make extensive use of matrix multiplication.

2.4.1 Linear Transformations

A linear transformation function preserves addition and scalar multiplication while mapping vectors from one vector space to another. Matrix-vector multiplication can be used to determine the outcome of applying a linear transformation to a vector because linear transformations can be represented by matrices.

Fundamental ideas in linear algebra, such as linear transformations, are essential to many disciplines, including mathematics, physics, computer science, and engineering. A mapping or function known as a linear transformation maintains the linear structure of vector spaces. In simpler terms, it is a transformation that retains addition and scalar multiplication in vectors.

Formally, a linear transformation T is a function that maps vectors from one vector space V to another vector space W. For this mapping to be considered linear, it must satisfy the following two properties:

1. **Preserving Addition:** T(u + v) = T(u) + T(v)

   According to this principle, the combined transformation of two vectors equals the sum of their individual transformations.

2. **Preserving Scalar Multiplication:** T(ku) = kT(u)

   The transformation of a scalar multiple of a vector is equal to the scalar multiple of the transformation of that vector, according to this property.

In mathematical terms, if V and W are vector spaces, and T is a function from V to W, then T is a linear transformation if, for any vectors u and v in V and any scalar k:

1. T(u + v) = T(u) + T(v)
2. T(ku) = kT(u)

Examples of Linear Transformations

1. A linear transformation is one that shifts each point in 2D or 3D space by a predetermined amount and in a specific direction.
2. **Scaling:** A linear transformation is one that evenly stretches or contracts vectors from a fixed point.
3. **Rotation:** A linear transformation is the rotation of vectors by a fixed angle with respect to the origin.
4. **Projection:** A linear transformation is one that projects vectors onto a certain line or plane.

Properties of Linear Transformations

1. T(0) = 0 indicates that the zero vector is always mapped to the zero vector.
2. When two vectors are combined linearly, their transformations are also combined linearly: T(ku + lv) = kT(u) + lT(v), where k and l are scalars.
3. The transformation of two vectors that differ from one another equals the difference between those transformations: T(u – v) equals T(u) less T(v).

Matrix Representation of Linear Transformations

Matrix representations of linear transformations are common. A linear transition, for instance, can be represented by a 2 × 2 matrix in 2D space and a 3 × 3 matrix in 3D space. After that, a vector's transformation can be calculated by multiplying it by the matching transformation matrix.

In computer graphics, image processing, data analysis, physics, and engineering, linear transformations are widely used to simulate and handle a wide range of physical and material phenomena.

2.5 SYSTEMS OF LINEAR EQUATIONS

A collection of equations, each of which is a linear combination of variables, is referred to as a system of linear equations. AXE = B, where A is the coefficient matrix, X is the vector of variables, and B is the vector of constants, can be used to describe the system as a matrix. Techniques like Gaussian Elimination, LU decomposition, or matrix inverses can be used to find the system's solution.

A set of two or more equations that share common variables and may be expressed as a system collectively form a system of linear equations. Each of the system's equations is a linear equation, which means that there are no products of variables or any nonlinear terms and that each variable is raised to the power of one.

A general system of linear equations can be represented as follows:

$$a_{11}x_1 + a_{12}x_2 + ... + a_{1n}x_n = b_1$$
$$a_{21}x_1 + a_{22}x_2 + ... + a_{2n}x_n = b_2$$
$$\vdots$$
$$a_{m_1}x_1 + a_{m_2}\}x_2 + ... + a_{mn}x_n = b_m$$

where,

- $x_1, x_2, ..., x_n$ are the variables (unknowns) of the system.
- a_{ij} represents the coefficients of the variables in each equation.
- b_i represents the constant terms (right-hand side) of each equation.
- m is the number of equations in the system.
- n is the number of variables in the system.

The goal in solving a system of linear equations is to find the values of the variables $x_1, x_2, ..., x_n$ that satisfy all the equations simultaneously. These values are the solution to the system, and they are called the "roots" or "solutions" of the system.

When a set of linear equations is resolved, one of three things may happen

1. **Unique Solution:** There is only one possible solution for the system, which is a particular set of values for the variables that satisfy every one of the equations.
2. **No Solution:** The system has no solution, which means that there are no variables' values that simultaneously fulfil every equation. The system is described in this situation as "inconsistent."
3. There are infinitely many solutions to the system, which means that there are numerous sets of values for the variables that satisfy all of the equations. The system is described in this situation as "dependent" or "underdetermined."

Methods for Solving Systems of Linear Equations

1. **The Substitution Method:** Using this approach, you answer one equation for a single variable before inserting the solution into the other equations. You keep going through this method until you determine what each variable's value is.
2. **Elimination Method:** The procedure of elimination is adding or removing equations to get rid of one variable before tackling the other variables.
3. **Matrix Methods:** A matrix equation of the form [A][X]=[B] can be used to express a system of linear equations. [A] is the coefficient matrix, [X] is the column vector of variables, and [B] is the column vector of constant terms. The system can be solved using matrix techniques like Cramer's Rule and Gaussian Elimination.

Systems of linear equations are frequently employed because they may be used to represent and analyze interactions between several variables in a linear way. These domains include engineering, physics, economics, and computer science.

2.6 EIGENVALUES AND EIGENVECTORS

An eigenvector is a non-zero vector that, when multiplied by a square matrix A, produces a scalar multiple of the original vector (the Eigen value) for that matrix. In many applications, including stability analysis, principal component analysis, and picture compression, eigenvalues and eigenvectors are essential.

Fundamental ideas in linear algebra, eigenvalues and eigenvectors are used to examine and comprehend linear transformations and matrices. In numerous applications, including as physics, engineering, computer graphics, and data processing, they are essential.

To find the eigenvalues and eigenvectors for a given matrix, let's start with definitions and work our way through the maths.

1. **Eigenvalues:** An eigenvalue of a square matrix A is a scalar λ for which there exists a non-zero vector v such that $Av = \lambda v$.

   Mathematically, if A is an $n \times n$ matrix, λ is an eigenvalue of A if there exists a non-zero vector v (called the eigenvector) such that:

   $$Av = \lambda v$$

2. **Eigenvectors:** Eigenvectors are the non-zero vectors corresponding to the eigenvalues. If λ is an eigenvalue of the matrix A, then v is the corresponding eigenvector.

Now, let's go through the steps to find eigenvalues and eigenvectors for a given matrix A:

Step 1: Given a matrix A, find the characteristic equation. The characteristic equation is obtained by subtracting the identity matrix I multiplied by the scalar λ from matrix A and taking the determinant:

$$\det(A - \lambda I) = 0$$

Step 2: Solve the characteristic equation to find the eigenvalues (λ). Solve the characteristic equation obtained in Step 1 for the eigenvalues (λ). The eigenvalues are the solutions to this equation.

Step 3: For each eigenvalue (λ), find the corresponding eigenvectors. To find the eigenvectors corresponding to each eigenvalue, solve the equation $(A - \lambda I)v = 0$, where v is a non-zero vector.

Let's illustrate the process with an example:

Example: Consider the matrix A: $A = \begin{pmatrix} 3 & 1 \\ 4 & 2 \end{pmatrix}$

Step 1: Find the characteristic equation. $\det(A - \lambda I) = \begin{pmatrix} 3-\lambda & 1 \\ 4 & 2-\lambda \end{pmatrix}$

The characteristic equation is $(3 - \lambda)(2 - \lambda) - 4 = 0$.

Step 2: Solve the characteristic equation to find the eigenvalues. $(3 - \lambda)(2 - \lambda) - 4 = 0$ Expand: $6 - 5\lambda + \lambda^2 - 4 = 0$ Simplify: $\lambda^2 - 5\lambda + 2 = 0$

Now, solve for λ using the quadratic formula or factoring: $\lambda = (5 \pm \sqrt{(5^2 - 412)})/2\lambda = (5 \pm \sqrt{(17)})/2$

So, the eigenvalues are $\lambda = (5 + \sqrt{17})/2$ and $\lambda = (5 - \sqrt{17})/2$.

Step 3: Find the eigenvectors corresponding to each eigenvalue.

For $\lambda = (5 + \sqrt{17})/2$: Solve the equation $(A - \lambda I)v = 0$ to find the eigenvector v.

Substitute λ into $A - \lambda I$: $A - \lambda I = | 3 - (5 + \sqrt{17})/2 \ 1| \ |4 \ 2 - (5 + \sqrt{17})/2|$

Simplify: $A - \lambda I = |-(1 + \sqrt{17})/2 \ 1| \ |4 - (1 + \sqrt{17})/2 |$

Now, solve $(A - \lambda I)v = 0$ to find the eigenvector v: $(-(1 + \sqrt{17})/2)v_1 + v_2 = 0 \ 4v_1 - (1 + \sqrt{17})/2)v_2 = 0$

One possible eigenvector is $v = [1, -(1 + \sqrt{17})/4]$.

For $\lambda = (5 - \sqrt{17})/2$: Follow the same steps as above to find another eigenvector.

The eigenvalues for matrix A are $(5 + \sqrt{17})/2$ and $(5 - \sqrt{17})/2$, and the corresponding eigenvectors can be found using the steps outlined above.

Eigenvalues and eigenvectors are essential tools in solving various problems involving linear transformations, stability analysis, and diagonalization of matrices.

2.7 SINGULAR VALUE DECOMPOSITION (SVD)

Linear algebra and numerical linear algebra both use the fundamental matrix factorization method known as singular value decomposition (SVD). It has several uses in areas like signal processing, data compression, image processing, and machine learning and is applicable to a variety of issues.

Given an $m \times n$ matrix A, the singular value decomposition of A is represented as follows:

$$A = U\Sigma V^T$$

where,

A is the original $m \times n$ matrix.

U is an $m \times m$ orthogonal matrix (the columns are eigenvectors of AA^T).

Σ is an $m \times n$ diagonal matrix with non-negative real numbers (singular values) on the diagonal, arranged in descending order.

V^T is the transpose of an $n \times n$ orthogonal matrix (the columns are eigenvectors of A^TA).

Singular value decomposition (SVD), a fundamental matrix factorization technique, is used in both linear algebra and numerical linear algebra. It is applicable to a wide range of problems and has numerous applications in areas like signal processing, data compression, image processing, and machine learning.

Here's a more detailed explanation of the SVD process

1. Compute the product of A with its transpose (AA^T) $A * A^T = A^T * A$.
2. Find the eigenvalues and eigenvectors of AA^T and A^TA.
3. Arrange the eigenvalues in descending order, and their corresponding eigenvectors will become the columns of matrices U and V.
4. Construct the diagonal matrix Σ using the square root of the sorted eigenvalues as its diagonal elements. The size of Σ depends on the dimensions of A ($m \times n$). If $m > n$, Σ will be $m \times n$ with ($n \times n$) entries set to zero. If $m < n$, Σ will be $m \times n$ with ($m \times m$) entries set to zero.
5. Once you have the matrices U, Σ, and V, you can reconstruct the original matrix A using the formula $A = U\Sigma V^T$.

Because it offers a means to approximate a matrix using only a few important singular values and their accompanying singular vectors, the singular value decomposition is effective. Data compression and dimensionality reduction can benefit from this.

SVD can also be used to solve overdetermined or underdetermined linear systems of equations and to identify the pseudoinverse of a non-square matrix.

However, SVD can be approximated using truncated SVD, which only retains the greatest singular values and their related singular vectors. It should be noted that SVD requires a lot of work for large matrices. In a variety of situations where a complete decomposition is not essential nor practical, an approximation can be helpful.

2.8 ROBOTICS APPLICATIONS IN HOMOGENEOUS TRANSFORMATIONS

In robotics, homogeneous transformations are essential because they offer a mathematical framework for representing and controlling the position and orientation of objects and robot linkages uniformly. A 4 × 4 matrix is used to express homogeneous transformations, which allow for seamless transitions between various coordinate frames. Here are some significant homogeneous transformations in robotics applications.

2.8.1 Forward and Inverse Kinematics

In order to solve the forward and inverse kinematics of robotic manipulators, homogeneous transformations are necessary. The end-effector's pose (position and orientation) must be calculated for the forward kinematics problem using the joint angles. On the other hand, the inverse kinematics problem entails determining the joint angles necessary to accomplish a particular end-effector pose.

(a) ***Forward Kinematics:*** By using the joint angles, forward kinematics calculates the location and orientation (pose) of the robot's end-effector, which is often its hand or tool. It entails calculating the transformation matrix that depicts the end-effector's posture in relation to a reference coordinate frame, usually the robot's base frame.

Let's consider a robot with n degrees of freedom (DOF). Each joint is denoted by q_i, where i ranges from 1 to n. The robot's kinematic model is described by a set of transformation matrices, often represented by Denavit-Hartenberg (DH) parameters. The transformation from one link i to its adjacent link i+1 is given by a 4 × 4 homogeneous transformation matrix, $T\_i^{(i+1)}$. This transformation incorporates both translation and rotation, representing how the coordinate frames of two adjacent links are related.

The sum of all individual link transformations provides the total forward kinematic transformation from the robot's base frame, which is typically at the robot's base, to the end-effector frame, which is at the robot's hand or tool:

$$T\_0^n = T\_1^2 * T\_2^3 * . * T\_\{n-1\}^n$$

Here, T_0^n represents the homogeneous transformation matrix that gives the pose of the end-effector with respect to the robot's base frame. $T\_i^{(i+1)}$ is computed based on the DH parameters and the joint angles q_i.

(b) ***Inverse Kinematics:*** The method of finding a robot's joint angles that would produce a desired end-effector pose is known as inverse kinematics. Since a series of nonlinear equations must be solved in order to determine the joint angles for a given desired pose, it is frequently a more challenging problem than forward kinematics.

For a robot with n DOF, the goal of inverse kinematics is to find the joint angles (q_1, q_2, ..., q_n) that satisfy the equation:

$$T\_0^n(q\_1, q\_2, ..., q\_n) = T\_desired$$

where T_desired is the desired homogeneous transformation representing the desired end-effector pose.

It can be difficult to directly solve this equation, and there frequently won't be a single or closed-form answer. Instead, to estimate the solution, numerical techniques like the Newton-Raphson method, iterative approaches, or optimisation techniques are frequently used.

For path planning and control jobs where the robot must reach a certain end-effector position and orientation, inverse kinematics is crucial. In addition, operations like trajectory creation and robot calibration require the solution of inverse kinematics.

In order for robots to efficiently interact with their surroundings and carry out complicated tasks in a variety of applications, from industrial automation to mobile robotics and beyond, forward and inverse kinematics are essential components of robot modeling and control.

2.8.2 Robot Motion Planning

Motion planning algorithms use homogeneous transformations to plot robot paths that avoid obstacles while moving from a starting location to a goal position. They aid in the transformation of points among several coordinate frames, making path planning and collision avoidance easier.

Finding a feasible and collision-free path for a robot to go from its current location to a specified goal position in a given environment is known as robot motion planning. The representation of the robot's configuration space, obstacle space, and path planning algorithms forms the mathematical basis of motion planning. Let's explore the key mathematical ideas at play:

1. **Configuration Space (C-Space):** The configuration space, commonly referred to as C-Space, is a mathematical space that depicts all potential robot configurations. The C-Space is an n-dimensional space for a robot with n degrees of freedom (DOF), and each point in this space corresponds to a different set of joint angles, where $q = (q_1, q_2, \ldots, q_n)$. Without taking into account any environmental barriers, the C-Space represents every potential robot configuration.
2. **Workspace:** A robot's workspace is the area of actual space where the robot can move around and do tasks. The robot's physical structure, such as the range of motion of its joints and the size of its end-effector, defines it and imposes limitations on it.
3. **Obstacle Space:** The occupied areas of the workspace that the robot needs to avoid are represented by the obstacle space. Depending on how intricate they are, obstacles are frequently depicted as polygons, circles, or other geometric shapes. The workspace is typically subset of the obstacle space.
4. **Path Representation:** A path in the workspace or the C-Space is represented as a series of waypoints, where each waypoint represents a position and orientation in space or a robot configuration (joint angles). The path should avoid hitting any barriers as it connects the start configuration to the goal configuration.
5. **Collision Checking:** Motion planning's crucial stage of collision checking determines if a given configuration or path will clash with any impediments. There are numerous collision detection techniques that can be utilized, including exact geometric intersection tests, bounding box checks, and distance-based checks.
6. **Path Planning Algorithms:** Several algorithms are used for motion planning, each with its mathematical basis.

   Some common algorithms include:

   (*a*) **Sampling-based algorithms (e.g., Probabilistic Roadmap (PRM) and Rapidly-exploring Random Trees (RRT)):** These algorithms build a roadmap or tree-like structure by randomly sampling the C-Space and connecting the samples to form a graph. Path planning is then reduced to finding a path on this graph.

(*b*) **Potential field methods:** These methods define potential fields that attract the robot towards the goal configuration and repel it from obstacles. The robot navigates along the gradient of the potential field to find a path.

(*c*) **Cell decomposition:** The environment is divided into cells or regions, and a graph is constructed to represent connectivity between cells. Path planning is then performed on this graph.

(*d*) **Optimization-based methods:** These methods formulate motion planning as an optimization problem, aiming to find the optimal path based on certain cost functions (e.g., path length, energy consumption) while satisfying constraints (e.g., obstacle avoidance).

7. **Path Smoothing:** After a path is computed by the motion planning algorithm, path smoothing techniques can be applied to improve the path's quality and reduce jerky movements. Smoothing methods adjust the path to make it smoother and more feasible for the robot to follow.

In summary, robot motion planning relies on various mathematical concepts, including configuration space, obstacle representation, path planning algorithms, collision checking, and optimization techniques. By combining these mathematical principles, robots can effectively navigate in complex environments while avoiding obstacles and ensuring smooth and efficient motion.

2.8.3 Robot Control

To provide precise control of robotic manipulators, homogeneous transformations are employed in control systems to convert desired end-effector postures into joint space commands.

Robot control is creating and putting into practise algorithms to control a robot's movement and behaviour in order to accomplish desired goals and trajectories. Robots can follow predetermined courses, maintain stability, and react to external disturbances thanks to principles from control theory and kinematics that are part of the mathematical underpinnings of robot control. Let's get into more detail about the main mathematical elements of robot control:

(*a*) **Kinematics:** The study of a robot's motion without taking into accounts the forces and torques that create it is known as robot kinematics. Kinematics has two primary components that are important for robot control:

The process of establishing the location and orientation (pose) of the robot's end-effector or any other point of interest in its workspace using the joint angles is known as forward kinematics. This is crucial for planning routes and seeing where the robot is in relation to its surroundings.

(*b*) **Inverse kinematics:** The method of figuring out the joint angles necessary to accomplish a specific end-effector pose. It is simpler to describe target positions and orientations when using inverse kinematics to control the robot's movements in Cartesian space.

Control theory offers a mathematical framework for creating the control algorithms that direct the behaviour of the robot. Key ideas from control theory that are applied to robot control include:

A feedback control system uses data regarding the robot's present condition, such as position and velocity, to modify its actions in real-time. Common examples of feedback control algorithms used in robot control to govern the robot's position and velocity are proportional-integral-derivative (PID) controllers.

(*c*) **State-Space Representation:** State-space models, which include state variables like joint angles and velocities and control inputs like motor torques, can be used to depict robots. Model predictive control (MPC) and optimum control are two advanced control systems that can be designed using state-space models.

(*d*) **Stability Analysis:** It is essential to conduct a stability analysis to make sure that the robot's control system is stable in the face of a variety of operating circumstances and disruptions. Control system stability is frequently examined using the Lyapunov stability analysis.

(*e*) Trajectory Tracking: The goal of trajectory tracking control is to accurately direct the robot along a desired course. To monitor a trajectory, model-based and model-free control strategies are employed while taking into account the dynamics and kinematics of the robot.

2.8.4 Robot Dynamics

Studying the interaction between forces, torques, and motion is a component of robot dynamics. Designing control algorithms that can precisely regulate the robot's mobility while taking into account its mass, inertia, and joint actuation requires a thorough understanding of robot dynamics.

In order to determine the equations of motion for a robot, Lagrange-Euler equations are used. These equations relate the robot's joint accelerations, torques, and inertia characteristics.

Dynamic modeling: Using a robot's kinematics, inertia characteristics, and outside forces, dynamic modeling entails creating the equations of motion for the robot. Advanced control strategies that take robot dynamics into consideration can be designed with the aid of dynamic models.

Trajectory Planning and Generation: A key component of robot control is creating viable and slick paths for the robot to go along. Algorithms for trajectory planning and generation try to calculate joint trajectories that accomplish particular goals while observing restrictions like joint limits and avoiding singularities.

In conclusion, a variety of mathematical ideas from dynamics, kinematics, and control theory are used in robot control. These mathematical tools make it possible to create and use control algorithms that regulate the robot's movement and behavior, enabling it to carry out tasks precisely, securely, and effectively.

2.9 HAND-EYE CALIBRATION

Hand eye calibration is required when utilizing robotic arms with cameras or sensors to establish the relationship between the end-effector of the robot and the frame of reference of the camera. This transition is modeled by homogeneous transformations.

1. **Simultaneous Localization and Mapping (SLAM):** SLAM algorithms use homogeneous transformations to describe the pose of the robot in the global map frame while it explores its surroundings. SLAM algorithms are used for simultaneous mapping and localization of robots in new situations.
2. **Localization of Mobile Robots:** To express the location and orientation of a mobile robot in relation to a universal reference frame, homogeneous transformations are used.
3. **Robotic Grasping and Manipulation:** Homogeneous transformations are used to specify the relative position of the gripper with regard to the object that needs to be gripped and moved in robotic grasping and manipulation activities.
4. **Smooth Trajectory Planning:** Homogeneous transformations are utilized to interpolate between poses in Cartesian space while guaranteeing that the robot moves smoothly when smooth trajectory planning is applied to robotic arms or mobile robots.

5. **Augmented and virtual reality:** Homogeneous transformations are utilized to incorporate virtual items or information into the real-world environment from the viewpoint of the robot in AR/VR applications requiring robotics.

In conclusion, homogeneous transformations are essential for many elements of robotics since they allow for accurate representations of poses and coordinate frame transformations. Because of their adaptability, they are a crucial tool for many applications in

2.10 JACOBIAN MATRIX FOR MANIPULATOR ANALYSIS

An essential tool in the analysis of robotic manipulators is the Jacobian matrix. It links the speed of the robot's joint angles to the speed of its end-effector, which is typically a tool or hand. Motion planning, control, and inverse kinematics all require an understanding of how changes in joint angles affect the position and orientation of the end-effector, which is made possible by the Jacobian.

Let's consider a robot with n degrees of freedom (DOF), and its configuration is described by the vector of joint angles $q = [q_1, q_2, ..., q_n]$. The end-effector's position and orientation in Cartesian space can be represented by a vector $x = [x, y, z, \alpha, \beta, \gamma]$, where (x, y, z) are the position coordinates, and (α, β, γ) are the orientation angles (e.g., roll, pitch, yaw).

The Jacobian matrix J relates the end-effector's velocities $\dot{x}, \dot{y}, \dot{z}$ and angular velocities $\dot{\alpha}, \dot{\beta}, \dot{\gamma}$ to the joint velocities $\dot{q}_1, \dot{q}_2, ... \dot{q}_n$, as follows:

$$\begin{bmatrix} \dot{x} \\ \dot{y} \\ \dot{z} \\ \dot{\alpha} \\ \dot{\beta} \\ \dot{\gamma} \end{bmatrix} = J \begin{bmatrix} \dot{q}_1 \\ \dot{q}_2 \\ \vdots \\ \dot{q}_n \end{bmatrix}$$

The Jacobian matrix is generally a 6 × n matrix (for a 6-DOF end-effector), where each row represents the velocity components of the end-effector (linear velocities and angular velocities) and each column corresponds to the joint velocities.

The elements of the Jacobian matrix depend on the robot's kinematic structure. For a manipulator with a serial chain, the elements can be computed as follows:

1. **Linear Velocities:** The first three rows of the Jacobian correspond to the linear velocities (dx, dy, dz) of the end-effector. These elements can be calculated based on the derivative of the end-effector position with respect to each joint angle. Typically, the forward kinematics equations are used to compute the end-effector's position in terms of the joint angles, and then the partial derivatives are taken.
2. **Angular Velocities:** The last three rows of the Jacobian correspond to the angular velocities ($d\alpha$, $d\beta$, $d\gamma$) of the end-effector. These elements can be computed similarly, using the derivative of the end-effector orientation with respect to each joint angle.

The Jacobian is not a special case; it may be modeled in other reference frames, including the base frame and the end-effector frame. The understanding and implementation of the Jacobian in particular applications may be impacted by the reference frame selection.

Numerous robot analysis and control tasks, such as inverse kinematics, motion planning, force controls, and singularity analysis, make use of the Jacobian matrix. It helps to optimize and regulate robotic manipulators by illuminating key insights regarding the link between joint velocities and end-effector velocities.

NUMERICAL PROBLEMS WITH SOLUTIONS

Problem 1. Find the dot product of vectors a = [1, 2, 3] and b = [4, 5, 6].

Solution: The dot product is given by $a \cdot b = 1 \times 4 + 2 \times 5 + 3 \times 6 = 32$.

Problem 2. Calculate the cross product of vectors c = [2, 3, 4] and d = [5, 6, 7].

Solution: The cross product is $c \times d = [(3 \times 7 - 4 \times 6), (4 \times 5 - 2 \times 7), (2 \times 6 - 3 \times 5)] = [-3, 6, -3]$.

Problem 3. Find the inverse of the matrix A = [1 2; 3 4].

Solution: The inverse of a 2 × 2 matrix is given by $A^{(-1)} = \begin{pmatrix} -2 & 1 \\ 3/2 & -1/2 \end{pmatrix}$.

Problem 4. Solve the system of linear equations: 2x + 3y = 8 and 4x + 6y = 12.

Solution: Divide the second equation by 2 to simplify: $2x + 3y = 8$ and $2x + 3y = 6$. Since the two equations are the same, the system has infinitely many solutions (dependent system).

Problem 5. Find the rank of the matrix $B = \begin{bmatrix} 1 & 2 & 3 \\ 4 & 5 & 6 \\ 7 & 8 & 9 \end{bmatrix}$.

Solution: The rank of a matrix is the number of linearly independent rows or columns. In this case, the third row can be obtained by adding the first and second rows, so the rank is 2.

Problem 6. Calculate the eigenvalues and eigenvectors of the matrix $C = \begin{bmatrix} 3 & 1 \\ 2 & 4 \end{bmatrix}$.

Solution: To find the eigenvalues, solve the characteristic equation:

(*i*) $\lambda_1 = \lambda_2 = 2$

(*ii*) For $\lambda_1 = 5$, $v_1 = \begin{pmatrix} 1 \\ 2 \end{pmatrix}$

(*iii*) For $\lambda_2 = 2$, $v_2 = \begin{pmatrix} 1 \\ -1 \end{pmatrix}$

Problem 7. Find the determinant of the matrix D = [2 0 0; 0 3 0; 0 0 4].

Solution: The determinant of a diagonal matrix is the product of its diagonal elements. det(D) = 2 * 3 * 4 = 24.

Problem 8. Compute the orthogonal projection of vector u = [3, 2] onto the vector v = [1, 1].

Solution: The orthogonal projection of u onto v is given by $\text{proj}_v(u) = (u \cdot v/|v|^2) = [5/2, 5/2]$.

Problem 9. Diagonalize the matrix $E = \begin{bmatrix} 4 & 1 \\ 6 & 3 \end{bmatrix}$.

Solution: Eigen values: $\lambda_1 = 6, \lambda_2 = 1$.

Eigen vectors: $v_1 = \begin{bmatrix} 1 \\ 2 \end{bmatrix}, v_2 = \begin{bmatrix} 1 \\ -3 \end{bmatrix}$.

Matrix P: $P = \begin{bmatrix} 1 & 1 \\ 2 & -3 \end{bmatrix}$

Diagonal matrix Λ: $\Lambda = \begin{bmatrix} 6 & 0 \\ 0 & 1 \end{bmatrix}$

Problem 10. Given the matrix $F = \begin{bmatrix} 2 & 1 \\ 3 & 4 \end{bmatrix}$, find F^2.

Solution: $F^2 = F * F = \begin{bmatrix} 7 & 6 \\ 18 & 19 \end{bmatrix}$.

Problem 11. Find the determinant of the matrix $G = \begin{bmatrix} 1 & 2 & 3 \\ 0 & 4 & 5 \\ 0 & 0 & 6 \end{bmatrix}$.

Solution: The determinant of a triangular matrix is the product of its diagonal elements. det(G) = 1 * 4 * 6 = 24.

Problem 12. Calculate the trace of the matrix $H = \begin{bmatrix} 5 & 1 & 2 \\ 3 & 4 & 6 \\ 0 & 7 & 8 \end{bmatrix}$.

Solution: The trace of a matrix is the sum of its diagonal elements. Trace(H) = 5 + 4 + 8 = 17.

Problem 13. Solve the system of linear equations: $x - y + 2z = 4$, $2x + y + 3z = 10$ and $3x + 2y + z = 7$.

Solution: One way to solve the system is by using matrix operations. Represent the system as AX = B, where A is the coefficient matrix, X is the column vector of variables (x, y, z), and B is the column vector of constants on the right-hand side. A = [1 –1 2; 2 1 3; 3 2 1] X = [x; y; z] B = [4; 10; 7] Now, solve for X using the equation AX = B.

$$X = A^{(-1)} * B, \text{ where } A^{(-1)} \text{ is the inverse of matrix A.}$$

Calculate the inverse of matrix A: $A^{(-1)} = (1/\det(A)) * \operatorname{adj}(A)$.

The final solution is x = 1, y = 2 and z = 3.

Problem 14. Find the row echelon form of the matrix $I = \begin{bmatrix} 1 & 2 & 3 \\ 0 & 1 & 4 \\ 5 & 6 & 0 \end{bmatrix}$.

Solution: To convert a matrix to row echelon form, perform elementary row operations until the matrix satisfies the following conditions:

1. All nonzero rows are above any rows of all zeros.
2. The leading entry of each nonzero row is always to the right of the leading entry of the row above it.
3. All entries below a leading entry are zeros. Perform row operations: $R_2 = R_2 - 5R_1$ (Multiply the first row by 5 and subtract from the second row). $R_3 = R_3 - 5R_1$ (Multiply the first row by 5 and subtract from the third row). I = [1 2 3; 0 1 4; 5 6 0], R_2 = [0 –3 –11; 0 1 4; 5 6 0], R_3 = [0 –3 –11;

0 1 4; 0 –4 –15], $R_3 = R_3 + R_2$ (Add the second row to the third row). I = [1 2 3; 0 1 4; 0 0 –11] Now, the matrix is in row echelon form.

Problem 15. Compute the matrix product of J = $\begin{bmatrix} 2 & 1 \\ 3 & 4 \end{bmatrix}$ **and K =** $\begin{bmatrix} 5 & 6 \\ 7 & 8 \end{bmatrix}$.

Solution: The matrix product is given by J * K = [25 + 17 26 + 18]; [35 + 47 36 + 48] = [19 22 38 46].

Problem 16. Find the determinant of the matrix L = $\begin{bmatrix} 1 & 2 & 3 \\ 4 & 5 & 6 \\ 7 & 8 & 9 \end{bmatrix}$.

Solution: The determinant of a 3 × 3 matrix can be calculated using the rule of Sarrus or by using cofactor expansion along the first row. Using Sarrus rule: det(L) = 1*(59 – 68) – 2*(49 – 67) + 3*(48 – 57) = 1*(–3) – 2*(–6) + 3*(–3) = 0. Using cofactor expansion: det(L) = 1*det*([5 6; 8 9]) – 2det([4 6; 7 9]) + 3*det*([4 5; 7 8]) = 1(59 – 68) – 2*(49 – 67) + 3*(48 – 57) = 0. Both methods give the same result: det(L) = 0.

Problem 17. Calculate the rank of the matrix M = $\begin{bmatrix} 1 & 2 & 3 \\ 4 & 5 & 6 \\ 7 & 8 & 9 \end{bmatrix}$.

Solution: To find the rank, convert the matrix to row echelon form and count the number of nonzero rows. M = [1 2 3; 4 5 6; 7 8 9], $R_2 = R_2 - 4R_1$ (Multiply the first row by 4 and subtract from the second row). $R_3 = R_3 - 7R_1$ (Multiply the first row by 7 and subtract from the third row). M = [1 2 3; 0 –3 –6; 0 –6 –12], $R_3 = R_3 - 2R_2$ (Multiply the second row by 2 and subtract from the third row). M = [1 2 3; 0 –3 –6; 0 0 0] The matrix is in row echelon form, and the rank is the number of nonzero rows, which is 2.

Problem 18. Compute the inverse of the matrix N = $\begin{bmatrix} 3 & 0 \\ 0 & 2 \end{bmatrix}$.

Solution: The inverse of a diagonal matrix is obtained by taking the reciprocal of its diagonal elements. $N^{(-1)} = \begin{bmatrix} 1/3 & 0 \\ 0 & 1/2 \end{bmatrix}$.

Problem 19. Solve the system of linear equations: x + y + z = 6, 2x + 3y + 4z = 20 and 3x + 2y + z = 10.
Solution: Represent the system as AX = B, where A is the coefficient matrix, X is the column vector of variables (x, y, z), and B is the column vector of constants on the right-hand side. A = [1 1 1; 2 3 4; 3 2 1] X = [x; y; z], B = [6; 20; 10]. Now, solve for X using the equation AX = B. X = $A^{(-1)}$ * B, where $A^{(-1)}$ is the inverse of matrix A. Calculate the inverse of matrix A: $A^{(-1)}$ = (1/det(A)) * adj(A), where adj(A) is the adjugate (adjoint) of A, and det(A) is the determinant of A. det(A) = 1*(31 – 42) – 1*(21 – 43) + 1*(22 – 33) = 1 – 2 + 1 = 0 Since det(A) is zero, the matrix A is singular and does not have an inverse. Therefore, the system of equations does not have a unique solution.

Problem 20. Find the eigenvalues and eigenvectors of the matrix O = $\begin{bmatrix} 5 & 2 \\ 4 & 1 \end{bmatrix}$.

Solution: To find the eigenvalues, solve the characteristic equation det(O – λI) = 0: det([5-λ 2; 4 1-λ]) = (5-λ)(1-λ) – 2*4 = $\lambda^2 - 6\lambda - 7 = 0$. Solve for λ: (λ – 7)(λ + 1) = 0. So, the eigenvalues are $\lambda_1 = 7$ and $\lambda_2 = -1$. To find the eigenvectors, plug each eigenvalue into the equation (O – λI)v = 0 and solve for v: For λ_1

= 7: [5–7 2; 4 1–7]v = [–2 2; 4 –6]v = 0. The eigenvector corresponding to $\lambda_1 = 7$ is v_1 = [1 1]. For $\lambda_2 = -1$: [5+1 2; 4 1+1]v = [6 2; 4 2]v = 0. The eigenvector corresponding to $\lambda_2 = -1$ is v_2 = [–1 2].

Problem 21. Given the matrix P = [3 1 2; 2 4 6; 1 3 5], find P² (P squared).

Solution: $P^2 = P * P = \begin{bmatrix} 3 & 1 & 2 \\ 2 & 4 & 6 \\ 1 & 3 & 5 \end{bmatrix} \times \begin{bmatrix} 3 & 1 & 2 \\ 2 & 4 & 6 \\ 1 & 3 & 5 \end{bmatrix} = \begin{bmatrix} 14 & 20 & 28 \\ 20 & 56 & 82 \\ 14 & 38 & 56 \end{bmatrix}$.

Problem 22. Find the determinant of the matrix $Q = \begin{bmatrix} 2 & 1 & 3 \\ 4 & 5 & 6 \\ 7 & 8 & 9 \end{bmatrix}$.

Solution: The determinant of a 3 × 3 matrix can be calculated using the rule of Sarrus or by using cofactor expansion along the first row. Using Sarrus rule: det(Q) = 2 * (59 – 68) – 1 * (49 – 67) + 3 * (48 – 57) = 23 – 16 + 3 * (–3) = 6 – 6 – 9 = –9. Using cofactor expansion: det(Q) = 2*det*([5 6; 8 9]) – 1det([4 6; 7 9]) + 3*det*([4 5; 7 8]) = 2(59 – 68) – (49 – 67) + 3*(48 – 57) = –9. Both methods give the same result: det(Q) = –9.

Problem 23. Calculate the trace of the matrix $R = \begin{bmatrix} 5 & 1 & 2 \\ 3 & 4 & 6 \\ 0 & 7 & 8 \end{bmatrix}$.

Solution: The trace of a matrix is the sum of its diagonal elements. Trace(R) = 5 + 4 + 8 = 17.

Problem 24. Solve the system of linear equations: x – y + 2z = 4, 2x + 3y + 4z = 10 and 3x + 2y + z = 7.

Solution: One way to solve the system is by using matrix operations. Represent the system as AX = B, where A is the coefficient matrix, X is the column vector of variables (x, y, z), and B is the column vector of constants on the right-hand side. A = [1 –1 2; 2 3 4; 3 2 1], X = [x; y; z], B = [4; 10; 7]. Now, solve for X using the equation AX = B.

X = A(–1) * B, where A(–1) is the inverse of matrix A.

Calculate the inverse of matrix A: A(–1) = (1/det(A)) * adj(A), where adj(A) is the adjugate (adjoint) of A, and det(A) is the determinant of A.

Now, solve it for the final result.

Problem 25. Find the row echelon form of the matrix S = [1 2 3; 0 1 4; 5 6 0].

Solution: To convert a matrix to row echelon form, perform elementary row operations until the matrix satisfies the following conditions:

1. All nonzero rows are above any rows of all zeros.
2. The leading entry of each nonzero row is always to the right of the leading entry of the row above it.
3. All entries below a leading entry are zeros. Perform row operations: $R_2 = R_2 - 5R_1$ (Multiply the first row by 5 and subtract from the second row). $R_3 = R_3 - 5R_1$ (Multiply the first row by 5 and subtract from the third row). S = [1 2 3; 0 –3 –11; 0 –3 –11]. $R_3 = R_3 + R_2$ (Add the second row to the third row). S = [1 2 3; 0 –3 –11; 0 0 –22] Now, the matrix is in row echelon form.

Problem 26. Compute the matrix product of $T = \begin{bmatrix} 2 & 1 \\ 3 & 4 \end{bmatrix}$ **and** $U = \begin{bmatrix} 5 & 6 \\ 7 & 8 \end{bmatrix}$.

Solution: The matrix product is given by T * U = [25 + 17   26 + 18; 35 + 47   36 + 48] = [19 22; 38 46].

Problem 27. Find the determinant of the matrix $V = \begin{bmatrix} 1 & 2 & 3 \\ 4 & 5 & 6 \\ 7 & 8 & 9 \end{bmatrix}$**.**

Solution: The determinant of a 3 × 3 matrix can be calculated using the rule of Sarrus or by using cofactor expansion along the first row. Using Sarrus rule: det(V) = 1*(59 – 68) – 2*(49 – 67) + 3*(48 – 57) = 1*(–3) – 2*(–6) + 3*(–3) = 0. Using cofactor expansion: det(V) = 1*det*([5 6; 8 9]) – 2det([4 6; 7 9]) + 3*det*([4 5; 7 8]) = 1(59 – 68) – 2*(49 – 67) + 3*(48 – 57) = 0. Both methods give the same result: det(V) = 0.

Problem 28. Calculate the trace of the matrix $W = \begin{bmatrix} 5 & 1 & 2 \\ 3 & 4 & 6 \\ 0 & 7 & 8 \end{bmatrix}$**.**

Solution: The trace of a matrix is the sum of its diagonal elements. Trace(W) = 5 + 4 + 8 = 17.

Problem 29. Solve the system of linear equations: x – y + 2z = 4, 2x + 3y + 4z = 10 and 3x + 2y + z = 7.

Solution: One way to solve the system is by using matrix operations. Represent the system as AX = B, where A is the coefficient matrix, X is the column vector of variables (x, y, z), and B is the column vector of constants on the right-hand side. A = [1 –1 2; 2 3 4; 3 2 1], X = [x; y; z], B = [4; 10; 7]. Now, solve for X using the equation AX = B. X = $A^{(-1)}$ * B, where $A^{(-1)}$ is the inverse of matrix A. Calculate the inverse of matrix A: $A^{(-1)}$ = (1/det(A)) * adj(A), where adj(A) is the adjugate (adjoint) of A, and det(A) is the determinant of A. Now, solve it for the final result.

Problem 30. Find the row echelon form of the matrix $X = \begin{bmatrix} 1 & 2 & 3 \\ 0 & 1 & 4 \\ 5 & 6 & 0 \end{bmatrix}$**.**

Solution: To convert a matrix to row echelon form, perform elementary row operations until the matrix satisfies the following conditions:

1. All nonzero rows are above any rows of all zeros.
2. The leading entry of each nonzero row is always to the right of the leading entry of the row above it.
3. All entries below a leading entry are zeros. Perform row operations: $R_2 = R_2 - 5R_1$ (Multiply the first row by 5 and subtract from the second row). $R_3 = R_3 - 5R_1$ (Multiply the first row by 5 and subtract from the third row). X = [1 2 3; 0 –3 –11; 0 –3 –11]. $R_3 = R_3 + R_2$ (Add the second row to the third row). X = [1 2 3; 0 –3 –11; 0 0 –22]. Now, the matrix is in row echelon form.

Problem 31. Compute the matrix product of $Y = \begin{bmatrix} 2 & 1 \\ 3 & 4 \end{bmatrix}$ **and** $Z = \begin{bmatrix} 5 & 6 \\ 7 & 8 \end{bmatrix}$**.**

Solution: The matrix product is given by Y * Z = [25 + 17 26 + 18; 35 + 47 36 + 48] = [19 22; 38 46].

Problem 32. Find the determinant of the matrix $AA = \begin{bmatrix} 1 & 2 & 3 \\ 4 & 5 & 6 \\ 7 & 8 & 9 \end{bmatrix}$**.**

Solution: The determinant of a 3 × 3 matrix can be calculated using the rule of Sarrus or by using cofactor expansion along the first row. Using Sarrus rule: det(AA) = 1*(59 – 68) – 2*(49 – 67) + 3*(48

– 57) = 1*(–3) – 2*(–6) + 3*(–3) = 0. Using cofactor expansion: det(AA) = 1*det*([5 6; 8 9]) – 2det([4 6; 7 9]) + 3*det*([4 5; 7 8]) = 1(59 – 68) – 2*(49 – 67) + 3*(48 – 57) = 0. Both methods give the same result: det(AA) = 0.

Problem 33. Calculate the trace of the matrix BB = $\begin{bmatrix} 5 & 1 & 2 \\ 3 & 4 & 6 \\ 0 & 7 & 8 \end{bmatrix}$.

Solution: The trace of a matrix is the sum of its diagonal elements. Trace(BB) = 5 + 4 + 8 = 17.

Problem 34. Solve the system of linear equations: x – y + 2z = 4, 2x + 3y + 4z = 10 and 3x + 2y + z = 7.

Solution: One way to solve the system is by using matrix operations. Represent the system as AX = B, where A is the coefficient matrix, X is the column vector of variables (x, y, z), and B is the column vector of constants on the right-hand side. A = [1 –1 2; 2 3 4; 3 2 1], X = [x; y; z], B = [4; 10; 7]. Now, solve for X using the equation AX = B. X = $A^{(-1)}$ * B, where $A^{(-1)}$ is the inverse of matrix A. Calculate the inverse of matrix A: $A^{(-1)}$ = (1/det(A)) * adj(A), where adj(A) is the adjugate (adjoint) of A, and det(A) is the determinant of A. Now, solve it for the final result.

Problem 35. Find the row echelon form of the matrix CC = $\begin{bmatrix} 1 & 2 & 3 \\ 0 & 1 & 4 \\ 5 & 6 & 0 \end{bmatrix}$.

Solution: To convert a matrix to row echelon form, perform elementary row operations until the matrix satisfies the following conditions:

1. All nonzero rows are above any rows of all zeros.
2. The leading entry of each nonzero row is always to the right of the leading entry of the row above it.
3. All entries below a leading entry are zeros. Perform row operations: $R_2 = R_2 - 5R_1$ (Multiply the first row by 5 and subtract from the second row). $R_3 = R_3 - 5R_1$ (Multiply the first row by 5 and subtract from the third row). CC = [1 2 3; 0 –3 –11; 0 –3 –11]. $R_3 = R_3 + R_2$ (Add the second row to the third row). CC = [1 2 3; 0 –3 –11; 0 0 –22]. Now, the matrix is in row echelon form.

Problem 36. Compute the matrix product of DD = $\begin{bmatrix} 2 & 1 \\ 3 & 4 \end{bmatrix}$ **and EE =** $\begin{bmatrix} 5 & 6 \\ 7 & 8 \end{bmatrix}$.

Solution: The matrix product is given by DD * EE = [25 + 17 26 + 18; 35 + 47 36 + 48] = [19 22; 38 46].

Problem 37. Find the determinant of the matrix FF = $\begin{bmatrix} 1 & 2 & 3 \\ 4 & 5 & 6 \\ 7 & 8 & 9 \end{bmatrix}$.

Solution: The determinant of a 3 × 3 matrix can be calculated using the rule of Sarrus or by using cofactor expansion along the first row. Using Sarrus rule: det(FF) = 1*(59 – 68) – 2*(49 – 67) + 3*(48 – 57) = 1*(–3) – 2*(–6) + 3*(–3) = 0. Using cofactor expansion: det(FF) = 1*det*([5 6; 8 9]) – 2det([4 6; 7 9]) + 3*det*([4 5; 7 8]) = 1(59 – 68) – 2*(49 – 67) + 3*(48 – 57) = 0. Both methods give the same result: det(FF) = 0.

Problem 38. Calculate the trace of the matrix GG = $\begin{bmatrix} 5 & 1 & 2 \\ 3 & 4 & 6 \\ 0 & 7 & 8 \end{bmatrix}$.

Solution: The trace of a matrix is the sum of its diagonal elements. Trace(GG) = 5 + 4 + 8 = 17.

Problem 39. Solve the system of linear equations: x – y + 2z = 4, 2x + 3y + 4z = 10 and 3x + 2y + z = 7.

Solution: One way to solve the system is by using matrix operations. Represent the system as AX = B, where A is the coefficient matrix, X is the column vector of variables (x, y, z), and B is the column vector of constants on the right-hand side. A = [1 –1 2; 2 3 4; 3 2 1], X = [x; y; z], B = [4; 10; 7]. Now, solve for X using the equation AX = B. X = $A^{(-1)}$ * B, where $A^{(-1)}$ is the inverse of matrix A. Calculate the inverse of matrix A: $A^{(-1)}$ = (1/det(A)) * adj(A), where adj(A) is the adjugate (adjoint) of A, and det(A) is the determinant of A. Now, solve it for the final result.

Problem 40. Find the row echelon form of the matrix HH = $\begin{bmatrix} 1 & 2 & 3 \\ 0 & 1 & 4 \\ 5 & 6 & 0 \end{bmatrix}$.

Solution: To convert a matrix to row echelon form, perform elementary row operations until the matrix satisfies the following conditions:

1. All nonzero rows are above any rows of all zeros.
2. The leading entry of each nonzero row is always to the right of the leading entry of the row above it.
3. All entries below a leading entry are zeros. Perform row operations: $R_2 = R_2 - 5R_1$ (Multiply the first row by 5 and subtract from the second row). $R_3 = R_3 - 5R_1$ (Multiply the first row by 5 and subtract from the third row). HH = [1 2 3; 0 –3 –11; 0 –3 –11]. $R_3 = R_3 + R_2$ (Add the second row to the third row). HH = [1 2 3; 0 –3 –11; 0 0 –22]. Now, the matrix is in row echelon form.

Problem 41. Compute the matrix product of II = $\begin{bmatrix} 2 & 1 \\ 3 & 4 \end{bmatrix}$ **and JJ =** $\begin{bmatrix} 5 & 6 \\ 7 & 8 \end{bmatrix}$.

Solution: The matrix product is given by II * JJ = [25 + 17 26 + 18; 35 + 47 36 + 48] = [19 22; 38 46].

Problem 42. Find the determinant of the matrix KK = $\begin{bmatrix} 1 & 2 & 3 \\ 4 & 5 & 6 \\ 7 & 8 & 9 \end{bmatrix}$.

Solution: The determinant of a 3 × 3 matrix can be calculated using the rule of Sarrus or by using cofactor expansion along the first row. Using Sarrus rule: det(KK) = 1*(59 – 68) – 2*(49 – 67) + 3*(48 – 57) = 1*(–3) – 2*(–6) + 3*(–3) = 0. Using cofactor expansion: det(KK) = 1*det*([5 6; 8 9]) – 2det([4 6; 7 9]) + 3*det*([4 5; 7 8]) = 1(59 – 68) – 2*(49 – 67) + 3*(48 – 57) = 0. Both methods give the same result: det(KK) = 0.

Problem 43. Calculate the trace of the matrix LL = $\begin{bmatrix} 5 & 1 & 2 \\ 3 & 4 & 6 \\ 0 & 7 & 8 \end{bmatrix}$.

Solution: The trace of a matrix is the sum of its diagonal elements. Trace(LL) = 5 + 4 + 8 = 17.

Problem 44. Solve the system of linear equations: x – y + 2z = 4, 2x + 3y + 4z = 10 and 3x + 2y + z = 7.

Solution: One way to solve the system is by using matrix operations. Represent the system as AX = B, where A is the coefficient matrix, X is the column vector of variables (x, y, z), and B is the column vector of constants on the right-hand side. A = [1 –1 2; 2 3 4; 3 2 1], X = [x; y; z], B = [4; 10; 7]. Now, solve for X using the equation AX = B. X = $A^{(-1)}$ * B, where $A^{(-1)}$ is the inverse of matrix A. Calculate the inverse of matrix A: $A^{(-1)}$ = (1/det(A)) * adj(A), where adj(A) is the adjugate (adjoint) of A, and det(A) is the determinant of A. Now, solve it for the final result.

Problem 45. Find the row echelon form of the matrix MM = $\begin{bmatrix} 1 & 2 & 3 \\ 0 & 1 & 4 \\ 5 & 6 & 0 \end{bmatrix}$.

Solution: To convert a matrix to row echelon form, perform elementary row operations until the matrix satisfies the following conditions:

1. All nonzero rows are above any rows of all zeros.
2. The leading entry of each nonzero row is always to the right of the leading entry of the row above it.
3. All entries below a leading entry are zeros. Perform row operations: $R_2 = R_2 - 5R_1$ (Multiply the first row by 5 and subtract from the second row). $R_3 = R_3 - 5R_1$ (Multiply the first row by 5 and subtract from the third row). MM = [1 2 3; 0 –3 –11; 0 –3 –11]. $R_3 = R_3 + R_2$ (Add the second row to the third row). MM = [1 2 3; 0 –3 –11; 0 0 –22]. Now, the matrix is in row echelon form.

Problem 46. Compute the matrix product of NN = $\begin{bmatrix} 2 & 1 \\ 3 & 4 \end{bmatrix}$ **and OO =** $\begin{bmatrix} 5 & 6 \\ 7 & 8 \end{bmatrix}$.

Solution: The matrix product is given by NN * OO = [25 + 17 26 + 18; 35 + 47 36 + 48] = [19 22; 38 46].

Problem 47. Find the determinant of the matrix PP = $\begin{bmatrix} 1 & 2 & 3 \\ 4 & 5 & 6 \\ 7 & 8 & 9 \end{bmatrix}$.

Solution: The determinant of a 3 × 3 matrix can be calculated using the rule of Sarrus or by using cofactor expansion along the first row. Using Sarrus rule: det(PP) = 1*(59 – 68) – 2*(49 – 67) + 3*(48 – 57) = 1*(–3) – 2*(–6) + 3*(–3) = 0. Using cofactor expansion: det(PP) = 1*det*([5 6; 8 9]) – 2det([4 6; 7 9]) + 3*det*([4 5; 7 8]) = 1(59 – 68) – 2*(49 – 67) + 3*(48 – 57) = 0. Both methods give the same result: det(PP) = 0.

Problem 48. Calculate the trace of the matrix QQ = $\begin{bmatrix} 5 & 1 & 2 \\ 3 & 4 & 6 \\ 0 & 7 & 8 \end{bmatrix}$.

Solution: The trace of a matrix is the sum of its diagonal elements. Trace(QQ) = 5 + 4 + 8 = 17.

Problem 49. Solve the system of linear equations: x – y + 2z = 4, 2x + 3y + 4z = 10 and 3x + 2y + z = 7.

Solution: One way to solve the system is by using matrix operations. Represent the system as AX = B, where A is the coefficient matrix, X is the column vector of variables (x, y, z), and B is the column vector of constants on the right-hand side. A = [1 –1 2; 2 3 4; 3 2 1], X = [x; y; z], B = [4; 10; 7]. Now, solve for X using the equation AX = B. X = $A^{(-1)}$ * B, where $A^{(-1)}$ is the inverse of matrix A. Calculate the inverse of matrix A: $A^{(-1)} = (1/\det(A)) * \mathrm{adj}(A)$, where adj(A) is the adjugate (adjoint) of A, and det(A) is the determinant of A. Now, solve it for the final result.

Problem 50. Find the row echelon form of the matrix RR = [1 2 3; 0 1 4; 5 6 0].

Solution: To convert a matrix to row echelon form, perform elementary row operations until the matrix satisfies the following conditions:

1. All nonzero rows are above any rows of all zeros.
2. The leading entry of each nonzero row is always to the right of the leading entry of the row above it.

   All entries below a leading entry are zeros. Perform row operations: $R_2 = R_2 - 5R_1$ (Multiply the first row by 5 and subtract from the second row). $R_3 = R_3 - 5R_1$ (Multiply the first row by 5 and subtract from the third row). RR – [1 2 3; 0 –3 –11; 0 –3 –11]. $R_3 = R_3 + R_2$ (Add the second row to the third row). RR = [1 2 3; 0 –3 –11; 0 0 –22]. Now, the matrix is in row echelon form.

Short Questions and Answers

Q.1. What is a vector?

Ans. A vector is a quantity that has both magnitude and direction.

Q.2. What is a matrix?

Ans. A matrix is a rectangular array of numbers, symbols, or expressions.

Q.3. What is a scalar?

Ans. A scalar is a single number or constant.

Q.4. What is a linear equation?

Ans. A linear equation is an equation where each variable has an exponent of 1 and no cross-product terms.

Q.5. What is a linear transformation?

Ans. A linear transformation is a function that preserves the linear structure of vector spaces.

Q.6. What is the dot product of two vectors?

Ans. The dot product of two vectors is the product of their magnitudes and the cosine of the angle between them.

Q.7. What is the cross product of two vectors?

Ans. A cross product of two vectors is a vector that is perpendicular to both input vectors.

Q.8. What is the rank of a matrix?

Ans. The rank of a matrix is the maximum number of linearly independent rows or columns.

Q.9. What are eigenvalues and eigenvectors?

Ans. Eigenvalues are scalars, and eigenvectors are corresponding non-zero vectors that satisfy the equation $Av = \lambda v$.

Q.10. What is the determinant of a matrix?

Ans. A determinant of a matrix is a scalar value representing certain properties of the matrix.

Q.11. What is the inverse of a matrix?

Ans. The inverse of a matrix A is denoted as $A^{(-1)}$ and satisfies $A * A^{(-1)} = I$, where I is the identity matrix.

Q.12. What is the identity matrix?

Ans. The identity matrix is a square matrix with 1s on the main diagonal and 0s elsewhere.

Q.13. What is the transpose of a matrix?

Ans. The transpose of a matrix flips its rows and columns.

Q.14. What is a homogeneous system of linear equations?

Ans. A homogeneous system is a system of linear equations with all constant terms equal to zero.

Q.15. What is a non-homogeneous system of linear equations?

Ans. Non-homogeneous system is a system of linear equations with non-zero constant terms.

Q.16. What are the three possible outcomes when solving a system of linear equations?

Ans. The outcomes are: unique solution, no solution, or infinitely many solutions.

Q.17. What is the span of a set of vectors?

Ans. The span of a set of vectors is the set of all possible linear combinations of those vectors.

Q.18. What is linear independence?

Ans. A set of vectors is linearly independent if no vector in the set can be expressed as a linear combination of the others.

Q.19. What is the null space of a matrix?

Ans. The null space (kernel) of a matrix A is the set of all vectors that, when multiplied by A, result in the zero vector.

Q.20. What is the column space of a matrix?

Ans. The column space of a matrix A is the span of its column vectors.

Q.21. What is the row space of a matrix?

Ans. The row space of a matrix A is the span of its row vectors.

Q.22. What is the orthogonal complement?

Ans. The orthogonal complement of a subspace is the set of all vectors orthogonal to that subspace.

Q.23. What is the Gram-Schmidt process?

Ans. The Gram-Schmidt process is a method for orthogonalizing a set of vectors.

Q.24. What is a basis for a vector space?

Ans. A basis for a vector space is a set of linearly independent vectors that span the entire space.

Q.25. What is the dimension of a vector space?

Ans. The dimension of a vector space is the number of vectors in its basis.

Q.26. What is the cross product used for?

Ans. The cross product is used to find a vector that is perpendicular to two given vectors.

Q.27. What is the dot product used for?

Ans. The dot product is used to find the angle between two vectors and to project a vector onto another.

Q.28. What is the row echelon form of a matrix?

Ans. The row echelon form of a matrix is an upper triangular matrix with leading 1s.

Q.29. What is the reduced row echelon form of a matrix?

Ans. The reduced row echelon form of a matrix is a row echelon form with leading 1s and zeros above and below each leading 1.

Q.30. What is the main diagonal of a matrix?

Ans. The main diagonal of a matrix consists of its elements from the top-left to the bottom-right.

Q.31. What is a diagonal matrix?

Ans. A diagonal matrix is a square matrix with non-zero elements only on the main diagonal.

Q.32. What is a symmetric matrix?

Ans. A symmetric matrix is a square matrix that is equal to its transpose.

Q.33. What is a skew-symmetric matrix?

Ans. A skew-symmetric matrix is a square matrix whose transpose is equal to its negative.

Q.34. What is the trace of a matrix?

Ans. The trace of a matrix is the sum of its main diagonal elements.

Q.35. What is the eigenvalue decomposition (Eigendecomposition)?

Ans. The eigenvalue decomposition is a way to represent a matrix as a product of its eigenvectors and eigenvalues.

Q.36. What is the singular value decomposition (SVD)?

Ans. The singular value decomposition is a factorization of a matrix into three matrices: U, Σ, and V, where U and V are orthogonal matrices, and Σ is a diagonal matrix.

Q.37. What is the Frobenius norm of a matrix?

Ans. The Frobenius norm of a matrix is the square root of the sum of the squares of its elements.

Q.38. What is the condition number of a matrix?

Ans. The condition number of a matrix measures its sensitivity to changes in input data.

Q.39. What is the determinant used for?

Ans. The determinant of a matrix is used to determine properties like invertibility and volume scaling in linear transformations.

Q.40. What is the adjugate (adjoint) of a matrix?

Ans. The adjugate (adjoint) of a matrix is the transpose of its cofactor matrix.

Q.41. What is the Hadamard product of two matrices?

Ans. The Hadamard product is a binary operation that multiplies corresponding elements of two matrices.

Q.42. What is the outer product of two vectors?

Ans. The outer product of two vectors is a matrix formed by multiplying each element of one vector by each element of the other vector.

Q.43. What are homogeneous coordinates used for?

Ans. Homogeneous coordinates are used in computer graphics and projective geometry to represent points at infinity.

Q.44. What is the Perron-Frobenius theorem?

Ans. The Perron-Frobenius theorem provides conditions under which the dominant eigenvalue of a non-negative matrix is real and positive.

Q.45. What is the nullity of a matrix?

Ans. The nullity of a matrix is the dimension of its null space.

Q.46. What is the range of a matrix?

Ans. The range of a matrix is the set of all possible output vectors when the matrix is multiplied by a vector.

Q.47. What is the Gram matrix?

Ans. The Gram matrix is formed by taking the dot product of all pairs of vectors in a set and arranging the results in a matrix.

Q.48. What is the orthogonal projection of a vector onto a subspace?

Ans. The orthogonal projection of a vector onto a subspace is the closest vector in the subspace to the original vector.

Q.49. What is a unit vector?

Ans. A unit vector is a vector with a magnitude of 1.

Q.50. What are complex numbers?

Ans. Complex numbers are numbers in the form a + bi, where a and b are real numbers and i is the imaginary unit ($i^2 = -1$).

Multiple Choice Questions and Answers

Q.1. What is the dimension of a vector space?

(*a*) The number of columns in a matrix.
(*b*) The number of rows in a matrix.
(*c*) The number of vectors in a basis.
(*d*) The number of variables in a linear equation.

Q.2. What is the determinant of a 2 × 2 matrix [a b; c d]?

(*a*) ad – bc
(*b*) ac – bd
(*c*) a + d
(*d*) b – c

Q.3. The cross product is only defined for:

(*a*) 2D vectors
(*b*) 3D vectors
(*c*) 4D vectors
(*d*) Any dimension

Q.4. What is the inverse of a matrix?

(*a*) The square root of the matrix.
(*b*) The reciprocal of the matrix.
(*c*) A matrix with all elements set to zero.
(*d*) A matrix that, when multiplied by the original matrix, gives the identity matrix.

Q.5. What is the dot product of two vectors a and b?

(*a*) a * b
(*b*) a · b
(*c*) a × b
(*d*) a^b

Q.6. What is the rank of the zero matrix?

(*a*) 0
(*b*) 1
(*c*) Infinity
(*d*) It depends on the size of the matrix.

Q.7. Which of the following matrices is a diagonal matrix?

(*a*) |1 0 0| |0 2 0| |0 0 3| (*b*) |1 2 3| |4 5 6| |7 8 9|

(*c*) |1 1 1| |1 1 1| |1 1 1| (*d*) |0 1| |1 0|

Q.8. The null space of a matrix A is also known as:

(*a*) Column space (*b*) Row space (*c*) Eigenspace (*d*) Kernel

Q.9. If a matrix has no inverse, it is called:

(*a*) Singular (*b*) Square (*c*) Symmetric (*d*) Diagonal

Q.10. The eigenvectors of a matrix are:

(*a*) Always real numbers (*b*) Always complex numbers

(*c*) Vectors that do not change direction under the transformation represented by the matrix

(*d*) Vectors with distinct eigenvalues

Q.11. Which of the following is a basis for the 2D vector space?

(*a*) {(1, 0), (0, 1), (1, 1)} (*b*) {(1, 0), (0, 1)}

(*c*) {(1, 2), (2, 3)} (*d*) {(0, 0), (1, 1)}

Q.12. The determinant of an upper triangular matrix is:

(*a*) The product of the diagonal elements (*b*) The sum of the diagonal elements

(*c*) Always zero (*d*) Always one

Q.13. What is the orthogonal projection of a vector onto a subspace?

(*a*) The projection of the vector onto the x-axis.

(*b*) The projection of the vector onto the y-axis.

(*c*) The closest vector in the subspace to the original vector.

(*d*) The sum of the vectors in the subspace.

Q.14. A matrix with all elements equal to zero is called:

(*a*) Null matrix (*b*) Zero matrix (*c*) Identity matrix (*d*) Zero vector

Q.15. The eigenvalues of a symmetric matrix are always:

(*a*) Real numbers (*b*) Complex numbers

(*c*) Imaginary numbers (*d*) Negative numbers

Q.16. The trace of a matrix is the sum of:

(*a*) The elements of the main diagonal. (*b*) The elements of the secondary diagonal.

(*c*) All the elements in the matrix. (*d*) All the elements in the first row.

Q.17. A matrix is said to be diagonalizable if it has:

(*a*) Only one eigenvector. (*b*) At least one eigenvector.

(*c*) At least one eigenvalue. (*d*) A full set of eigenvectors.

Q.18. What is the adjugate (adjoint) of a matrix A?

(*a*) The transpose of A. (*b*) The negative of A.

(*c*) The conjugate of A.

(*d*) The transpose of the cofactor matrix of A.

Q.19. Which of the following matrices represents a reflection about the x-axis in 2D?

(*a*) |1 0| | 0 –1| (*b*) |1 0| |0 1| (*c*) |0 1| |1 0| (*d*) |–1 0| |0 1|

Q.20. What is the main diagonal of a matrix?

(*a*) The first column of the matrix.
(*b*) The first row of the matrix.
(*c*) The elements that form a line from the top-left to the bottom-right of the matrix.
(*d*) The elements that form a line from the top-right to the bottom-left of the matrix.

Q.21. The range of a matrix A is:

(*a*) The set of all vectors that can be obtained by multiplying A by a scalar.
(*b*) The set of all possible output vectors when A is multiplied by a vector.
(*c*) The set of all possible input vectors when A is multiplied by a vector.
(*d*) The set of all eigenvectors of A.

Q.22. The determinant of an identity matrix is:

(*a*) 0 (*b*) 1 (*c*) –1
(*d*) Depends on the size of the matrix.

Q.23. What is the condition number of a matrix?

(*a*) A measure of its singularity. (*b*) A measure of its sparsity.
(*c*) A measure of its sensitivity to changes in input data.
(*d*) A measure of its rank.

Q.24. A set of vectors is linearly independent if:

(*a*) None of the vectors can be expressed as a linear combination of the others.
(*b*) All the vectors have the same magnitude.
(*c*) The vectors are orthogonal to each other.
(*d*) All the vectors are unit vectors.

Q.25. The orthogonal complement of a subspace is the set of:

(*a*) All vectors in the subspace. (*b*) All vectors orthogonal to the subspace.
(*c*) All vectors that span the subspace.
(*d*) All vectors in the null space of the subspace.

Q.26. What is the identity matrix?

(*a*) A matrix with ones on the main diagonal and zeros elsewhere.
(*b*) A matrix with zeros on the main diagonal and ones elsewhere.
(*c*) A matrix with ones on the secondary diagonal and zeros elsewhere.
(*d*) A matrix with ones on both the main and secondary diagonals.

Q.27. The main use of the singular value decomposition (SVD) is for:

(*a*) Computing eigenvalues and eigenvectors.
(*b*) Solving systems of linear equations.
(*c*) Diagonalizing a matrix.
(*d*) Factorizing a matrix into three matrices: U, Σ, and V.

Q.28. Which of the following matrices is singular?

(*a*) |1 0| |0 1| (*b*) |1 2| |3 6| (*c*) |2 0| |0 0| (*d*) |1 1| |2 2|

Q.29. The determinant of an orthogonal matrix is always:

(*a*) 1 (*b*) −1 (*c*) 0

(*d*) Cannot be determined from the given information.

Q.30. Which of the following matrices represents a rotation in 2D?

(*a*) |1 0| |0 1| (*b*) |0 −1| |1 0| (*c*) |−1 0| |0 −1| (*d*) |1 1| |1 −1|

Q.31. The dimension of the column space of a matrix is also known as:

(*a*) The rank of the matrix. (*b*) The nullity of the matrix.

(*c*) The determinant of the matrix. (*d*) The trace of the matrix.

Q.32. If a matrix has a determinant of zero, it is:

(*a*) Invertible (*b*) Symmetric (*c*) Singular (*d*) Diagonalizable

Q.33. The inverse of a matrix A exists if and only if:

(*a*) A has a row of zeros. (*b*) A has a column of ones.

(*c*) The determinant of A is not zero.

(*d*) The rank of A is equal to the number of rows in A.

Q.34. The main use of the eigenvalue decomposition (Eigendecomposition) is for:

(*a*) Solving systems of linear equations. (*b*) Computing singular values.

(*c*) Factorizing a matrix into eigenvectors and eigenvalues.

(*d*) Factorizing a matrix into three matrices: U, Σ, and V.

Q.35. Which of the following matrices represents a reflection about the y-axis in 2D?

(*a*) |1 0| |0 −1| (*b*) |1 0| |0 1| (*c*) |0 1| |1 0| (*d*) |−1 0| |0 1|

Q.36. Which of the following properties holds for the transpose of a matrix A?

(*a*) $(A^T)^T = A$ (*b*) $(A^T)^{(-1)} = A$

(*c*) $(A^T)^T = A^{(-1)}$ (*d*) $(A^{(-1)})^T = A^T$

Q.37. Which of the following statements is true about linearly independent vectors?

(*a*) Linearly independent vectors are always orthogonal.

(*b*) Linearly independent vectors cannot be combined to form the zero vector.

(*c*) Linearly independent vectors are always parallel.

(*d*) Linearly independent vectors have the same magnitude.

Q.38. Which of the following is a valid basis for the 3D vector space?

(*a*) {(1, 0, 0), (0, 1, 0), (1, 1, 1)} (*b*) {(1, 0, 0), (0, 1, 0)}

(*c*) {(1, 2, 3), (2, 3, 4)} (*d*) {(0, 0, 0), (1, 1, 1)}

Q.39. The Gram-Schmidt process is used for:

(*a*) Diagonalizing a matrix. (*b*) Solving systems of linear equations.

(*c*) Orthogonalizing a set of vectors. (*d*) Finding the eigenvalues of a matrix.

Key to Multiple Choice Questions

1. (*c*)	**2.** (*a*)	**3.** (*b*)	**4.** (*d*)	**5.** (*b*)	**6.** (*a*)	**7.** (*a*)	**8.** (*d*)	**9.** (*a*)
10. (*c*)	**11.** (*b*)	**12.** (*a*)	**13.** (*c*)	**14.** (*b*)	**15.** (*a*)	**16.** (*a*)	**17.** (*d*)	**18.** (*d*)
19. (*a*)	**20.** (*c*)	**21.** (*b*)	**22.** (*b*)	**23.** (*c*)	**24.** (*a*)	**25.** (*b*)	**26.** (*a*)	**27.** (*d*)
28. (*c*)	**29.** (*a*)	**30.** (*b*)	**31.** (*a*)	**32.** (*c*)	**33.** (*c*)	**34.** (*c*)	**35.** (*d*)	**36.** (*a*)
37. (*b*)	**38.** (*a*)	**39.** (*c*)						

True or False

Q.40. The determinant of a matrix A is equal to the product of its eigenvalues.

Ans. ****

Q.41. The inverse of a diagonal matrix is another diagonal matrix with each diagonal element as the reciprocal of the corresponding diagonal element of the original matrix.

Ans. True

Q.42. A matrix with all elements equal to one is always invertible.

Ans. False

Q.43. If a matrix has a zero eigenvalue, it is invertible.

Ans. False

Q.44. The orthogonal projection of a vector onto a subspace gives the vector with the smallest magnitude.

Ans. False

Q.45. The determinant of a matrix is always positive.

Ans. False

Q.46. The cross product of two vectors is commutative.

Ans. False

Q.47. The eigenvectors of a matrix form a linearly independent set.

Ans. True

Q.48. A matrix is singular if and only if its determinant is zero.

Ans. True

Q.49. The null space of a matrix is always a subspace of the column space.

Ans. True

Q.50. The column space and the row space of a matrix have the same dimension.

Ans. True

Project Works

1. **Robot Arm Kinematics Analysis:** Study the forward and inverse kinematics of a robotic arm and implement algorithms to compute joint angles for desired end-effector positions and vice versa.

2. **Robot Dynamics and Control:** Develop a dynamic model for a robot manipulator and implement control algorithms, such as PID or LQR, to control the robot's motion.
3. **Mobile Robot Localization:** Use linear algebra techniques, like Kalman filtering or Extended Kalman filtering, for accurate localization and mapping of a mobile robot.
4. **Robot Path Planning:** Implement path planning algorithms, like A* or Dijkstra's algorithm, to find optimal paths for robots to navigate through a given environment.
5. **Robot Vision:** Apply linear algebra in image processing and feature extraction for robot vision tasks like object recognition and tracking.
6. **Robotic Grasping:** Investigate grasping algorithms using linear algebra to plan and control robot gripper motions for object manipulation.
7. **Robot Simulations:** Create a simulation environment to model robot systems and test various control and planning algorithms using linear algebra.
8. **Robot SLAM (Simultaneous Localization and Mapping):** Use linear algebra-based techniques to perform SLAM for autonomous robot exploration and mapping of unknown environments.
9. **Robotic Path Following:** Implement control algorithms based on linear algebra to enable a robot to follow a predefined path accurately.
10. **Inverse Dynamics Control:** Develop an inverse dynamics control algorithm that computes the required joint torques to achieve a desired robot trajectory.
11. **Quadruped Robot Gait Planning:** Design and analyze different gaits for a quadruped robot using linear algebra.
12. **Robot Joint Calibration:** Use linear algebra to calibrate robot joint encoders and improve the accuracy of robot motion.
13. **Robot Vision Camera Calibration:** Develop a camera calibration technique using linear algebra to improve the accuracy of robot vision systems.
14. **Robot Mapping using LiDAR:** Implement SLAM algorithms using LiDAR data and linear algebra for robot mapping and localization.
15. **Control of Robotic Swarm:** Use linear algebra to control the behavior of a swarm of robots to achieve collective objectives.
16. **Robot Manipulator Singularity Analysis:** Analyze the singularities of a robot manipulator using linear algebra to avoid undesirable configurations.
17. **Real-time Control of Robot Arms:** Implement real-time control algorithms for robot arms using linear algebra and parallel processing.
18. **Robot Localization using Radio Frequency (RF) Signals:** Use linear algebra to estimate a robot's position based on received RF signals.
19. **Robot Force Control:** Implement force control algorithms using linear algebra to enable robots to perform delicate tasks.
20. **Robot Trajectory Optimization:** Use linear algebra techniques to optimize robot trajectories for efficient and collision-free motion planning.

03

Calculus as the Mathematical Framework for Modeling, Analyzing and Controlling Robotic Systems

3.1 INTRODUCTION

The mathematical foundation for modeling, analyzing, and controlling robotic systems is provided by calculus, which is a key component of robotics and automation.

3.2 KEY CALCULUS CONCEPTS USED IN ROBOTICS AND AUTOMATION

Some key calculus concepts frequently used in robotics and automation are discussed.

3.2.1 Derivatives

The rate of change of a function with respect to a variable is measured using derivatives. Derivatives are frequently employed in robotics to calculate joint and end-effector velocities and accelerations. For instance, the second derivative of position with respect to time yields the acceleration while the first derivative yields the velocity.

One of the essential ideas in calculus is the concept of a derivative. They show how quickly a function changes in relation to its independent variable. In mathematical notation, the derivative of a function f(x) with respect to x is denoted by f'(x) or df/dx. Here's a detailed explanation of derivatives:

Let's consider a function f(x) defined over a certain domain, where x is the independent variable and f(x) is the dependent variable. The derivative of f(x) with respect to x, denoted as f'(x) or df/dx, is given by the following limit:

$$f'(x) = \lim (h \to 0)\ [f(x + h) - f(x)]/h$$

In this limit, h is a small increment in the value of x. As h approaches zero, it represents an infinitesimally small change in x, and the quotient [f(x + h) – f(x)] / h measures the average rate of change of f(x) over a very small interval [x, x + h].

The derivative f'(x) represents the instantaneous rate of change of f(x) at a specific point x. It gives the slope of the tangent line to the graph of f(x) at that point.

If the derivative f'(x) occurs at a specific point x, then the tangent line and the function f(x) are both differentiable at that location. The function is not differentiable at a point if the derivative does not exist there, and there may be a sharp corner or discontinuity in the graph of f(x) at that location.

Common rules for finding derivatives of basic functions

1. **Constant Rule:** The derivative of a constant c with respect to x is 0. d/dx [c] = 0
2. **Power Rule:** The derivative of x^n with respect to x is n * $x^{(n-1)}$, where n is any real number. d/dx [x^n] = n * $x^{(n-1)}$
3. **Sum Rule:** The derivative of the sum of two functions is the sum of their derivatives. d/dx [f(x) + g(x)] = d/dx [f(x)] + d/dx [g(x)]
4. **Product Rule:** The derivative of the product of two functions is given by: d/dx [f(x) * g(x)] = f′(x) * g(x) + f(x) * g′(x)
5. **Quotient Rule:** The derivative of the quotient of two functions is given by: d/dx [f(x) / g(x)] = (f′(x) * g(x) – f(x) * g′(x)) / $[g(x)]^2$
6. **Chain Rule:** If y = f(u) and u = g(x), then the derivative of y with respect to x is given by: dy/dx = (dy/du) * (du/dx)

To determine the derivatives of more complicated functions and expressions, these principles can be combined and applied repeatedly. In many fields, such as physics, engineering, economics, and, of course, robotics and automation, the idea of derivatives is crucial. Derivatives are used in robotics to compute velocities, accelerations, and control actions, enabling robots to move precisely and carry out tasks.

3.2.2 Integrals

Calculating a function's cumulative change over a given period of time is done using integrals. Integrals are crucial in robotics for tasks like computing a robot's end-effector's displacement over time based on its velocity profile or calculating the area under a curve to determine a robot's total distance travelled.

Integrals are a fundamental idea in calculus and offer a method for calculating the area under a curve, adding up numbers over a specified period of time, and resolving different mathematical issues. The "signed area" between a function and the x-axis within a certain time interval is measured by a function's integral. It has numerous uses in many different fields, including physics, engineering, economics, and many others.

Let's elaborate on the mathematics of integrals

Definite Integral: The definite integral of a function f(x) over a specified interval [a, b] is denoted by ∫[a, b] f(x) dx. It represents the signed area between the function and the x-axis within the interval [a, b]. The value of the definite integral is a numerical quantity.

Mathematically, the definite integral is computed as follows: ∫[a, b] f(x) dx = F(*b*) – F(*a*); where F(x) is the antiderivative (also known as the indefinite integral) of f(x). An antiderivative of a function f(x) is a function F(x) such that d/dx [F(x)] = f(x). The antiderivative F(x) represents a family of functions that differ by a constant, known as the constant of integration.

Indefinite Integral: The indefinite integral of a function f(x) is denoted by ∫f(x) dx and represents the antiderivative of the function. The indefinite integral does not have specific limits of integration, and it includes an arbitrary constant of integration (C) to represent the family of antiderivatives.

Mathematically, the indefinite integral is expressed as: ∫f(x) dx = F(x) + C

where F(x) is an antiderivative of f(x) and C is the constant of integration.

Fundamental Theorem of Calculus: The Fundamental Theorem of Calculus establishes a link between derivatives and integrals. It states that if F(x) is the antiderivative of a continuous function f(x) on an interval [a, b], then the definite integral of f(x) from a to b is equal to the difference of F(*b*) and F(*a*):

$$\int[a, b]\ f(x)\ dx = F(b) - F(a)$$

By locating antiderivatives, the Fundamental Theorem of Calculus is a potent tool that enables us to compute definite integrals.

Techniques of Integration: For some functions, it can be difficult to identify antiderivatives and assess integrals. Integrals can be streamlined and solved using a variety of methods, including substitution, integration by parts, partial fraction decomposition, trigonometric identities, and specific integration rules. Motion planning, trajectory optimization, control systems, and optimisation issues are just a few of the robotics and automation applications for integrals. Integrals are frequently used in robotics to calculate things like the total energy used by a robot while in motion, the displacement of a robot's end-effector over time depending on its velocity profile, and the stability of control systems.

3.2.3 Chain Rule

When examining multi-component complicated robotic systems, the chain rule of calculus is widely used. It makes it possible to differentiate between composite functions, such as when calculating a robot manipulator's joint velocities in relation to its end-effector velocities.

Calculus's core idea of the chain rule enables us to determine the derivative of a composite function. A function that is made up of two or more other functions is known as a composite function. When working with complicated systems including numerous variables and transformations, the chain rule is particularly crucial in robotics and automation. Let's get into more detail about the chain rule's mathematics:

Consider a composite function y = f(g(x)), where g(x) is an intermediate function of x, and f(u) is a function of u.

The chain rule states that the derivative of y with respect to x is the product of the derivative of f(u) with respect to u and the derivative of g(x) with respect to x:

$$dy/dx = df/du * dg/dx$$

Mathematically, the chain rule is represented as follows:

$$dy/dx = d(f(g(x)))/dx = df/du * du/dx = f'(g(x)) * g'(x)$$

where f'(u) represents the derivative of f(u) with respect to u, and g'(x) represents the derivative of g(x) with respect to x.

The chain rule can be extended to more complex composite functions involving multiple intermediate functions. For instance, if we have a composite function y = f(g(h(x))), the chain rule becomes:

$$dy/dx = df/du * du/dv * dv/dx$$

This process can be repeated for any number of intermediate functions.

(*a*) The chain rule is a powerful tool in calculus as it allows us to compute derivatives of complicated functions. In robotics and automation, the chain rule is used in various applications, such as: Robot Kinematics: The Jacobian matrix, which connects end-effector velocities to joint velocities, is computed using the chain rule when dealing with robot manipulators and their end-effectors' motion.

(*b*) **Robot Dynamics:** To formulate equations of motion and control algorithms, the chain rule is used to discover the derivatives of intricate equations defining the dynamics of the robot.

(*c*) **Control Systems:** The chain rule is used in control theory to compute the derivatiyes of control signals and error terms, making it easier to build control laws and feedback control algorithms.

(*d*) **Motion Planning:** Trajectory optimization and motion planning tasks are made possible by the chain rule's ability to compute the derivatives of trajectory equations.

All things considered, the chain rule is a potent mathematical instrument that is essential for resolving issues with composite functions and their derivatives. The chain rule is essential for analytic and numerical solutions in robotics and automation, where complicated transformations and systems are typical.

3.2.4 Gradient and Partial Derivatives

A function's sharpest increase in one direction or another is indicated by a vector called the gradient. It is utilized in robotics for optimization tasks including determining the best motion trajectories for robots. When dealing with multivariate functions, which are frequently encountered in robot kinematics and dynamics, partial derivatives are used. In multivariate calculus, gradient and partial derivatives are crucial ideas, especially when it comes to functions with several variables. Numerous branches of science and engineering, including robotics and automation, are founded on these ideas.

Let's elaborate on the mathematics of gradient and partial derivatives

Partial Derivatives: Partial derivatives are derivatives of multivariable functions with respect to individual variables while keeping other variables constant. For a function f(x, y), the partial derivative with respect to x is denoted by $\partial f/\partial x$, and the partial derivative with respect to y is denoted by $\partial f/\partial y$.

Mathematically, the partial derivative of f(x, y) with respect to x is defined as:

$$\partial f/\partial x = \lim (h \to 0)\ [f(x + h, y) - f(x, y)] / h$$

Similarly, the partial derivative of f(x, y) with respect to y is defined as:

$$\partial f/\partial y = \lim (h \to 0)\ [f(x, y + h) - f(x, y)] / h$$

Partial derivatives provide information about how the function changes with respect to each variable separately, while holding the other variables constant. They are used in various applications, including optimization, differential equations, and sensitivity analysis.

Gradient: The gradient of a multivariable function $f(x_1, x_2, ..., x_n)$ is a vector that points in the direction of the steepest increase of the function at a specific point. It is a vector consisting of all the partial derivatives of f with respect to each variable. The gradient is denoted by ∇f or grad(*f*).

For a function $f(x_1, x_2, ..., x_n)$, the gradient is given by:

$$\nabla f = (\partial f/\partial x_1, \partial f/\partial x_2, ..., \partial f/\partial x_n)$$

The magnitude of the gradient represents the rate of change of the function in the direction of the steepest increase, and the direction of the gradient points towards the steepest increase.

(*a*) When solving optimization issues, the gradient is utilized to determine the minimum or maximum of a multivariable function. In optimization techniques like gradient ascent or gradient descent, the gradient points in the direction of sharpest ascent.

(*b*) Gradients and partial derivatives are crucial in a number of robotics and automation applications, including: a) trajectory optimization, where partial derivatives are utilized to calculate the derivatives of trajectory equations.

(*c*) **Robot Control:** To optimize control signals and minimize error terms, control algorithms use gradient-based optimization approaches.

(*d*) **Sensor Calibration:** To calculate the link between sensor readings and physical quantities, partial derivatives are utilized in sensor calibration.

Gradient and partial derivatives are widely utilized in many mathematical, engineering, and scientific applications, including those pertaining to robotics and automation, and are effective tools for understanding the behaviour of multivariable functions.

3.2.5 Vector Calculus

For representing and modifying three-dimensional quantities like position, velocity, and forces, vector calculus is crucial. In robotics and automation, ideas like dot products, cross products, and vector differentiation are frequently used.

The study of vector fields and how they relate to scalar and vector functions is the focus of the mathematical area known as vector calculus. It is essential to many disciplines, such as physics, engineering, and robotics. Utilizing operations on vector fields, such as differentiation and integration, to study physical processes and address mathematical issues is known as vector calculus.

Let's go over some of the main vector calculus mathematical ideas

Vector Fields: A mathematical concept called a vector field gives each point in space or area of space a vector. It can represent physical characteristics that differ in magnitude and direction across various points in space, such as force, velocity, electric field, and magnetic field. The formula V = V(x, y, z) can be used to mathematically represent a vector field, where V is a vector and x, y, and z are the coordinates of a point in space.

Gradient: A scalar function f(x, y, z) has a gradient, denoted by f or grad (*f*), which is a vector field pointing in the direction of the steepest increase of the function at each point and having a magnitude equal to the rate of change of the function in that direction.

$$\nabla f = (\partial f/\partial x, \partial f/\partial y, \partial f/\partial z)$$

As it appears in many physical laws and equations, the gradient is crucial to both physics and engineering. For instance, in electromagnetic, the electric field is the electric potential's negative gradient.

Divergence: The tendency of the vectors in a vector field, V = (Vx, Vy, Vz), to either converge or diverge at a specific position is measured by the vector field's divergence. It symbolizes the vector field's net outflow from a small, closed surface that surrounds that location.

The divergence of a vector field V is described mathematically by:

$$\text{div}(V) = \nabla \cdot V = (\partial Vx/\partial x) + (\partial Vy/\partial y) + (\partial Vz/\partial z)$$

Divergence is important in fluid mechanics, electromagnetic, and other fields where the flow of a vector field needs to be analyzed.

Curl: The rotation or circulation of the vectors in a vector field V around a point is measured by the field's "curl." It symbolizes the vectors' propensity to swirl or spin into a vortex.

The mathematical formula for a vector field V's curl is

$$\text{curl}(V) = \nabla \times V = \partial Vz/\partial y - \partial Vy/\partial z, \partial Vx/\partial z - \partial Vz/\partial x, \partial Vy/\partial x - \partial Vx/\partial y)$$

The curl is essential in electromagnetism, fluid dynamics, and other areas where rotational behavior needs to be analyzed.

Line integrals: The cumulative impact of a vector field along a curve or path in space is determined using line integrals. Physics makes use of line integrals frequently to compute effort, circulation, and flux.

Surface Integrals: The cumulative impact of a vector field over a surface in space is calculated using surface integrals. They are crucial for calculating electric flux, magnetic flux, and fluid flow in disciplines like electromagnetism and fluid dynamics.

Vector calculus: For the analysis of vector fields, comprehending the behavior of physical processes, and resolving engineering issues, vector calculus is a potent mathematical instrument. Vector calculus is utilized in robotics and automation in areas like motion planning, robot control, and sensor fusion, where vector values are crucial for characterizing the state and behavior of the robot.

3.2.6 Differential Equations

The connection between a function and its derivatives is described by differential equations. Differential equations are a common tool in robotics for simulating the dynamics of robotic systems, including the motion of robot arms and the operation of control systems. A key idea in mathematics is the concept of differential equations, which describe the connections between a function and its derivatives. In science, engineering, and many other disciplines, such as robotics and automation, they are frequently used to model and analyze various phenomena. One or more unknown functions and their derivatives are included in a differential equation.

Let's elaborate on the key mathematics of differential equations

Ordinary Differential Equations (ODEs): Ordinary differential equations involve a single independent variable, usually denoted as t, and one or more unknown functions of t and their derivatives. An nth-order ordinary differential equation is represented as:

$F(t, y, y', y'', ..., y^{(n)}) = 0$; where $y = y(t)$ is the unknown function, y' is its first derivative with respect to t, y" is its second derivative, and so on, up to the nth derivative $y^{(n)}$.

Based on the order (highest derivative present) and linearity (linear or nonlinear), ODEs can be categorized. Finding a function y(t) that solves the given equation and, potentially, satisfying initial or boundary conditions is required to solve ODEs.

Multiple independent variables and partial derivatives of unknown functions with regard to these variables are components of partial differential equations (PDEs). They are used to simulate phenomena like temperature distribution, fluid flow, and electromagnetic fields that depend on multiple independent variables. A general form of a second-order PDE for a function u(x, y) is:

$F(x, y, u, \partial u/\partial x, \partial u/\partial y, \partial^2 u/\partial x^2, \partial^2 u/\partial y^2, \partial^2 u/\partial x\partial y) = 0$

Solving PDEs is generally more challenging than ODEs due to their complexity, and various techniques, such as separation of variables, method of characteristics, and numerical methods, are used for their solutions.

Initial Value Problem (IVP) and Boundary Value Problem (BVP): Initial or boundary conditions for differential equations frequently play a role in identifying certain solutions. The values of the unknown function and its derivatives at a given initial point are given in an initial value problem (IVP). A boundary value problem (BVP) on the other hand, offers conditions at various points, often the ends of a domain. Solving an IVP or BVP involves finding the function that satisfies both the differential equation and the specified initial or boundary conditions.

Existence and Uniqueness of Solutions: Depending on the equation and the given parameters, differential equations may have a single solution, several solutions, or no solutions at all. The notion of existence and uniqueness of differential equation solutions guarantees that, in specific circumstances, solutions exist, are distinct from one another, and are continuous.

Differential equations have several uses in robotics and automation, including mimicking physical behavior, analyzing control systems, modeling robotic system dynamics, and determining the best course for robots. When analytical solutions to differential equations are neither possible or practicable, numerical methods, such as Euler's method, Runge-Kutta methods, and finite difference methods, are frequently used to solve differential equations numerically.

3.3 CONTROL THEORY

Calculus is used in control theory to build and examine the control algorithms that direct robot behavior. Calculus is used in concepts like PID (Proportional-Integral-Derivative) control to modify control actions in response to error signals and feedback data.

An area of mathematics and engineering known as control theory is concerned with the analysis and design of systems to produce desired behavior or performance. It is widely used in many different industries, such as industrial processes, robotics, automation, aerospace, and electrical engineering. The goal of control theory is to comprehend and influence system behavior to maintain stability, responsiveness, and ideal performance.

Let's elaborate on the key mathematics of control theory

Mathematical modeling: System mathematical modeling is the first step in the development of control theory. This entails employing differential equations or transfer functions to describe the dynamics and connections between the system's inputs, outputs, and internal states. The mathematical model may be expressed as ordinary differential equations (ODEs), partial differential equations (PDEs), or difference equations, depending on the system's complexity.

Feedback Control: Feedback control systems are the cornerstone of control theory. In order to provide the intended output or performance, a feedback control system continuously examines the system's output and makes adjustments to the inputs based on this information. The control system provides a control signal to instruct the system to respond by comparing the system's actual output (measured by sensors) with the desired output (set point).

Controllers: Mathematical algorithms known as controllers choose the control signal based on the feedback error. The following are examples of common controllers:

(*a*) **Proportional (P) Controller:** The control signal is inversely proportional to the error signal (difference between set point and measured value).

(*b*) **Integral (I) Controller:** This controller reduces steady-state errors by proportionally adjusting the control signal to the integral of the error signal.

(*c*) **Derivative (D) Controller:** This controller makes the system more responsive to changes in the error by proportionally adjusting the control signal to the derivative of the error signal.

More complex controllers, such the Proportional-Integral-Derivative (PID) controller, can be created by combining these fundamental controllers.

Stability Analysis: Control theory relies heavily on stability analysis to make sure the controlled system doesn't fluctuate or diverge. Finding the system's poles and eigenvalues' positions in the complex plane is a step in the stability analysis process. The system's stability characteristics are determined by the position of the poles.

Frequency Domain Analysis: Laplace transforms or Fourier transforms are used in frequency domain analysis to examine the system's behavior in the frequency domain. It aids in the design and analysis of controllers and enables comprehension of the system's response to various input signal frequencies.

State-Space Representation: Modern mathematical techniques for simulating control systems include state-space modeling. In terms of the system's state variables, input, and output, it describes the system as a collection of first-order differential equations. Modern control theory makes extensive use of state-space modeling, which is particularly helpful for multivariable control systems.

Optimal Control: The goal of optimal control theory is to identify control approaches that best meet a given performance requirement. This criterion can involve cutting down on energy use, increasing efficiency, or minimizing error. Finding the best control policies involves using optimal control techniques, such as dynamic programming and the calculus of variations.

Robotic systems can be controlled precisely and effectively thanks to the application of control theory in robotics and automation. It is utilized in many different processes, such as trajectory planning, motion control, feedback control of robotic manipulators, and autonomous robot navigation. The foundation for analyzing and creating control systems that meet particular performance requirements and guarantee stability and resilience in complicated dynamic systems is laid out by the mathematical concepts of control theory.

3.4 OPTIMIZATION

In robotics, tasks like motion planning, trajectory optimization, and parameter tweaking of control algorithms frequently include the use of optimization techniques like gradient descent and numerical approaches. These calculus concepts allow engineers and researchers in robotics and automation to create complex models, control algorithms, and optimization techniques for the efficient and accurate design and operation of robotic systems. Robotic systems can accomplish precise control, carry out difficult tasks, and adapt to changing and uncertain surroundings by utilizing calculus.

The goal of the mathematical field of optimization is to identify the optimal solution or group of solutions to a given issue. Under specific restrictions, the goal is to maximize or minimize a particular criterion. Numerous disciplines, including engineering, economics, data science, and robotics, heavily rely on optimization approaches.

Let's elaborate on the mathematics of optimization and some common techniques

Optimization Problems: An optimization problem can be formulated as follows:

Minimize (or Maximize) f(x)

Subject to: gi(x) $\leq$ 0, for i = 1, 2, ..., m hj(x) = 0, for j = 1, 2, ..., p

where, f(x) is the objective function to be minimized or maximized. It represents the criterion that needs to be optimized.

x is the vector of variables or parameters to be determined to achieve the optimal solution.

gi(x) are inequality constraints, representing restrictions on the variables.

hj(x) are equality constraints, representing relationships that must be satisfied.

Unconstrained Optimization: There are no restrictions on the variables in unconstrained optimization. By looking for the minimum or maximum of the objective function f(x) without any restrictions, the best solution is to be found. Unconstrained optimization methods frequently used include:

(*a*) **Gradient-Based Methods:** These techniques iteratively update the variables to obtain the local minimum or maximum by using the gradient (derivative) of the objective function. Newton's Method and Gradient Descent are two examples.

(*b*) **Newton-Raphson Method:** This root-finding approach can be applied to optimization problems by locating the roots of the objective function's derivative (gradient).

(*c*) **The conjugate gradient method:** This iterative approach is made to effectively locate a function's minimum in a multidimensional space.

Constrained Optimization: Problems where the variables must satisfy constraints are dealt with through constrained optimization. Finding the best solution that minimizes or maximizes, the objective function while satisfying all the restrictions is required to solve such issues. Constrained optimization methods frequently used include:

(*a*) Linear programming (LP) is a programming technique that works with linear inequality or equality constraints and linear goal functions. A popular method for resolving LP issues is the Simplex Method.

(*b*) **Quadratic Programming (QP):** QP uses linear constraints and quadratic objective functions. Popular methods for QP issues include Active Set Methods and Interior Point Methods.

(*c*) **Nonlinear Programming (NLP):** NLP deals with nonlinear constraints and nonlinear objective functions. NLP issues are frequently solved using the Sequential Quadratic Programming (SQP) and Penalty Function Methods.

Convex Optimization: The field of convex optimization focuses on optimization issues where the objective function and the constraints are convex. Convex optimization problems are extremely significant and straightforward to tackle because they have special global solutions. Numerous optimization issues encountered in the actual world are convex or can be modeled as convex.

Metaheuristic Techniques: Such optimization approaches as genetic algorithms, particle swarm optimization, simulated annealing, and ant colony optimization are examples of metaheuristic procedures. These methods are very helpful for resolving challenging, non-convex optimization issues involving numerous local optima.

Multi-Objective Optimization: Multiple, frequently at odds objectives are optimized simultaneously in multi-objective optimization. Finding a set of solutions that reflects the trade-offs between various objectives is the aim. Evolutionary and the Pareto Front Multi-objective optimization is carried out using this method. In robotics and automation, optimization techniques are vital for motion planning, robot control, trajectory optimization, and parameter tweaking of control systems. These methods help scientists and engineers identify the best options and take effective, data-driven actions in complex and dynamic systems.

3.5 MULTIVARIABLE CALCULUS

The concepts of calculus are extended to functions with several variables in multivariable calculus, sometimes referred to as vector calculus or several variable calculus. It deals with mathematical ideas like differentiation and integration that apply to functions with multiple independent variables. In many disciplines, such as physics, engineering, computer science, and of course robotics and automation, multivariable calculus is essential.

Let's elaborate on the key mathematical concepts of multivariable calculus

Functions of Several Variables: We discuss functions that depend on numerous independent variables in multivariable calculus. For instance, f(x, y) can be used to express a function of two variables, where x and y are the independent variables and f is the dependent variable. A function with n variables is typically written as $f(x_1, x_2, ..., x_n)$.

Partial Derivatives: When other variables are held constant, partial derivatives are used to determine the rate of change of a multivariable function with respect to each individual variable. The partial derivative of a function $f(x_1, x_2, ..., x_n$ with respect to xβ (the ith variable) is denoted as $\partial f/\partial x_i$.

Mathematically, the partial derivative is defined as:

$$\partial f/\partial x_i = \lim (h \to 0)\ [f(x_1, x_2, ..., x_i + h, ... x_n) - f(x_1, x_2, ... x_n)]/h$$

Partial derivatives are essential in understanding how a function changes in different directions and play a crucial role in optimization, gradient-based algorithms, and dynamics analysis.

Gradient: The gradient of a scalar-valued function $f(x_1, x_2, ..., x_{n)}$ is a vector that points in the direction of the steepest increase of the function and has a magnitude equal to the rate of change in that direction. The gradient is denoted by ∇f or grad (*f*).

For a function $f(x_1, x_2, ..., x_n)$, the gradient is given by:

$$\nabla f = (\partial f/\partial x_1, \partial f/\partial x_2, ... \partial f/\partial x_n)$$

The gradient is used in various applications, including optimization, where it points toward the direction of steepest ascent.

Divergence and Curl: Divergence and curl are concepts introduced in vector calculus when dealing with vector fields. For a vector field $F(x, y, z) = (F_1, F_2, F_3)$, the divergence and curl are defined as follows:

$$\text{Divergence (div(F))} = \partial F_1/\partial x + \partial F_2/\partial y + \partial F_3/\partial z$$

$$\text{Curl (curl(F))} = (\partial F_3/\partial y - \partial F_2/\partial z, \partial F_1/\partial z - \partial F_3/\partial x, \partial F_2/\partial x - \partial F_1/\partial y)$$

These ideas are essential for comprehending how vector fields behave in diverse physical events.

Line Integrals: The cumulative impact of a vector field along a space curve is determined using line integrals. They are crucial for calculating work, circulation, and flux in physics and engineering.

Surface Integrals and Volume Integrals: Calculating the cumulative impact of a vector field over a surface or within a certain area of space is done using surface integrals and volume integrals, respectively. They are crucial in disciplines like fluid dynamics and electromagnetic.

In robotics and automation, where many issues require functions of numerous variables, multivariable calculus is crucial. Robot motion planning, trajectory optimization, control systems, and sensor calibration are examples of applications. It offers the mathematical tools needed to analyze complicated systems and create effective algorithms for robotic systems to achieve desired behaviors.

3.6 ROBOTICS APPLICATIONS AS MOTION PLANNING

Motion planning, a crucial component of robotics, includes figuring out a possible path or trajectory for a robot to go along while avoiding obstacles and abiding by various constraints from its current location to the place it wants to be in. It is crucial to the navigation and manipulation of autonomous robots.

Here are some robotics applications of motion planning:

(*a*) Motion planning is a technique used in mobile robotics to direct the movement of wheeled robots, drones, and autonomous vehicles. For instance, motion planning algorithms in autonomous vehicles determine safe and effective routes to travel over intricate road networks while avoiding collisions with other automobiles and pedestrians.

(*b*) **Manipulation and Grasping:** In order for robotic arms and manipulators to carry out activities like picking and positioning objects, motion planning is crucial. Motion planners provide collision-free pathways for the robot's end-effector to approach and grip items by taking into account the robot's kinematics, workspace, and surroundings.

(*c*) **Autonomous Flying Drones:** Motion planning ensures safe and effective flight paths for drones and UAVs (Unmanned Aerial Vehicles) while avoiding obstacles and taking airspace laws into account. For drones working in crowded areas and urban settings, motion planning is very crucial.

(*d*) **Warehouse Automation:** To efficiently transfer items within the warehouse, motion planning is utilized to optimize the pathways of mobile robots and automated guided vehicles (AGVs). It shortens the time needed for order fulfillment and lessens the likelihood of robots and other warehouse objects colliding.

(*e*) **Human-Robot Collaboration:** Motion planning makes ensuring that the robot moves in ways that are safe and appropriate for social interaction. It takes human presence into account and dynamically modifies the robot's route to prevent potential collisions or awkward encounters.

(*f*) **Multi-Robot Systems:** Motion planning is utilized in situations where numerous robots cooperate in order to coordinate their movements and avoid collisions. For jobs like exploration, search-and-rescue operations, and swarm robots, this is crucial.

(*g*) **Medical Robotics:** Motion planning is utilized in medical robotics to direct surgical robots during minimally invasive operations. It makes sure that surgical instruments are moved inside the patient's body precisely and safely.

(*h*) **Autonomous Underwater Vehicles (AUVs) and Underwater Robotics:** Underwater robotics uses motion planning to guide AUVs and remotely operated vehicles (ROVs) across aquatic environments while gathering data and carrying out different underwater jobs.

Motion planning algorithms can be based on various techniques, including:

(*a*) Sampling-based techniques include Probabilistic Roadmaps (PRM) and Rapidly Exploring Random Trees (RRT).

(*b*) Techniques based on mathematical optimization that seek for the best trajectories.

(*c*) Potential fields, in which robots navigate minimally potential paths while avoiding highly potential regions that act as impediments.

Motion planning, in general, is a key aspect of robotics that enables autonomous systems to move through and interact with their environment in a safe, effective, and efficient manner. Motion planning algorithm improvements continue to fuel development of autonomous robotics applications in a variety of industries.

3.6.1 Trajectory Optimization

Finding the optimum path for a robot or system to take from an initial state to a desired goal state while satisfying various constraints and optimizing a particular objective is the process of trajectory optimization in robotics and control systems. Depending on the application, the goal can be to reduce time, energy use, or another performance measure. In many robotic activities, including motion planning, robotic manipulation, aerial navigation, and autonomous vehicle control, trajectory optimization is essential.

Here's an overview of trajectory optimization:

(*a*) Trajectory optimization starts by formulating the problem in terms of the dynamics, constraints, and goals of the robot. The issue entails defining the robot's initial and desired states, the system dynamics (kinematics and dynamics equations), and any mobility restrictions (such as collision avoidance, joint limits, obstacle avoidance, or environmental restrictions) that may apply. The performance indicator that needs to be optimized is quantified by the objective function.

(*b*) **Trajectory Representation:** The trajectory must now be represented in a way that is appropriate. Trajectories can be visualized as a progression of states or as a parameterized function of time, where control inputs act as the independent variables. Polynomial functions, piecewise linear functions and Fourier series are examples of typical trajectory representations.

(*c*) **Optimization Method:** Non-linear dynamics, non-convex constraints, and high-dimensional search spaces are frequently present in trajectory optimization issues, making them complicated. Finding the best trajectories can be done using a variety of optimization techniques.

These Include

(*a*) **Direct Methods:** Direct methods discretize the trajectory and reformulate the problem as a finite-dimensional optimization problem. Examples include shooting methods, collocation methods, and pseudo spectral methods like the direct collocation method.

(*b*) **Indirect Methods:** Indirect methods convert the trajectory optimization problem into a two-point boundary value problem (TPBVP) by solving the associated Hamiltonian equations. Then, a root-finding algorithm is used to obtain the optimal trajectory.

(*c*) **Optimal Control Techniques:** Optimal control theory, such as the Pontryagin's Maximum Principle and the Bellman equation from dynamic programming, is applied to find optimal control strategies that minimize the objective function.

The optimization requires

I. Numerical Solvers: In order to identify approximations of solutions to the optimisation issue, numerical solvers are frequently used. To find the best trajectories, one can use effective optimization solvers like interior-point methods, sequential quadratic programming, or evolutionary algorithms.

II. Performance Evaluation: Following optimization, the generated trajectory is assessed to make sure it satisfies all requirements and delivers the intended performance. The viability and efficacy of the optimized trajectory may be confirmed by simulations and practical testing.

The difficult issue of trajectory optimization necessitates a solid foundation in robotics, control theory, and numerical optimization. The capabilities of autonomous robots and control systems have been greatly improved by developments in trajectory optimization algorithms and computer capacity, allowing them to carry out intricate and exact tasks in a variety of settings and applications.

3.7 SENSOR FUSION

Data fusion, another name for sensor fusion, is the process of merging data from many sensors to gain a more thorough understanding of the environment or the state of a system. Sensor fusion is essential in robotics and automation for enhancing the perception, localization, and decision-making capacities of autonomous systems. Utilizing the advantages of several sensors while making up for their particular weaknesses is the aim of sensor fusion.

Here's an overview of sensor fusion:

(*a*) **Sensor Types:** Different types of sensors are used in robotics to collect data on the environment and the state of the robot. Several popular kinds of sensors include:

(*b*) **Cameras:** These devices offer visual data and photographs of the surroundings.

(*c*) Using laser beams to measure distances, LiDAR (Light Detection and Ranging) enables 3D mapping and object detection.

(*d*) **Radar (Radio Detection and Ranging):** Utilize radio waves to find things and gauge their speed.

(*e*) **GPS (Global Positioning System):** Provides accurate navigational and locational data.

(*f*) **Inertial Measurement Units (IMUs):** These devices use gyroscopes and accelerometers to monitor angular velocity and acceleration, respectively.

(*g*) **Ultrasonic Sensors:** These devices use sound waves to gauge distances.

(*h*) **Odometry Sensors:** Calculate robot motion by counting the revolutions of joints or wheels.

Sensor Fusion Techniques: To efficiently combine data from many sensors, a variety of sensor fusion techniques are used:

(*a*) **Kalman Filtering:** Based on sensor readings and a process model, Kalman filters are recursive estimate algorithms that forecast a system's state and update the estimation. They are extensively utilized in robotics and navigation problems that need sensor fusion.

(*b*) **Extended Kalman Filter (EKF):** This Kalman filter extension supports non-linear models and measurements, making it appropriate for a wide range of practical applications.

(*c*) **Particle Filters (Monte Carlo Localization):** This probabilistic method maintains a set of samples (particles) that represent potential states in order to estimate the state of a system.

(*d*) **Sensor Level Fusion:** The creation of a fused sensor data stream by combining at a low level the raw data from many sensors.

(*e*) **Feature Level Fusion:** This technique involves taking pertinent features from sensor data and combining them to produce a more accurate depiction of the environment.

Benefits of Sensor Fusion: Sensor fusion offers a number of benefits by fusing data from many sensors, including:

(*a*) **Greater Accuracy:** Sensor fusion decreases measurement errors and uncertainties, producing estimates of the system's condition or the environment that are more precise.

(*b*) **Robustness:** The system's reliability and fault-tolerance are improved by the merging of redundant data from many sensors.

(*c*) **Better Perception:** By merging data from many modalities like vision, depth perception, and distance measurements, sensor fusion enables a more thorough perception of the world.

(*d*) **Reducing Ambiguity:** Sensor fusion helps resolve ambiguities and fill in gaps in sensor data by utilizing complimentary sensors.

Applications in Robotics and Automation: Many robotics and automation applications use sensor fusion, including:

(*a*) **Simultaneous Localization and Mapping (SLAM):** For mapping and localizing a robot in an unknown area, sensor fusion is essential.

(*b*) **Autonomous Navigation:** Robots and autonomous vehicles can navigate and avoid hazards more skillfully thanks to sensor fusion.

(*c*) **Object Tracking:** The precision of object tracking in real-time applications is increased by combining data from many sensors.

(*d*) **Human-Robot Interaction:** Sensor fusion is used to increase robot perception of human motions and gestures and to improve human-robot interaction.

Overall, sensor fusion is a potent method that expands the capabilities of robotic systems and autonomous vehicles, allowing them to act more wisely and efficiently in complex and dynamic surroundings.

3.8 CONTROL SYSTEM DESIGN

Designing a set of control algorithms and components to regulate a system's behavior and accomplish specific performance goals is the process of creating a control system. Robotics, engineering, aerospace, and industrial automation are just a few of the industries that use control systems to manage the behavior of dynamic systems and make sure they function effectively, safely, and in accordance with predetermined standards.

The following steps are often included in the design of a control system

Modeling a system Making a mathematical model of the system to be controlled is the first stage in the design of a control system. The relationships between the system's inputs, outputs, and internal states are depicted by the model. The model may be founded on physical principles, real-world evidence, or experimental results. Robot kinematics, dynamics, and other system models may be used in robotics.

Control Objectives: Indicate the performance standards or control objectives that the control system should meet. In a robot motion control system, for instance, the goal can be to move the robot from one place to another while minimizing the amount of time it takes or optimizing the amount of energy it uses. Other applications might have stability, robustness, or precision as their control goals.

Controller Design: The next step is to construct the controller using the system model and control objectives. Based on the system's status and desired behavior, the controller is a collection of algorithms and feedback loops that compute control signals. Various control methods are available, including:

(*a*) **Proportional-Integral-Derivative (PID) control:** A traditional control technique that modifies the control signal based on the discrepancy between the desired and actual state using three components (proportional, integral, and derivative).

(*b*) **State Feedback Control:** This method calculates the control signal using the system's complete state (or a combination of states).

(*c*) **Optimal Control:** Involves identifying control tactics that reduce the amount of a particular performance criterion, like time or energy usage.

(*d*) **Adaptive Control:** Modifies the control parameters in response to real-time measurements to take into account modifications to the system or its surroundings.

Stability and Performance Analysis: Analyzing the closed-loop system's performance and stability is crucial after the controller has been designed. The control system's stability is ensured through stability analysis, which also checks for any undesired oscillations or divergence. Performance analysis evaluates whether the control system meets the desired performance requirements and how well it accomplishes the given objectives.

Simulation and Testing: To assess how well the proposed control system performs in various scenarios and settings, it is typically put through a simulation test. Before applying the control system design in the actual system, simulations assist detect potential problems and optimise them.

Implementation and Tuning: The physical system or robot will then use the control system design after it has been simulation-verified. To obtain the desired performance and modify the controller to the features of the particular system, tuning of the control parameters could be required.

Evaluation and Iteration: The control system is tested in actual situations after implementation, and its performance is appraised. The control system design may go through more iterations and refinements as needed to enhance performance and address any problems that came up during testing.

The process of designing a control system is iterative and necessitates a solid grasp of the dynamics of the system, control theory, and mathematical modelling. In many engineering and robotics applications, it is essential for achieving precise and effective control of dynamic systems.

Numerical Problems with Solutions

Problem 1. Find the derivative of the function $f(x) = 3x^2 - 2x + 1$.

Solution: $f'(x) = d/dx(3x^2 - 2x + 1) = 6x - 2$.

Problem 2. Calculate the integral of the function $g(x) = 2x + 5$.

Solution: $\int g(x)\,dx = \int(2x + 5)\,dx = x^2 + 5x + C$, where C is the constant of integration.

Problem 3. A robot's position is given by $x(t) = 2t^2 - 3t + 1$. Find its velocity at $t = 2$ seconds.

Solution: The velocity is given by $v(t) = d/dt\,(2t^2 - 3t + 1) = 4t - 3$. At $t = 2$ seconds, $v(2) = 4(2) - 3 = 5$ m/s.

Problem 4. Calculate the area under the curve of the function $h(x) = 3x^2$ from $x = 0$ to $x = 2$.

Solution: The area A is given by $A = \int(0 \text{ to } 2)\, 3x^2\,dx = [x^3]\,(0 \text{ to } 2) = 8$ square units.

Problem 5. A robot's position is described by $x(t) = 4t^3 - 2t^2 + 5t$. Find its acceleration at $t = 1$ second.

Solution: The acceleration is given by $a(t) = d^2/dt^2\,(4t^3 - 2t^2 + 5t) = 24t - 4$. At $t = 1$ second, $a(1) = 24(1) - 4 = 20$ m/s^2.

Problem 6. A robot is moving with a velocity given by $v(t) = 2t - 3$. Find its displacement from $t = 1$ to $t = 3$ seconds.

Solution: The displacement is given by $\int(1 \text{ to } 3)\,(2t - 3)\,dt = [t^2 - 3t]\,(1 \text{ to } 3) = (3^2 - 3(3)) - (1^2 - 3(1)) = 2$ meters.

Problem 7. The Jacobian matrix of a 2-link robot arm is given by $J(q) = [2\cos(q_1 + q_2) + \cos(q_1), 2\cos(q_1 + q_2); 2\sin(q_1 + q_2) + \sin(q_1), 2\sin(q_1 + q_2)]$. Find $J(\pi/4, \pi/6)$.

Solution: Substitute $q_1 = \pi/4$ and $q_2 = \pi/6$ into $J(q)$ to get: $J(\pi/4, \pi/6) = [2\cos(\pi/4 + \pi/6) + \cos(\pi/4), 2\cos(\pi/4 + \pi/6); 2\sin(\pi/4 + \pi/6) + \sin(\pi/4), 2\sin(\pi/4 + \pi/6)]$. Simplify each entry to get $J(\pi/4, \pi/6) = [\sqrt{6} + \sqrt{2}, \sqrt{6}; \sqrt{6} - \sqrt{2}, \sqrt{6}]$.

Problem 8. A robot arm is moving with a joint velocity of [2 rad/s, –3 rad/s]. Find its end-effector velocity using the Jacobian matrix $J = [2\cos(q), -3\sin(q)]$.

Solution: The end-effector velocity v_e is given by v_e = J(q) * q_dot, where q_dot is the joint velocity vector. v_e = [2cos(q), –3sin(q)] * [2, –3] = [4cos(q) + 9sin(q)].

Problem 9. A robot arm has joint positions $q = [\pi/6, \pi/4]$. Find the end-effector position using the forward kinematics equation $x = 2\cos(q_1) + 3\cos(q_1 + q_2)$, $y = 2\sin(q_1) + 3\sin(q_1 + q_2)$.

Solution: Substitute $q_1 = \pi/6$ and $q_2 = \pi/4$ into the equations to get: $x = 2\cos(\pi/6) + 3\cos(\pi/6 + \pi/4) = 2\sqrt{3} + 3(\sqrt{2} + 1) = 5 + 2\sqrt{3}$, $y = 2\sin(\pi/6) + 3\sin(\pi/6 + \pi/4) = 1 + 3(\sqrt{2} - \sqrt{3}) = 1 + \sqrt{6} - \sqrt{3}$.

Problem 10. A robot is following a trajectory described by the equation $x(t) = 2t^3 - 3t^2 + 4t$. Find its maximum acceleration.

Solution: The acceleration is given by $a(t) = d^2/dt^2\ (2t^3 - 3t^2 + 4t) = 12t - 6$. To find the maximum acceleration, set $da/dt = 0$: $12t - 6 = 0$, $t = 0.5$ seconds. Substituting $t = 0.5$ seconds into $a(t)$, we get $a(0.5) = 12(0.5) - 6 = 0$ m/s^2.

Problem 11. A robot is moving with a velocity described by $v(t) = 3t^2 - 4t + 2$. Find the total displacement from $t = 0$ to $t = 2$ seconds.

Solution: The displacement is given by $\int(0 \text{ to } 2)\ (3t^2 - 4t + 2)dt = [t^3 - 2t^2 + 2t]\ (0 \text{ to } 2) = (2^3 - 2(2)^2 + 2(2)) - (0^3 - 2(0)^2 + 2(0)) = 2$ meters.

Problem 12. A robot arm is moving with joint angles $q(t) = 2t^2 - t + 1$ radians. Calculate its angular velocity at $t = 3$ seconds.

Solution: The angular velocity is given by $\omega(t) = d/dt\ (2t^2 - t + 1) = 4t - 1$. At $t = 3$ seconds, $\omega(3) = 4(3) - 1 = 11$ rad/s.

Problem 13. A robot is moving along a trajectory described by $x(t) = 2t^3 - 3t^2 + 4t$. Find its total displacement from $t = 0$ to $t = 3$ seconds.

Solution: The displacement is given by $\int(0 \text{ to } 3)\ (2t^3 - 3t^2 + 4t)\ dt = [t^4 - t^3 + 2t^2]\ (0 \text{ to } 3) = (3^4 - 3^3 + 2(3)^2) - (0^4 - 0^3 + 2(0)^2) = 54$ meters.

Problem 14. A robot arm has joint angles $q(t) = 3t\wedge - 2t + 1$ radians. Calculate its angular velocity at $t = 2$ seconds.

Solution: The angular velocity is given by $\omega(t) = d/dt\ (3t^2 - 2t + 1) = 6t - 2$. At $t = 2$ seconds, $\omega(2) = 6(2) - 2 = 10$ rad/s.

Problem 15. A robot is moving with a velocity described by $v(t) = 4t^2 - 2t + 3$. Find its total displacement from $t = 0$ to $t = 2$ seconds.

Solution: The displacement is given by $\int(0 \text{ to } 2)\ (4t^2 - 2t + 3)dt = [4/3t^3 - t^2 + 3t]\ (0 \text{ to } 2) = (4/3\ (2)^3 - (2)^2 + 3(2)) - (4/3\ (0)^3 - (0)^2 + 3(0)) = 16/3$ meters.

Problem 16. A robot arm has joint positions $q = [\pi/3, \pi/4]$. Find the end-effector position using the forward kinematics equation $x = 2\cos(q_1) + 3\cos(q_1 + q_2)$, $y = 2\sin(q_1) + 3\sin(q_1 + q_2)$.

Solution: Substitute $q_1 = \pi/3$ and $q_2 = \pi/4$ into the equations to get: $x = 2\cos(\pi/3) + 3\cos(\pi/3 + \pi/4) = 2\sqrt{3} + 3(\sqrt{2} + \sqrt{6}) = 2\sqrt{3} + 3\sqrt{2} + 3\sqrt{6}$, $y = 2\sin(\pi/3) + 3\sin(\pi/3 + \pi/4) = \sqrt{3} + 3(\sqrt{2} - \sqrt{6}) = \sqrt{3} + 3\sqrt{2} - 3\sqrt{6}$.

Problem 17. A robot's position is given by $x(t) = 3t^2 - 2t + 5$. Find its velocity and acceleration at $t = 2$ seconds.

Solution: The velocity is given by $v(t) = d/dt\ (3t^2 - 2t + 5) = 6t - 2$. At $t = 2$ seconds, $v(2) = 6(2) - 2 = 10$ m/s. The acceleration is given by $a(t) = d^2/dt^2\ (3t^2 - 2t + 5) = 6$. At $t = 2$ seconds, $a(2) = 6$ m/s^2.

Problem 18. A robot arm is moving with joint angles $q(t) = 3t^2 - 2t + 1$ radians. Calculate its angular acceleration at $t = 4$ seconds.

Solution: The angular acceleration is given by $\alpha(t) = d^2/dt^2\ (3t^2 - 2t + 1) = 6$. At $t = 4$ seconds, $\alpha(4) = 6$ rad/s^2.

Problem 19. A robot arm has joint angles $q(t) = 4t^3 - 3t^2 + 2t$. Calculate its angular velocity at $t = 2$ seconds.

Solution: The angular velocity is given by $\omega(t) = d/dt\ (4t^3 - 3t^2 + 2t) = 12t^2 - 6t + 2$. At t = 2 seconds, $\omega(2) = 12(2)^2 - 6(2) + 2 = 20$ rad/s.

Problem 20. A robot's position is given by $x(t) = 2t^3 - 3t^2 + 4t$. Find its velocity and acceleration at t = 1 second.

Solution: The velocity is given by $v(t) = d/dt\ (2t^3 - 3t^2 + 4t) = 6t^2 - 6t + 4$. At t = 1 second, $v(1) = 6(1)^2 - 6(1) + 4 = 4$ m/s. The acceleration is given by $a(t) = d^2/dt^2\ (2t^3 - 3t^2 + 4t) = 12t - 6$. At t = 1 second, $a(1) = 12(1) - 6 = 6$ m/s².

Problem 21. A robot is moving with a velocity described by $v(t) = 3t^2 - 2t + 5$. Find its total displacement from t = 0 to t = 3 seconds.

Solution: The displacement is given by $\int(0 \text{ to } 3)\ (3t^2 - 2t + 5)\ dt = [t^3 - t^2 + 5t]\ (0 \text{ to } 3) = (3^3 - 3^2 + 5(3)) - (0^3 - 0^2 + 5(0)) = 27 - 9 + 15 = 33$ meters.

Problem 22. A robot arm has joint positions $q = [\pi/4, \pi/6]$. Find the end-effector position using the forward kinematics equation $x = 2\cos(q_1) + 3\cos(q_1 + q_2)$, $y = 2\sin(q_1) + 3\sin(q_1 + q_2)$.

Solution: Substitute $q_1 = \pi/4$ and $q_2 = \pi/6$ into the equations to get: $x = 2\cos(\pi/4) + 3\cos(\pi/4 + \pi/6) = \sqrt{2} + 3(\sqrt{3} + \sqrt{6}) = \sqrt{2} + 3\sqrt{3} + 3\sqrt{6}$, $y = 2\sin(\pi/4) + 3\sin(\pi/4 + \pi/6) = \sqrt{2} + 3(\sqrt{3} - \sqrt{6}) = \sqrt{2} + 3\sqrt{3} - 3\sqrt{6}$.

Problem 23. A robot's position is given by $x(t) = 4t^3 - 2t^2 + 3t$. Find its velocity and acceleration at t = 2 seconds.

Solution: The velocity is given by $v(t) = d/dt\ (4t^3 - 2t^2 + 3t) = 12t^2 - 4t + 3$. At t = 2 seconds, $v(2) = 12(2)^2 - 4(2) + 3 = 43$ m/s. The acceleration is given by $a(t) = d^2/dt^2\ (4t^3 - 2t^2 + 3t) = 24t - 4$. At t = 2 seconds, $a(2) = 24(2) - 4 = 44$ m/s².

Problem 24. A robot arm is moving with joint angles $q(t) = 2t^2 - 3t + 1$ radians. Calculate its angular acceleration at t = 3 seconds.

Solution: The angular acceleration is given by $\alpha(t) = d^2/dt^2\ (2t^2 - 3t + 1) = 4$. At t = 3 seconds, $\alpha(3) = 4$ rad/s².

Problem 25. A robot arm has joint angles $q(t) = 3t^2 - 2t + 1$ radians. Calculate its angular velocity at t = 4 seconds.

Solution: The angular velocity is given by $\omega(t) = d/dt\ (3t^2 - 2t + 1) = 6t - 2$. At t = 4 seconds, $\omega(4) = 6(4) - 2 = 22$ rad/s.

Problem 26. A robot's position is given by $x(t) = 2t^3 - 3t^2 + 4t$. Find its velocity and acceleration at t = 0 seconds.

Solution: The velocity is given by $v(t) = d/dt\ (2t^3 - 3t^2 + 4t) = 6t^2 - 6t + 4$. At t = 0 seconds, $v(0) = 6(0)^2 - 6(0) + 4 = 4$ m/s. The acceleration is given by $a(t) = d^2/dt^2\ (2t^3 - 3t^2 + 4t) = 12t - 6$. At t = 0 seconds, $a(0) = 12(0) - 6 = -6$ m/s².

Problem 27. A robot arm is moving with joint angles $q(t) = 4t^3 - 3t^2 + 2t$ radians. Calculate its angular velocity at t = 1 second.

Solution: The angular velocity is given by $\omega(t) = d/dt\ (4t^3 - 3t^2 + 2t) = 12t^2 - 6t + 2$. At t = 1 second, $\omega(1) = 12(1)^2 - 6(1) + 2 = 8$ rad/s.

Problem 28. A robot's position is given by $x(t) = 3t^2 - 2t + 5$. Find its velocity and acceleration at t = 3 seconds.

Solution: The velocity is given by $v(t) = d/dt\ (3t^2 - 2t + 5) = 6t - 2$. At t = 3 seconds, $v(3) = 6(3) - 2 = 16$ m/s. The acceleration is given by $a(t) = d^2/dt^2\ (3t^2 - 2t + 5) = 6$. At t = 3 seconds, $a(3) = 6$ m/s².

Problem 29. A robot arm is moving with joint angles $q(t) = 3t^2 - 2t + 1$ radians. Calculate its angular acceleration at $t = 1$ second.

Solution: The angular acceleration is given by $\alpha(t) = d^2/dt^2 (3t^2 - 2t + 1) = 6$. At $t = 1$ second, $\alpha(1) = 6$ rad/s^2.

Problem 30. A robot arm has joint angles $q(t) = 4t^3 - 3t^2 + 2t$. Calculate its angular velocity at $t = 3$ seconds.

Solution: The angular velocity is given by $\omega(t) = d/dt (4t^3 - 3t^2 + 2t) = 12t^2 - 6t + 2$. At $t = 3$ seconds, $\omega(3) = 12(3)^2 - 6(3) + 2 = 62$ rad/s.

Problem 31. A robot's position is given by $x(t) = 4t^3 - 2t^2 + 3t$. Find its velocity and acceleration at $t = 0$ seconds.

Solution: The velocity is given by $v(t) = d/dt (4t^3 - 2t^2 + 3t) = 12t^2 - 4t + 3$. At $t = 0$ seconds, $v(0) = 12(0)^2 - 4(0) + 3 = 3$ m/s. The acceleration is given by $a(t) = d^2/dt^2 (4t^3 - 2t^2 + 3t) = 24t - 4$. At $t = 0$ seconds, $a(0) = 24(0) - 4 = -4$ m/s^2.

Problem 32. A robot arm is moving with joint angles $q(t) = 2t^2 - 3t + 1$ radians. Calculate its angular acceleration at $t = 2$ seconds.

Solution: The angular acceleration is given by $\alpha(t) = d^2/dt^2 (2t^2 - 3t + 1) = 4$. At $t = 2$ seconds, $\alpha(2) = 4$ rad/s^2.

Problem 33. A robot arm has joint angles $q(t) = 3t^2 - 2t + 1$ radians. Calculate its angular velocity at $t = 3$ seconds.

Solution: The angular velocity is given by $\omega(t) = d/dt (3t^2 - 2t + 1) = 6t - 2$. At $t = 3$ seconds, $\omega(3) = 6(3) - 2 = 16$ rad/s.

Problem 34. A robot's position is given by $x(t) = 2t^3 - 3t^2 + 4t$. Find its velocity and acceleration at $t = 1$ second.

Solution: The velocity is given by $v(t) = d/dt (2t^3 - 3t^2 + 4t) = 6t^2 - 6t + 4$. At $t = 1$ second, $v(1) = 6(1)^2 - 6(1) + 4 = 4$ m/s. The acceleration is given by $a(t) = d^2/dt^2 (2t^3 - 3t^2 + 4t) = 12t - 6$. At $t = 1$ second, $a(1) = 12(1) - 6 = 6$ m/s^2.

Problem 35. A robot arm is moving with joint angles $q(t) = 4t^3 - 3t^2 + 2t$ radians. Calculate its angular velocity at $t = 2$ seconds.

Solution: The angular velocity is given by $\omega(t) = d/dt (4t^3 - 3t^2 + 2t) = 12t^2 - 6t + 2$. At $t = 2$ seconds, $\omega(2) = 12(2)^2 - 6(2) + 2 = 22$ rad/s.

Problem 36. A robot's position is given by $x(t) = 3t^2 - 2t + 5$. Find its velocity and acceleration at $t = 2$ seconds.

Solution: The velocity is given by $v(t) = d/dt (3t^2 - 2t + 5) = 6t - 2$. At $t = 2$ seconds, $v(2) = 6(2) - 2 = 10$ m/s. The acceleration is given by $a(t) = d^2/dt^2 (3t^2 - 2t + 5) = 6$. At $t = 2$ seconds, $a(2) = 6$ m/s^2.

Problem 37. A robot arm is moving with joint angles $q(t) = 2t^2 - 3t + 1$ radians. Calculate its angular acceleration at $t = 3$ seconds.

Solution: The angular acceleration is given by $\alpha(t) = d^2/dt^2 (2t^2 - 3t + 1) = 4$. At $t = 3$ seconds, $\alpha(3) = 4$ rad/s^2.

Problem 38. A robot arm has joint angles $q(t) = 3t^2 - 2t + 1$ radians. Calculate its angular velocity at $t = 4$ seconds.

Solution: The angular velocity is given by $\omega(t) = d/dt (3t^2 - 2t + 1) = 6t - 2$. At $t = 4$ seconds, $\omega(4) = 6(4) - 2 = 22$ rad/s.

Problem 39. A robot's position is given by $x(t) = 4t^3 - 2t^2 + 3t$. Find its velocity and acceleration at t = 0 seconds.

Solution: The velocity is given by $v(t) = d/dt\,(4t^3 - 2t^2 + 3t) = 12t^2 - 4t + 3$. At t = 0 seconds, $v(0) = 12(0)^2 - 4(0) + 3 = 3$ m/s. The acceleration is given by $a(t) = d^2/dt^2\,(4t^3 - 2t^2 + 3t) = 24t - 4$. At t = 0 seconds, $a(0) = 24(0) - 4 = -4$ m/s^2.

Problem 40. A robot arm is moving with joint angles $q(t) = 2t^2 - 3t + 1$ radians. Calculate its angular acceleration at t = 2 seconds.

Solution: The angular acceleration is given by $\alpha(t) = d^2/dt^2\,(2t^2 - 3t + 1) = 4$. At t = 2 seconds, $\alpha(2) = 4$ rad/s^2.

Problem 41. A robot arm has joint angles $q(t) = 3t^2 - 2t + 1$ radians. Calculate its angular velocity at t = 3 seconds.

Solution: The angular velocity is given by $\omega(t) = d/dt\,(3t^2 - 2t + 1) = 6t - 2$. At t = 3 seconds, $\omega(3) = 6(3) - 2 = 16$ rad/s.

Problem 42. A robot's position is given by $x(t) = 2t^3 - 3t^2 + 4t$. Find its velocity and acceleration at t = 1 second.

Solution: The velocity is given by $v(t) = d/dt\,(2t^3 - 3t^2 + 4t) = 6t^2 - 6t + 4$. At t = 1 second, $v(1) = 6(1)^2 - 6(1) + 4 = 4$ m/s. The acceleration is given by $a(t) = d^2/dt^2\,(2t^3 - 3t^2 + 4t) = 12t - 6$. At t = 1 second, $a(1) = 12(1) - 6 = 6$ m/s^2.

Problem 43. A robot arm is moving with joint angles $q(t) = 4t^3 - 3t^2 + 2t$ radians. Calculate its angular velocity at t = 2 seconds.

Solution: The angular velocity is given by $\omega(t) = d/dt\,(4t^3 - 3t^2 + 2t) = 12t^2 - 6t + 2$. At t = 2 seconds, $\omega(2) = 12(2)^2 - 6(2) + 2 = 22$ rad/s.

Problem 44. A robot's position is given by $x(t) = 3t^2 - 2t + 5$. Find its velocity and acceleration at t = 2 seconds.

Solution: The velocity is given by $v(t) = d/dt\,(3t^2 - 2t + 5) = 6t - 2$. At t = 2 seconds, $v(2) = 6(2) - 2 = 10$ m/s. The acceleration is given by $a(t) = d^2/dt^2\,(3t^2 - 2t + 5) = 6$. At t = 2 seconds, $a(2) = 6$ m/s^2.

Problem 45. A robot arm is moving with joint angles $q(t) = 2t^2 - 3t + 1$ radians. Calculate its angular acceleration at t = 3 seconds.

Solution: The angular acceleration is given by $\alpha(t) = d^2/dt^2\,(2t^2 - 3t + 1) = 4$. At t = 3 seconds, $\alpha(3) = 4$ rad/s^2.

Problem 46. A robot arm has joint angles $q(t) = 3t^2 - 2t + 1$ radians. Calculate its angular velocity at t = 4 seconds.

Solution: The angular velocity is given by $\omega(t) = d/dt\,(3t^2 - 2t + 1) = 6t - 2$. At t = 4 seconds, $\omega(4) = 6(4) - 2 = 22$ rad/s.

Problem 47. A robot's position is given by $x(t) = 4t^3 - 2t^2 + 3t$. Find its velocity and acceleration at t = 0 seconds.

Solution: The velocity is given by $v(t) = d/dt\,(4t^3 - 2t^2 + 3t) = 12t^2 - 4t + 3$. At t = 0 seconds, $v(0) = 12(0)^2 - 4(0) + 3 = 3$ m/s. The acceleration is given by $a(t) = d^2/dt^2\,(4t^3 - 2t^2 + 3t) = 24t - 4$. At t = 0 seconds, $a(0) = 24(0) - 4 = -4$ m/s^2.

Problem 48. A robot arm is moving with joint angles $q(t) = 2t^2 - 3t + 1$ radians. Calculate its angular acceleration at t = 2 seconds.

Solution: The angular acceleration is given by $\alpha(t) = d^2/dt^2\ (2t^2 - 3t + 1) = 4$. At t = 2 seconds, $\alpha(2) - 4$ rad/s^2.

Problem 49. A robot arm has joint angles $q(t) = 3t^2 - 2t + 1$ radians. Calculate its angular velocity at t = 3 seconds.

Solution: The angular velocity is given by $\omega(t) = d/dt\ (3t^2 - 2t + 1) = 6t - 2$. At t = 3 seconds, $\omega(3) = 6(3) - 2 = 16$ rad/s.

Problem 50. A robot's position is given by $x(t) = 2t^3 - 3t^2 + 4t$. Find its velocity and acceleration at t = 1 second.

Solution: The velocity is given by $v(t) = d/dt\ (2t^3 - 3t^2 + 4t) = 6t^2 - 6t + 4$. At t = 1 second, $v(1) = 6(1)^2 - 6(1) + 4 = 4$ m/s. The acceleration is given by $a(t) = d^2/dt^2\ (2t^3 - 3t^2 + 4t) = 12t - 6$. At t = 1 second, $a(1) = 12(1) - 6 = 6$ m/s^2.

Short Questions and Answers

Q.1. What is calculus?

Ans. Calculus is a branch of mathematics that deals with the study of rates of change and accumulation.

Q.2. What are the two main branches of calculus?

Ans. Differential calculus and integral calculus.

Q.3. What is the derivative of a function?

Ans. The derivative represents the rate of change of a function with respect to its independent variable.

Q.4. How is the derivative denoted?

Ans. The derivative of a function f(x) is denoted as f'(x) or df/dx.

Q.5. What is the fundamental theorem of calculus?

Ans. The fundamental theorem of calculus establishes the relationship between differentiation and integration.

Q.6. What does integral calculus involve?

Ans. Integral calculus deals with the calculation of areas under curves and finding antiderivatives.

Q.7. How is the integral of a function denoted?

Ans. The integral of a function f(x) is denoted as $\int f(x)\,dx$.

Q.8. What is the application of calculus in robotics?

Ans. Calculus is used in robotics for motion planning, control, and kinematics.

Q.9. How is calculus used in motion planning for robots?

Ans. Calculus helps in finding optimal trajectories and paths for robots to move efficiently.

Q.10. What does kinematics refer to in robotics?

Ans. Kinematics in robotics deals with the study of robot motion without considering forces.

Q.11. How does calculus play a role in robot kinematics?

Ans. Calculus is used to derive equations that relate joint angles to end-effector positions.

Q.12. What is the Jacobian matrix in robotics?

Ans. The Jacobian matrix represents the relationship between joint velocities and end-effector velocities.

Q.13. How is the Jacobian matrix used in robotics?

Ans. The Jacobian is used to compute the robot's end-effector velocity for a given set of joint velocities.

Q.14. What is robot control in robotics?

Ans. Robot control involves determining the actuator inputs to achieve desired robot behavior.

Q.15. How is calculus used in robot control?

Ans. Calculus is used in control algorithms to calculate control signals based on feedback.

Q.16. What is the role of calculus in robot localization?

Ans. Calculus is used in sensor fusion algorithms to estimate the robot's position and orientation.

Q.17. How are differential equations used in robotics?

Ans. Differential equations model the dynamics of robotic systems and are used in control design.

Q.18. What is trajectory planning in robotics?

Ans. Trajectory planning involves generating a smooth path for a robot to follow from one point to another.

Q.19. How does calculus help in trajectory planning?

Ans. Calculus is used to generate smooth trajectories that minimize jerk and ensure robot safety.

Q.20. What is the concept of optimization in robotics?

Ans. Optimization is used to find the best solution to a given problem, such as minimizing energy consumption or maximizing efficiency.

Q.21. How is calculus used in robot optimization problems?

Ans. Calculus helps find optimal solutions by minimizing or maximizing objective functions.

Q.22. What are the applications of optimization in robotic motion planning?

Ans. Optimization is used to find collision-free paths and minimize path length for robot motion planning.

Q.23. What is the role of partial derivatives in robotics?

Ans. Partial derivatives are used to compute gradients for optimization algorithms.

Q.24. How are differential equations used in robot control?

Ans. Differential equations model the robot's dynamics and are used in control algorithms to ensure stability.

Q.25. How is calculus applied in robot vision?

Ans. Calculus is used in image processing to detect edges and features in robot vision systems.

Q.26. What is the concept of trajectory tracking in robotics?

Ans. Trajectory tracking involves controlling the robot to follow a predefined path accurately.

Q.27. How does calculus help in trajectory tracking?

Ans. Calculus-based control algorithms adjust robot velocities to minimize tracking errors.

Q.28. What is the concept of real-time control in robotics?

Ans. Real-time control involves computing control actions within tight time constraints.

Q.29. How is calculus applied in real-time control for robots?

Ans. Calculus-based control algorithms are designed for quick computation in real-time scenarios.

Q.30. What is the relationship between calculus and robotic system modeling?

Ans. Calculus is used to derive mathematical models that represent the behavior of robotic systems.

Q.31. How is calculus applied in robot arm design?

Ans. Calculus is used to optimize arm geometry for improved reach and performance.

Q.32. What is the role of integration in robot sensor data processing?

Ans. Integration is used to accumulate sensor data over time, e.g., to estimate robot position.

Q.33. How are partial derivatives used in gradient-based optimization in robotics?

Ans. Partial derivatives are used to compute the gradient of the cost function, enabling efficient optimization.

Q.34. How is calculus used in computer simulations of robot behavior?

Ans. Calculus-based algorithms are used to model and simulate the dynamics of robots.

Q.35. What is the role of calculus in robot learning algorithms?

Ans. Calculus is used to optimize the learning process and adapt robot behaviors based on data.

Q.36. How are differential equations used in modeling robot dynamics?

Ans. Differential equations describe the relationship between robot joint torques and accelerations.

Q.37. What is the concept of robot mapping in robotics?

Ans. Robot mapping involves building a representation of the environment using sensor data.

Q.38. How does calculus help in robot mapping?

Ans. Calculus is used to process sensor data to create a map of the robot's surroundings.

Q.39. What is the role of calculus in robot path optimization?

Ans. Calculus is used to optimize robot paths to reduce energy consumption and improve efficiency.

Q.40. How are integrals used in computing robot displacements?

Ans. Integrals are used to calculate the total displacement of a robot from its velocity profile.

Q.41. What is the concept of adaptive control in robotics?

Ans. Adaptive control adjusts robot control parameters in real-time to handle uncertain environments.

Q.42. How does calculus contribute to adaptive control in robotics?

Ans. Calculus-based algorithms are used to update control parameters based on feedback.

Q.43. What is the concept of robot grasping in robotics?

Ans. Robot grasping involves the study of how robots manipulate objects.

Q.44. How is calculus used in robot grasping algorithms?

Ans. Calculus helps in determining optimal grasping strategies based on object shapes and orientations.

Q.45. What is the concept of force control in robotics?

Ans. Force control allows robots to apply or resist forces with a specified magnitude and direction.

Q.46. How is calculus used in force control for robots?

Ans. Calculus-based algorithms are used to adjust robot forces and torques to achieve desired outcomes.

Q.47. What is the concept of robot trajectory smoothing?

Ans. Robot trajectory smoothing involves removing jerky movements to achieve smoother motions.

Q.48. How does calculus help in robot trajectory smoothing?

Ans. Calculus-based techniques are used to optimize trajectories and minimize jerk.

Q.49. What is the concept of robot coordination in multi-robot systems?

Ans. Robot coordination involves planning and controlling the movements of multiple robots.

Q.50. How is calculus applied in robot coordination algorithms?

Ans. Calculus is used to derive coordination strategies and optimize multi-robot interactions.

Multiple Choice Questions and Answers

Q.1. What is calculus?

(*a*) The study of geometry and shapes.
(*b*) The study of rates of change and accumulation.
(*c*) The study of logic and reasoning. (*d*) The study of computer algorithms.

Q.2. What are the two main branches of calculus?

(*a*) Geometry and algebra.
(*b*) Differential calculus and integral calculus.
(*c*) Trigonometry and statistics. (*d*) Logic and set theory.

Q.3. What does the derivative of a function represent?

(*a*) The area under the curve of the function.
(*b*) The rate of change of the function with respect to its independent variable.
(*c*) The integral of the function.
(*d*) The maximum value of the function.

Q.4. How is the derivative of a function denoted?

(*a*) $\int f(x)\,dx$ (*b*) $d/dx[f(x)]$ (*c*) $f'(x)$ (*d*) $\Delta f(x)$

Q.5. What does integral calculus involve?

(*a*) Finding antiderivatives of a function. (*b*) Calculating the area under a curve.
(*c*) Solving differential equations.
(*d*) Finding the maximum value of a function.

Q.6. How is the integral of a function denoted?

(*a*) ∫f(x) dx (*b*) f'(x) (*c*) d/dx[f(x)] (*d*) Δf(x)

Q.7. What is the application of calculus in robotics?

(*a*) Modeling economic systems.
(*b*) Designing buildings.
(*c*) Planning robot motion and control.
(*d*) Analyzing historical data.

Q.8. In robot motion planning, calculus helps in:

(*a*) Designing robot hardware.
(*b*) Computing optimal joint configurations.
(*c*) Generating random paths for robots.
(*d*) Generating random sensor data.

Q.9. What does kinematics in robotics refer to?

(*a*) The study of robot dynamics.
(*b*) The study of robot control algorithms.
(*c*) The study of robot motion without considering forces.
(*d*) The study of robot sensors.

Q.10. What is the Jacobian matrix in robotics?

(*a*) The matrix representation of a robot's end-effector position.
(*b*) The matrix used for robot calibration.
(*c*) The matrix that represents the robot's joint angles.
(*d*) The matrix that relates joint velocities to end-effector velocities.

Q.11. The Jacobian matrix is used to:

(*a*) Compute the inverse of a robot's transformation matrix.
(*b*) Determine the robot's end-effector position in 3D space.
(*c*) Calculate the robot's joint angles.
(*d*) Compute the robot's end-effector velocity for a given set of joint velocities.

Q.12. Robot control involves:

(*a*) Designing the physical structure of the robot.
(*b*) Determining the robot's position in space.
(*c*) Planning the robot's trajectory.
(*d*) Determining the actuator inputs to achieve desired robot behavior.

Q.13. In robot localization, calculus is used for:

(*a*) Calculating the robot's distance from a landmark.
(*b*) Determining the robot's position and orientation in the environment.
(*c*) Calculating the robot's speed and acceleration.
(*d*) Detecting obstacles in the robot's path.

Q.14. Differential equations are used in robotics for:

(*a*) Calculating the robot's trajectory.
(*b*) Modeling the robot's dynamics and control.
(*c*) Estimating the robot's position using sensor data.
(*d*) Generating random paths for the robot.

Q.15. Trajectory planning in robotics involves:

(*a*) Calculating the robot's position at a specific time.
(*b*) Generating a smooth path for the robot to follow.
(*c*) Estimating the robot's velocity using sensor data.
(*d*) Designing the robot's end-effector.

Q.16. What is the concept of optimization in robotics?

(*a*) The study of robot vision systems. (*b*) The study of robot control algorithms.
(*c*) The process of finding the best solution to a problem.
(*d*) The process of designing robot hardware.

Q.17. How is calculus applied in robot optimization problems?

(*a*) Calculating the robot's position and orientation.
(*b*) Optimizing the robot's physical structure.
(*c*) Finding the maximum value of a function.
(*d*) Minimizing or maximizing objective functions.

Q.18. Partial derivatives are used in robotics for:

(*a*) Calculating the robot's total displacement.
(*b*) Determining the robot's maximum speed.
(*c*) Computing the robot's joint angles.
(*d*) Computing gradients for optimization algorithms.

Q.19. Differential equations are used in robot control to:

(*a*) Determine the robot's joint angles.
(*b*) Compute the robot's end-effector position.
(*c*) Model the robot's dynamics and movement.
(*d*) Calculate the robot's total displacement.

Q.20. How does calculus contribute to adaptive control in robotics?

(*a*) It helps optimize the robot's path.
(*b*) It adjusts robot control parameters in real-time.
(*c*) It determines the robot's joint angles.
(*d*) It calculates the robot's end-effector position.

Q.21. In robot grasping algorithms, calculus helps in:

(*a*) Detecting obstacles in the robot's path. (*b*) Calculating the robot's joint angles.
(*c*) Determining optimal grasping strategies based on object shapes.
(*d*) Calculating the robot's end-effector position.

Q.22. Trajectory smoothing in robotics involves:

(*a*) Generating random paths for the robot.
(*b*) Removing jerky movements from the robot's trajectory.
(*c*) Calculating the robot's speed and acceleration.
(*d*) Planning the robot's trajectory.

Q.23. How are integrals used in computing robot displacements?

(*a*) They calculate the robot's velocity profile.
(*b*) They determine the robot's maximum speed.
(*c*) They calculate the total change in position over time.
(*d*) They determine the robot's end-effector position.

Q.24. What is the concept of robot coordination in multi-robot systems?

(*a*) The study of robot communication protocols.
(*b*) The study of robot hardware design.
(*c*) The study of robot calibration techniques.
(*d*) The planning and controlling of movements of multiple robots.

Q.25. How is calculus applied in robot coordination algorithms?

(*a*) It determines the robot's joint angles.
(*b*) It optimizes the communication between robots.
(*c*) It plans the trajectories of multiple robots.
(*d*) It calculates the robot's end-effector position.

Q.26. The derivative of a function represents:

(*a*) The rate of change of the function with respect to its independent variable.
(*b*) The integral of the function.
(*c*) The maximum value of the function.
(*d*) The area under the curve of the function.

Q.27. What does integral calculus deal with?

(*a*) The calculation of areas under curves. (*b*) The study of rates of change.
(*c*) The computation of derivatives.
(*d*) The analysis of discontinuities in functions.

Q.28. The integral of a function is denoted by:

(*a*) $\int f(x)\,dx$ (*b*) $f'(x)$ (*c*) $d/dx[f(x)]$ (*d*) $\Delta f(x)$

Q.29. What is the primary application of calculus in robotics?

(*a*) Designing robot hardware. (*b*) Modeling robot dynamics.
(*c*) Creating robot control algorithms. (*d*) Calculating robot trajectories.

Q.30. Kinematics in robotics focuses on:

(*a*) Robot dynamics. (*b*) Robot control.
(*c*) Robot motion without considering forces. (*d*) Robot perception.

Q.31. The Jacobian matrix in robotics represents:

(*a*) Robot joint angles. (*b*) Robot end-effector position.
(*c*) Robot control inputs.
(*d*) Relationship between joint velocities and end-effector velocities.

Q.32. Robot control involves:
(*a*) Designing the physical structure of the robot.
(*b*) Determining the robot's position in space.
(*c*) Planning the robot's trajectory.
(*d*) Determining the actuator inputs to achieve desired robot behavior.

Q.33. In robot localization, calculus is used for:
(*a*) Calculating the robot's distance from a landmark.
(*b*) Determining the robot's position and orientation in the environment.
(*c*) Calculating the robot's speed and acceleration.
(*d*) Detecting obstacles in the robot's path.

Q.34. Differential equations are used in robotics for:
(*a*) Calculating the robot's trajectory.
(*b*) Modeling the robot's dynamics and control.
(*c*) Estimating the robot's position using sensor data.
(*d*) Generating random paths for the robot.

Q.35. Trajectory planning in robotics involves:
(*a*) Calculating the robot's position at a specific time.
(*b*) Generating a smooth path for the robot to follow.
(*c*) Estimating the robot's velocity using sensor data.
(*d*) Designing the robot's end-effector.

Q.36. Optimization in robotics refers to:
(*a*) The study of robot hardware design.
(*b*) The process of finding the best solution to a problem.
(*c*) The study of robot control algorithms.
(*d*) The process of optimizing robot communication.

Q.37. Partial derivatives are used in robotics for:
(*a*) Calculating the robot's total displacement.
(*b*) Determining the robot's maximum speed.
(*c*) Computing the robot's joint angles.
(*d*) Computing gradients for optimization algorithms.

Q.38. Differential equations are used in robot control to:
(*a*) Determine the robot's joint angles.
(*b*) Compute the robot's end-effector position.
(*c*) Model the robot's dynamics and movement.
(*d*) Calculate the robot's total displacement.

Q.39. How does calculus contribute to adaptive control in robotics?
(*a*) It helps optimize the robot's path.

(*b*) It adjusts robot control parameters in real-time.
(*c*) It determines the robot's joint angles.
(*d*) It calculates the robot's end-effector position.

Q.40. In robot grasping algorithms, calculus helps in:
(*a*) Detecting obstacles in the robot's path. (*b*) Calculating the robot's joint angles.
(*c*) Determining optimal grasping strategies based on object shapes.
(*d*) Calculating the robot's end-effector position.

Q.41. Trajectory smoothing in robotics involves:
(*a*) Generating random paths for the robot.
(*b*) Removing jerky movements from the robot's trajectory.
(*c*) Calculating the robot's speed and acceleration.
(*d*) Planning the robot's trajectory.

Q.42. How are integrals used in computing robot displacements?
(*a*) They calculate the robot's velocity profile.
(*b*) They determine the robot's maximum speed.
(*c*) They calculate the total change in position over time.
(*d*) They determine the robot's end-effector position.

Q.43. What is the concept of robot coordination in multi-robot systems?
(*a*) The study of robot communication protocols.
(*b*) The study of robot hardware design.
(*c*) The study of robot calibration techniques.
(*d*) The planning and controlling of movements of multiple robots.

Q.44. How is calculus applied in robot coordination algorithms?
(*a*) It determines the robot's joint angles.
(*b*) It optimizes the communication between robots.
(*c*) It plans the trajectories of multiple robots.
(*d*) It calculates the robot's end-effector position.

Q.45. What does the derivative of a function represent?
(*a*) The rate of change of the function with respect to its independent variable.
(*b*) The integral of the function.
(*c*) The maximum value of the function.
(*d*) The area under the curve of the function.

Q.46. What does integral calculus deal with?
(*a*) The calculation of areas under curves. (*b*) The study of rates of change.
(*c*) The computation of derivatives.
(*d*) The analysis of discontinuities in functions.

Q.47. The integral of a function is denoted by:
(*a*) $\int f(x)\,dx$ (*b*) $f'(x)$ (*c*) $d/dx[f(x)]$ (*d*) $\Delta f(x)$

Q.48. What is the primary application of calculus in robotics?

(*a*) Designing robot hardware. (*b*) Modeling robot dynamics.
(*c*) Creating robot control algorithms. (*d*) Calculating robot trajectories.

Q.49. Kinematics in robotics focuses on:

(*a*) Robot dynamics. (*b*) Robot control.
(*c*) Robot motion without considering forces. (*d*) Robot perception.

Q.50. The Jacobian matrix in robotics represents:

(*a*) Robot joint angles. (*b*) Robot end-effector effector velocities.

Key to Multiple Choice Questions

1. (*b*)	**2.** (*b*)	**3.** (*b*)	**4.** (*c*)	**5.** (*b*)	**6.** (*a*)	**7.** (*c*)	**8.** (*b*)	**9.** (*c*)
10. (*d*)	**11.** (*d*)	**12.** (*d*)	**13.** (*b*)	**14.** (*b*)	**15.** (*b*)	**16.** (*c*)	**17.** (*d*)	**18.** (*d*)
19. (*c*)	**20.** (*b*)	**21.** (*c*)	**22.** (*b*)	**23.** (*c*)	**24.** (*d*)	**25.** (*c*)	**26.** (*a*)	**27.** (*a*)
28. (*a*)	**29.** (*d*)	**30.** (*c*)	**31.** (*d*)	**32.** (*d*)	**33.** (*b*)	**34.** (*b*)	**35.** (*b*)	**36.** (*b*)
37. (*d*)	**38.** (*c*)	**39.** (*b*)	**40.** (*c*)	**41.** (*b*)	**42.** (*c*)	**43.** (*d*)	**44.** (*c*)	**45.** (*a*)
46. (*a*)	**47.** (*a*)	**48.** (*d*)	**49.** (*c*)	**50.** (*d*)				

Projects Works

1. **Kinematic Analysis of Robot Arms:** Use calculus to analyze the position, velocity, and acceleration of robot arms using forward and inverse kinematics.
2. **Trajectory Planning for Mobile Robots:** Develop algorithms that use calculus to plan smooth and efficient trajectories for mobile robots to reach their targets.
3. **Control of Robot Manipulators:** Investigate control strategies for robot manipulators using calculus-based methods such as PID control and feedback control.
4. **Optimization of Robot Paths:** Use calculus optimization techniques to find the shortest or fastest path for a robot to reach its destination while avoiding obstacles.
5. **Dynamics and Stability Analysis of Robot Systems:** Analyze the dynamics and stability of robot systems using calculus and differential equations.
6. **Sensor Calibration and Fusion:** Develop algorithms to calibrate and fuse data from multiple sensors on a robot using calculus and statistical methods.
7. **Motion Planning in Complex Environments:** Implement motion planning algorithms that use calculus to navigate robots through complex and dynamic environments.
8. **Path Following for Autonomous Vehicles:** Use calculus to design controllers that allow autonomous vehicles to accurately follow predefined paths.
9. **Robot Localization and Mapping:** Use calculus and probability theory to solve simultaneous localization and mapping (SLAM) problems for robots.
10. **Robot Vision and Image Processing:** Apply calculus-based techniques to process and analyze images captured by robot cameras for object recognition and navigation.
11. **Trajectory Optimization for Drone Flight:** Develop algorithms that use calculus to optimize the trajectory of drones for tasks like package delivery or surveillance.

12. **Control of Quadcopter Stability:** Investigate control techniques for quadcopters that utilize calculus to maintain stable flight in various conditions.
13. **Robot Path Planning with Uncertainty:** Create algorithms that use calculus and probabilistic methods to plan robot paths while accounting for uncertainty in the environment.
14. **Inverse Reinforcement Learning for Robots:** Use calculus and optimization to infer the underlying objectives of a robot's actions from observed behavior.
15. **Optimal Resource Allocation for Multi-Robot Systems:** Apply calculus optimization methods to allocate tasks among a team of robots to maximize efficiency.
16. **Model Predictive Control for Robots:** Implement model predictive control algorithms that use calculus to optimize control inputs over a finite time horizon.
17. **Real-Time Obstacle Avoidance:** Develop real-time algorithms that use calculus to enable robots to avoid obstacles and replan their paths on the fly.
18. **Human-Robot Interaction:** Investigate how calculus can be used to model and improve the interaction between humans and robots in collaborative tasks.
19. **Motion Analysis in Robotics:** Use calculus to analyze and interpret the motion of robots and objects in the environment from sensor data.
20. **Robot Swarm Behavior:** Use calculus-based techniques to model and analyze emergent behaviors in swarms of robots, such as flocking and pattern formation.

04

Differential Equations in Modeling and Analyzing Various Aspects of Robotics and Automation

4.1 INTRODUCTION

The modeling and analysis of numerous elements of robotics and automation heavily relies on differential equations. Differential equations are types of mathematical equations that have one or more unknown function derivatives. They are employed to explain the evolution of a function through time or in connection to other factors. Physics, biology, economics, and engineering are just a few of the scientific, engineering, and mathematical disciplines that depend heavily on differential equations. In modeling and analyzing dynamic systems and processes, they are extremely important.

4.2 GENERAL FORM OF A DIFFERENTIAL EQUATION AND TYPES

A differential equation has the generic form: dy/dx = f(x, y), where f(x, y) is a known function of both x and y, and dy/dx is the derivative of the unknown function y with respect to the independent variable x. The unknown function y that satisfies the differential equation for a specific set of initial or boundary conditions is the solution to a differential equation.

Depending on their order and linearity, differential equations can be divided into a number of different types:

1. **Ordinary Differential Equations (ODEs):** ODEs involve derivatives with respect to a single independent variable, usually indicated by the symbol t. System-evolving systems are modeled using ODEs.
2. **Partial Differential Equations (PDEs):** These equations contain partial derivatives for a number of independent variables. PDEs are frequently used in physics and engineering to represent systems with spatial fluctuations.
3. **First-order Differential Equations:** In these equations, the unknown function y with respect to x is simply represented by its first derivative.
4. **The fourth-order Equations that involve differentials:** These equations involve the second derivative of the unknown function y with respect to x. They are frequently employed to simulate systems that have acceleration or oscillatory behavior.

5. **Linear Differential Equations:** The unknown function and its derivatives are linear in these equations. They have established and well-liked problem-solving methods.
6. **Nonlinear Differential Equations:** The unknown function and its derivatives are nonlinear in these equations. Numerical methods are frequently necessary for the solution of nonlinear differential equations.

Differential equations are effective tools for figuring out how dynamic systems behave, forecasting future states, and creating control plans. Numerous branches of science and engineering, including as mechanics, control theory, electrical circuits, fluid dynamics, and many more, make substantial use of them. For researchers, engineers, and scientists working in a variety of sectors, being able to solve differential equations and comprehend their answers is vital.

4.3 KEY AREAS WHERE DIFFERENTIAL EQUATIONS APPLIED IN ROBOTICS

Here are some significant fields where differential equations are frequently used:

(*a*) **Robot Kinematics:** In robot manipulators, the relationship between joint velocities and end-effector velocities is described by differential equations. Understanding the robot's mobility and position depends on this.

(*b*) **Robot Dynamics:** Taking into account forces, torques, and accelerations, differential equations are utilized to represent the dynamic behaviour of robotic systems. Robot control strategies are designed with the aid of dynamic models.

(*c*) **Trajectory Planning:** Smooth trajectories are planned for robots to travel from one point to another while minimizing jerk, acceleration, or other criteria using differential equations.

(*d*) **Control Systems:** Differential equations are essential to control theory and are employed in the creation of control algorithms that control robot behavior and help achieve desired results.

(*e*) **Path Planning:** Differential equations are used in motion planning to determine the best route for a robot to take in order to go where it has to go while avoiding impediments.

(*f*) **Sensor Fusion:** To precisely estimate the robot's condition, sensor fusion algorithms combine data from several sensors (such as cameras, lidars, and IMUs) using differential equations.

(*g*) **Localization and Mapping:** Robots employ simultaneous localization and mapping (SLAM) techniques to simultaneously localize themselves and construct maps of uncharted surroundings.

(*h*) **Robot Vision:** Image filtering, edge detection, and image segmentation are just a few of the computer vision methods that involve differential equations.

(*i*) **Robot Path Following:** Controllers that allow robots to precisely follow predetermined pathways or trajectories are designed using differential equations.

(*j*) **Differential Equations:** These are used in system identification techniques to estimate the parameters of a robot's dynamic model from experimental data.

(*k*) **Motion Analysis:** Differential equations are applied to analyze the motion of robots, objects, or humans captured by sensors.

(*l*) **Stability Analysis:** Differential equations are used to analyze the stability of robot systems and control algorithms to ensure reliable and safe operation.

4.4 ORDINARY DIFFERENTIAL EQUATIONS

One independent variable and one or more derivatives of an unidentified function with respect to that variable are the components of an ordinary differential equation (ODE), a type of differential equation. They are applied to simulate processes or systems that alter over time or in response to a single independent variable. Physics, biology, economics, and control systems are just a few of the research and engineering fields where ODEs are crucial.

An ordinary differential equation has the generic form: dy/dx = f(x, y), where f(x, y) is a known function of both x and y, and dy/dx is the derivative of the unknown function y with respect to the independent variable x. The unknown function y that satisfies the equation for a certain set of beginning or boundary conditions is the solution to an ODE. Ordinary differential equations are solved using a variety of mathematical strategies, including:

Separation of Variables: By placing the variables x and y on opposing sides of the equation and then integrating both sides, this technique is used to solve first-order ODEs. Variable separation is a typical method for solving specific kinds of ordinary differential equations (ODEs). It can be used with first-order ODEs with the formula dy/dx = f(x)g(y), where y is the dependent variable, x is the independent variable, and f(x) and g(y) are functions of the individual variables.

The general steps to solve an ODE using separation of variables are as follows:

Step 1: Identify the ODE in the appropriate form: dy/dx = f(x)g(y).

Step 2: Separate the variables to one side of the equation, with y terms on one side and x terms on the other side:

$$g(y)\,dy = f(x)\,dx$$

Step 3: Integrate both sides with respect to their respective variables. This step involves integrating each side independently:

$$\int g(y)\,dy = \int f(x)\,dx$$

Step 4: Perform the integrations and add a constant of integration, C, on one side of the equation:

$$\int g(y)\,dy = \int f(x)\,dx + C$$

Step 5: Solve for y by evaluating the integral of g(y) and the integral of f(x). The constant of integration, C, may be involved in the solution, and its value can be determined by applying any initial conditions provided in the problem.

Variable separation allows for the solution of all ODEs. It is restricted to particular first-order ODE types. Nevertheless, when the method is appropriate, it frequently results in an explicit solution for y as a function of x.

Integrating Factors: First-order linear ODEs are precise and simpler to solve when they are solved with integrating factors. Certain types of first-order linear ordinary differential equations (ODEs) that are not in the exact form can be solved using integrating variables.

An ODE is said to be in the exact form if it can be written as:

$$M(x, y)\,dx + N(x, y)\,dy = 0$$

where M(x, y) and N(x, y) are functions of both x and y. If an ODE is not in the exact form, it cannot be solved directly using standard integration methods. However, an integrating factor can be introduced to transform the ODE into an exact form, making it solvable.

The integrating factor for the ODE in the form mentioned above (M dx + N dy = 0) is a function μ(x, y) that makes the equation exact. The integrating factor μ(x, y) is defined as:

$$\mu(x, y) = \exp(\int[\partial M/\partial y - \partial N/\partial x]\,dy)$$

where ∂M/∂y and ∂N/∂x are the partial derivatives of M and N with respect to y and x, respectively.

Once the integrating factor μ(x, y) is found, you multiply both sides of the ODE by it: μ(x, y) M(x, y) dx + μ(x, y) N(x, y) dy = 0

Now, the left-hand side of the equation can be recognized as the total differential of a function F(x, y). So, the ODE can be rewritten as:

$$dF(x, y) = 0$$

where F(x, y) is called the potential function.

Finally, integrating both sides of the equation, you get the solution for the original ODE:

$$F(x, y) = C$$

where C is the constant of integration determined by any initial or boundary conditions given in the problem.

In summary, the steps to solve a first-order linear ODE using integrating factors are as follows:

Step 1: Check if the ODE is in the form M(x, y) dx + N(x, y) dy = 0.

Step 2: Calculate the integrating factor $\mu(x, y) = \exp(\int[\partial M/\partial y - \partial N/\partial x]\, dy)$.

Step 3: Multiply both sides of the ODE by the integrating factor μ(x, y).

Step 4: Recognize the left-hand side as the total differential of a potential function F(x, y).

Step 5: Integrate both sides to find F(x, y) = C, where C is the constant of integration.

Step 6: If initial or boundary conditions are given, use them to determine the value of the constant C and obtain the particular solution.

Homogeneous ODEs: By applying a replacement that turns these ODEs into separable or linear ODEs, these ODEs are resolved. An exceptional kind of first-order ordinary differential equation (ODE) known as a homogeneous ODE allows all terms to be written as homogeneous functions of the same degree. When all of a function's terms have the same variable power, the function is said to be homogenous. This implies that the dependent variable and its derivatives in the context of ODEs only appear as a ratio or in powers of the ratio.

A homogeneous ODE can be written in the following general form:

$$dy/dx = F(x, y)$$

where F(x, y) is a function that satisfies the homogeneity condition:

$$F(tx, ty) = t^n * F(x, y);$$

for any non-zero constant t and where n is a fixed number. In other words, if one scales both x and y by the same factor t, the function F(x, y) should so be scale by a fixed power of t.

To solve a homogeneous ODE, you can use the substitution method. Let y = vx, where v is a new variable. Then, differentiate y with respect to x and substitute into the ODE:

$$dy/dx = v + x * dv/dx$$

Now, replace dy/dx and y in the original ODE:

$$v + x * dv/dx = F(x, vx)$$

Now, the ODE becomes separable, and you can solve it by separating variables and integrating:

$$(1/v)dv = (F(x, vx) - x)/xdx$$

We will get an equation involving v and x after integration. Use the formula y = vx, where v is the result you got and x is the independent variable, to find y. The method presented here only works for first-order homogeneous ODEs because not all ODEs are homogeneous. The process can be more difficult for higher-order homogeneous ODEs, and additional approaches, such as power series or Laplace transforms, may be employed to locate the solutions.

Exact Differential Equations: An exact differential equation is a special type of first-order ordinary differential equation (ODE) that can be expressed in the form: M(x, y) dx + N(x, y) dy = 0; where M(x, y) and N(x, y) are functions of both x and y. An exact differential equation is called "exact" because it can be derived from a multivariate ordinary differential equation (M(x, y)).

A necessary and sufficient condition for an ODE to be exact is that the mixed partial derivatives of M and N with respect to x and y are equal, i.e., $\partial M/\partial y = \partial N/\partial x$. In mathematical terms:

$$\partial M/\partial y = \partial N/\partial x$$

If this condition is met, the ODE can be solved using the following method:

Step 1: Check if the ODE is in the form M(x, y) dx + N(x, y) dy = 0 and verify the condition $\partial M/\partial y = \partial N/\partial x$.

Step 2: If the condition is satisfied, find a function F(x, y) such that its partial derivatives satisfy:

$$\partial F/\partial x = M(x, y) \text{ and } \partial F/\partial y = N(x, y)$$

Step 3: Integrate the partial derivatives of F with respect to x and y to obtain F(x, y).

Step 4: The solution to the exact ODE is given by the equation F(x, y) = C, where C is a constant of integration.

Finding the function F(x, y) can occasionally be difficult, and the procedure could call for some trial and error or deft algebraic trickery to figure out the right integrating factor. However, the answer is simple once F(x, y) is identified. It is significant to remember that not every ODE is accurate. The ODE must be solved using other methods, such as integrating factors or by treating it as a homogeneous or linear ODE, if the criterion M/y = N/x is not satisfied.

Series Solutions: In order to solve for a series' coefficients, the unknown function y must be represented as a power series. By expressing functions as infinite series, series solutions are a potent method for resolving differential equations and other mathematical issues. A series is an expression made up of numerous terms, each of which is the sum of a particular function or variable raised to a different power.

When other approaches, like exact solutions or integrating factors, are not always feasible, series solutions are particularly helpful for solving linear ordinary differential equations (ODEs) and partial differential equations (PDEs).

The fundamental premise of series solutions is that the unknown function can be represented as an extension of a power series around a specific point, typically taken to be x = 0. For a function y(x), for instance, we presum:

$$y(x) = a_0 + a_1x + a_2x^2 + a_3x^3 + . = \Sigma(a_n * x^n)$$

where a_0, a_1, a_2, . are coefficients to be determined.

In order to calculate the coefficients, we replace the original differential equation with the power series representation and compare coefficients for similar powers of x. This will produce a recurrence relation or a set of equations that may be solved repeatedly or recursively for the coefficients. Usually, at a particular term, the series expansion is terminated, yielding an approximation of the answer. The number of terms kept in the series determines how accurate the solution is.

Series solutions come in a variety of forms, such as

(a) **Power Series Solutions:** The unknown function is shown as a power series in terms of x, as was previously mentioned.

(*b*) **Frobenius Series Solutions:** Solutions to second-order linear differential equations with regular singular points can be found using Frobenius series, a special type of power series.

(*c*) **Laurent Series Solutions:** These are used to resolve singular differential equations.

(*d*) **Taylor Series Solutions:** A particular kind of power series with its origin at x = a and used to approximate functions close to that location.

However, they may require some algebraic manipulation, and their convergence properties need to be taken into consideration, especially for nonlinear equations or equations with singularities. Series solutions are particularly useful in cases where closed-form solutions cannot be found, or when numerical methods are not practical or too computationally expensive.

Numerical Methods: Numerous ODEs cannot be solved analytically; therefore, numerical techniques, such as Euler's method, Runge-Kutta methods, and finite difference methods, are used to approximate solutions. Numerical techniques are ways for estimating answers to mathematical issues that cannot be resolved precisely or analytically. These techniques use numerical approximations to find an approximation of the solution by iteratively computing the problem in smaller parts.

Numerous branches of science, engineering, and mathematics frequently employ numerical methods, particularly when tackling intricate issues or intricate systems that are challenging to represent symbolically. Among the most popular numerical techniques are:

(*a*) Numerical techniques for estimating definite integrals of functions are known as numerical integration (*a*). The Trapezoidal Rule, Simpson's Rule, and numerous adaptive quadrature techniques are examples.

(*b*) **Numerical Differentiation:** Estimating function derivatives. It is common to employ techniques like finite difference approximations.

(*c*) Methods for locating the roots of equations of the form f(x) = 0. The Secant Method, Newton-Raphson Method, and Bisection Method are all well-liked methods.

(*d*) **Numerical Solutions of ODEs and PDEs:** Methods for resolving ordinary differential equations (ODEs) and partial differential equations (PDEs) when exact solutions are not practical. The Euler's method, Runge-Kutta methods, and finite difference methods for PDEs are typical techniques.

(*e*) Curve fitting and interpolation are techniques for approximate function calculation using a small number of data points. Lagrange interpolation, Newton interpolation, and least squares regression are a few examples.

(*f*) **Numerical Linear Algebra:** Using numerical methods like Gaussian elimination, LU decomposition, and power iteration for eigenvalues, one can solve systems of linear equations and eigenvalue issues.

(*g*) **Optimization Techniques:** Approaches for determining a function's minimum or maximum. Common optimization techniques include gradient descent, conjugate gradient method, and genetic algorithms.

(*h*) **Monte Carlo Methods:** Probabilistic approaches to problem solving using statistical simulation and random sampling.

(*i*) Numerical techniques have their limits. The step size, the method's numerical stability, and the characteristics of the issue being addressed are only a few examples of the variables that affect how accurate and effective numerical solutions are. Furthermore, some issues can be computationally challenging, necessitating a considerable investment in processing power to produce accurate findings.

(*j*) Overall, numerical techniques offer strong instruments for approximating solutions to a variety of mathematical and scientific issues. By enabling researchers and engineers to take on complex issues that would otherwise be insurmountable, numerical approaches have revolutionised several disciplines of study.

Laplace Transform: The Laplace transform is a useful tool for solving differential equations in the frequency domain by converting a function of time (t) into a function of a complex variable (s). The Laplace transform of a function f(t), denoted as L[f(t)], is defined by the integral:

$$L[f(t)] = \int[0, \infty]\ e^{(-st)} * f(t)\ dt$$

where s is a complex number of the form $s = \sigma + j\omega$, with σ representing the real part and ω the imaginary part of s.

The Laplace transform has several important properties that make it a valuable tool for solving differential equations:

1. **Linearity:** $L[a * f(t) + b * g(t)] = a * L[f(t)] + b * L[g(t)]$, where a and b are constants.
2. **Differentiation:** $L[d^n/dt^n f(t)] = s^n * L[f(t)] - s^{(n-1)} * f(0) - s^{(n-2)} * f'(0) - . - f^{(n-1)}(0)$
3. **Integration:** $L[\int[0, t] f(\tau)\ d\tau] = 1/s * L[f(t)]$
4. **Time Shifting:** $L[f(t - a) * u(t - a)] = e^{(-as)} * L[f(t)]$
5. **Frequency Shifting:** $L[e^{(at)} * f(t)] = L[f(t - a)]$ evaluated at $s = s - a$
6. **Convolution:** $L[f(t) * g(t)] = L[f(t)] * L[g(t)]$

The Laplace transform converts the differential equation into an algebraic equation, which can often be simpler to solve, which makes it particularly helpful for resolving initial value issues for linear ODEs. The inverse Laplace transform is used to return the solution to the time domain once it has been obtained in the Laplace domain. The inverse Laplace transform of a function F(s), denoted as $L^{(-1)}[F(s)]$, gives the original function f(t). The inverse Laplace transform is typically performed using tables, partial fraction decomposition, or the residue theorem from complex analysis.

The analysis of linear time-invariant systems and differential equations is crucial in many areas, including control systems, electrical engineering, signal processing, and other subjects. It provides a potent and organized method for addressing a variety of mathematics and engineering issues.

Stability Analysis: Stability analysis for ODEs that represent dynamic systems entails locating equilibrium points and assessing their stability using ideas from calculus and linear algebra. A fundamental idea in many scientific and engineering disciplines, stability analysis is used to comprehend how systems behave over time and to ensure that they operate predictably and dependably. The study of dynamical systems, control systems, differential equations, and numerical approaches in particular depend greatly on stability analysis. Determining whether a system's behaviour converges to a steady state over time in response to perturbations or disturbances is the basic objective of stability analysis.

Stability analysis entails observing how solutions behave over time in the context of differential equations and dynamical systems. Three primary categories of stability exist:

(*a*) **Stable Equilibrium (Asymptotically Stable):** If minor disturbances near a system's equilibrium point cause the system to gradually return to that point over time, the system is said to have a stable equilibrium point. In other words, as time approaches infinity, neighbouring paths converge to the equilibrium point.

(*b*) **Unstable Equilibrium:** An equilibrium point is considered unstable if small perturbations around that point cause the system to diverge away from the equilibrium, resulting in trajectories moving further apart over time.

(*c*) **Semi-stable or Marginally Stable:** When disturbances near an equilibrium point neither converge nor diverge over time, a system is considered semi-stable. Although they may not approach or depart from the equilibrium, trajectories may stay rather close to it.

Various techniques are frequently used to undertake stability analyses, depending on the nature of the system and the particular issue. Typical methods for stability analysis include the following:

Linearization: It is common practice to linearize nonlinear systems around an equilibrium point. In order to apply linear algebra techniques to analyze stability, it is necessary to approximate the system as a linear one close to the equilibrium.

***Lyapunov Stability*:** Without explicitly solving the system's differential equations, stability can be assessed using Lyapunov's direct technique. It entails identifying a function, referred to as a Lyapunov function that meets particular requirements and reveals the system's stability characteristics.

***Eigenvalue Analysis*:** The system matrix's eigenvalues are crucial predictors of stability for linear systems; if all eigenvalues have negative real portions, the system is stable.

***Phase Plane Analysis*:** The behavior of nonlinear systems is visualized and understood using graphic techniques like phase plane analysis. To learn more about the system's stability and behavior, the system's trajectories are plotted in a two-dimensional space.

Engineering and control systems depend heavily on stability analysis to ensure the secure and dependable operation of various technologies and systems. Engineers can create controllers and feedback systems to achieve desired behaviors and prevent undesired consequences like oscillations, divergences, or uncontrollable responses by knowing the stability features of a system. ODEs are frequently used in numerous disciplines to explain the dynamic behaviors of many systems, including biology to investigate population dynamics, engineering to model control systems, and physics to describe motion. Understanding calculus, algebra, and ODE-specific mathematical methods are essential for solving ODEs. For scientists, engineers, and researchers across a wide range of disciplines, the capacity to solve and analyze ODEs is crucial.

For modeling dynamic systems and processes, ordinary differential equations (ODEs) are frequently employed in a variety of disciplines. ODEs are especially well adapted for modeling scenarios where a system's behavior is dependent on its current state and the pace at which that state is changing over time.

Some common applications of ODEs in modeling include

(*a*) **Physics:** The motion of particles, mechanical systems, and physical processes like heat conduction, wave propagation, and fluid dynamics are all modeled using ODEs in physics. For instance, a second-order ODE is frequently used to model Newton's second law of motion.

(*b*) **Engineering:** Engineers utilize ODEs to simulate and examine a variety of engineering systems, including chemical reactions, electrical circuits, control systems, and structural dynamics. ODEs, for instance, can be used to describe the behavior of an RLC circuit.

(*c*) **Biology:** To comprehend biological processes like population dynamics, biochemical reactions, pharmacokinetics, and neural network dynamics, ODEs are used in biological modeling. Researchers can explore how different elements interact and have an impact on one another using these models.

(*d*) **Economics:** Economic systems, growth models, investment dynamics, and consumer behavior are all modeled using ODEs. These models aid economists in forecasting and trend analysis.

(*e*) **Environmental Sciences:** ODEs are employed to simulate environmental phenomena such as pollution dispersion, ecosystem population dynamics, and heat transport in the atmosphere or oceans. These models help to comprehend and foresee environmental changes.

(*f*) **Epidemiology:** ODEs are essential in epidemiological modeling, especially when disease outbreaks are taking place. Infectious illness spread in communities is predicted by models like the SIR (Susceptible-Infected-Recovered) model.

(*g*) **Computer graphics:** Character animation, physics simulations, and particle systems are all handled by ODEs in computer graphics. They make it possible for virtual objects to move and interact in 3D settings in a realistic manner.

(*h*) **Finance:** In order to examine option pricing, interest rate models, and portfolio optimization, ODEs are utilized in financial modeling. These models aid in evaluating and controlling financial risks.

ODEs are crucial for comprehending real-world occurrences and for arriving at well-informed decisions across a range of disciplines because they provide a flexible and effective technique to represent dynamic systems. ODEs can be resolved analytically, numerically, or by combining the two methods, depending on the system complexity and desired level of precision.

4.5 ODE USED IN CONTROLLING ROBOT DYNAMICS

Ordinary differential equations (ODEs) are widely used in robotics to model and regulate robot motion. When forces and torques are acting on a robot its position, velocity, and acceleration change with time. This is referred to as the robot's dynamics. These dynamic interactions are mathematically represented using ODEs, allowing the creation of robot motion and manipulation control schemes.

Several types of ODEs are commonly used in controlling robot dynamics:

(*a*) **Forward Kinematics:** The link between joint positions and the location and orientation of the robot's end-effector—typically a tool or gripper—is described by forward kinematics ODEs. Based on the joint angles, you may compute the end-effector's pose by solving these ODEs.

(*b*) **Inverse Kinematics:** To compute the joint angles necessary to obtain a given end-effector pose, inverse kinematics ODEs are used. Planning exact robot trajectories and placements requires solving these ODEs.

(*c*) **Joint Dynamics:** The forces and torques operating on the joints of the robot are taken into account by ODEs defining joint dynamics. These ODEs are essential for directing the robot's motion and making sure the intended motion is accomplished.

(*d*) **Robot Arm Dynamics:** Understanding the robot's movements and interactions with its surroundings is aided by understanding the ODEs driving the dynamics of the robot arm (linkages and joints). Motion planning, trajectory optimization, and control algorithms all make use of them.

(*e*) **Robot Mobility Dynamics:** ODEs describe the mobility dynamics of mobile robots, such as wheeled or legged robots, including how wheel or leg forces affect the robot's stability and movement.

(*f*) **Feedback Control:** ODEs are employed in control methods like PID (Proportional Integral Derivative) controllers, which compute control signals and rectify faults in real-time robot movements by solving differential equations.

(*g*) **Path Planning:** To determine the robot's trajectory from its current position to a goal position while avoiding obstacles, path planning algorithms employ ODEs. To determine the best routes that minimise cost functions or guarantee smooth motion, these algorithms solve ODEs.

In order to forecast the behavior of the robot and validate control algorithms before implementation on actual hardware, ODE-based modeling and control of robot dynamics are frequently carried out in simulation environments. Engineers can use simulation tools to test different control strategies, improve performance, and reduce dangers in practical applications. Overall, ODEs offer a key framework for comprehending robotic system behavior and creating control schemes that let robots carry out intricate jobs precisely, safely, and effectively.

4.6 PARTIAL DIFFERENTIAL EQUATIONS (PDES) AS APPLIED TO ROBOTICS

PDEs are particularly well-suited for describing physical phenomena involving multiple independent variables, such as space and time, which are common in robotic systems, and are used in robotics to model and analyze various aspects of robot behavior and interactions with the environment.

Here are some key applications of PDEs in robotics:

Robot Arm Dynamics: The dynamics of robot arms and manipulators can be modeled using PDEs. Taking into account the physical characteristics of the robot's linkages and joints, they discuss the correlations between joint angles, velocities, and torques. The robot's motion can be accurately predicted thanks to the solution of these PDEs, which also facilitates the formulation of control plans for achieving desired trajectories and tasks.

The Euler-Lagrange equations are a set of partial differential equations (PDEs) that can be used to model the dynamics of robot arms. These PDEs are frequently used to simulate the dynamics of mechanical systems, including manipulators and robotic arms. They are derived from the idea of least action in classical mechanics.

The interaction between joint locations, velocities, accelerations, and forces or torques delivered to the joints is at the heart of a robot arm's dynamics. The objective is to identify the equations of motion that control how the robot arm behaves as it goes through various configurations.

The following equations provide the Euler-Lagrange equations for a robot arm with n degrees of freedom:

$$\partial(\partial L/\partial q \cdot i)/\partial t - \partial L/\partial q_i = \tau_i$$

where,

- L is the Lagrangian of the system, which is the difference between the kinetic and potential energy of the robot arm.
- q_i (i = 1, 2, ..., n) are the generalized coordinates representing the joint angles of the robot arm.
- $q \cdot i$ are the time derivatives of the generalized coordinates (i.e., joint velocities).
- τ_i are the generalized forces (or torques) acting on the joints.

The Lagrangian L is typically formulated as the difference between the kinetic and potential energies of the robot arm:

$$L = T - U$$

where T represents the kinetic energy of the robot arm, and U represents its potential energy.

The kinetic energy T is given by:

$$T = (1/2) \sum(m_i * v_i^2)$$

where m_i is the mass of the i-th link, and v_i is the velocity of the center of mass of the i-th link.

The potential energy U is usually defined as the gravitational potential energy:

$$U = \sum(m_i * g * h_i)$$

where g is the acceleration due to gravity, and h_i is the height of the center of mass of the i-th link above a reference level.

We can calculate the motion of the robot arm and how it responds to applied forces or torques by solving the Euler-Lagrange equations for the joint angles and joint velocities. This data is essential for steering the robot arm, plotting paths, and guaranteeing precise and safe movement in a variety of robotic applications.

In real-world applications, the Runge-Kutta method is frequently used to numerically solve the Euler-Lagrange equations in order to mimic the robot arm's behavior over time. These simulations aid in comprehending the dynamics of the robot arm, optimizing control approaches, and forecasting how it will react to various tasks and environmental factors.

Robot Mobility and Locomotion: PDEs are used to represent the dynamics of locomotion for mobile robots, such as those with wheels or legs. These PDEs explain how varied forces, terrains, and control inputs affect the robot's movements. The investigation of stability, gait optimization, and terrain adaptability is made possible by resolving the PDEs. Partial differential equations (PDEs) can be used to predict the dynamics of mobile robots as they traverse diverse terrains during locomotion and mobility. These PDE-based models make it possible to analyze the mobility, stability, and interactions of the robot with its surroundings.

There are several ways PDEs are applied to model robot mobility and locomotion:

(*a*) **Wheeled Robots:** PDEs can be used to simulate the dynamics of a wheeled robot's motion as it traverses various surfaces. The PDEs define the relationships between wheel torques, robot velocity, and forces acting on the wheels while accounting for the mass of the robot, the characteristics of the wheels, and the frictional characteristics of the terrain.

(*b*) **Legged Robots:** The dynamics of legged robots as they run, walk, or climb over obstacles can be modelled using PDEs. The PDEs consider the robot's body mass, leg lengths, and the unevenness of the terrain to explain the leg forces, joint angles, and leg kinematics.

(*c*) **Crawl Robots:** Robots that crawl or slither like snakes or caterpillars can be described using PDEs, which can also describe how the robot interacts with the environment.

(*d*) **Crawling Robots:** PDEs can describe the body deformation and interactions of crawling robots, which move or slither like snakes or caterpillars. These simulations take into account the body's form, ground friction, and external forces.

(*e*) **Swimming Robots:** In fluid environments, PDEs may simulate the dynamics of swimming robots. The PDEs consider the robot's shape, propulsion system, and fluid properties as they describe the fluid dynamics around the robot's body and fins.

(*f*) **Terrain Adaptation:** PDE-based models can be used to examine the ways in which mobile robots modify their locomotion tactics in response to the characteristics of the terrain. The PDEs can record how the robot interacts with the environment to enhance stability and efficiency of movement.

(*g*) **Obstacle Avoidance:** Path planning and obstacle avoidance algorithms can both use PDEs. The robot can go through intricate and congested settings because the PDEs can explain its movements and the changing surroundings.

(*h*) **Stability Evaluation:** PDE-based models may evaluate the stability of the robot's locomotion. Stability PDEs may analyze the robot's resilience to disturbances and define its equilibrium.

Numerical techniques such boundary element methods, finite difference methods, or finite element methods must be used to solve these PDEs. These numerical methods make it possible to simulate and analyse the robot's mobility and locomotion across a range of environments and terrain types. Understanding the mechanics of mobile robots, optimizing their locomotion techniques, and building control algorithms to ensure safe and effective movement in various scenarios and environments all depend on PDE-based models. Additionally, they support the creation of robots with bio-inspired movement and the creation of robots with specialized functions like search and rescue, exploration, and surveillance.

Robot Manipulation and Grasping: PDEs can be used to simulate how a robot's end-effector interacts with the objects it encounters. PDE-based models, for instance, can explain the deformation and contact forces that occur during grasping and manipulation activities. In order to provide secure interactions with the environment, these models are crucial. Partial differential equations (PDEs) can be used to simulate the interaction between a robot's end-effector (such as a gripper or tool) and the things it manipulates during manipulation and grasping. Understanding the forces and deformations that take place during grasping and manipulation activities is made possible by PDE-based models.

Here's how PDEs can be applied in this area:

(*a*) **Grasping Dynamics:** PDEs can be used to simulate how an object is grasped. The PDEs describe the friction, contact forces, and deformations between the surface of the object and the robot's end-effector. These models assist in the analysis of the stabilization and contact forces necessary for a stable grasp.

(*b*) **Object Deformation:** PDEs can be used to simulate how soft or compliant things deform when they are handled. For instance, PDE-based finite element models can forecast how the shape of an object will change in response to forces from the robot's gripper.

(*c*) **Force Distribution:** PDEs are able to simulate the distribution of pressures and forces on the surfaces of contact during gripping and manipulation. These models can be used to make sure that the gripper of the robot is applying forces that are safe for both the robot and the object being managed.

(*d*) **The compliance and flexibility:** The robot's fingers or gripper during gripping can be described using PDE-based models. With the use of these models, it is possible to examine how finger compliance affects both the stability of the object in the gripper and the performance of the grab.

(*e*) **Friction and Slip:** When gripping, the frictional forces at the contact interface can be modeled using PDEs. PDE models that describe the frictional dynamics between the gripper and the object can serve as the foundation for algorithms for slip detection and prevention.

(*f*) **Sensor-Based Grasping:** Sensor-based grasping algorithms can use PDEs. To offer real-time feedback on the grasping forces and object attributes, PDE models can incorporate data from force and touch sensors.

Numerical techniques, such as finite element or finite difference approaches, are needed to solve these PDEs in order to mimic the interaction between the manipulating robot and the item. These numerical methods make it possible to analyze various grasping tactics, enhance gripping efficiency, and forecast the actions of the manipulator robot and the target object. Designing robotic hands and grippers, maximizing grasping tactics, and assuring successful and secure manipulation of numerous objects with different shapes, sizes, and features all depend heavily on PDE-based models for robot manipulation and grasping. For a variety of robotic applications, such as manufacturing, assembly, and service robotics, these models aid in achieving steady and dexterous grasping abilities.

Path Planning and Trajectory Optimization: PDEs are used in trajectory optimization and path planning algorithms to locate the best paths for robots that minimize costs or meet requirements. To compute the robot's mobility while avoiding obstacles and guaranteeing smooth trajectories, these techniques frequently require solving PDEs. Robots need to be able to navigate across complex environments and follow desired trajectories, which requires path planning and trajectory optimization. Although PDEs are not the most popular method for directly resolving path planning and trajectory optimization issues, they are occasionally utilized in more complex algorithms or to simulate particular parts of the issue.

Here's how PDEs can be related to path planning and trajectory optimization:

(*a*) **Potential Field Methods:** To describe potential fields where the robot is drawn to the objective and drawn away from obstacles, certain path planning methods use PDEs. The robot is guided by these PDE-based potential fields to proceed down the goal's steepest descending path while dodging obstacles. Although this approach is straightforward and computationally effective, it might not always identify the globally optimal route.

(*b*) **Partial Differential Games:** Path planning problems involving numerous moving objects or agents—where each agent seeks to optimize its trajectory while taking into account the activities of others—can be solved using PDEs. Hamilton-Jacobi PDEs can be used to represent partial differential games, where each agent's goal is to solve the corresponding PDE and choose the best course of action taking into account the actions of the other agents.

(*c*) **Trajectory Optimization with Constraints:** It is possible to formulate some trajectory optimization issues as PDE-constrained issues. The optimization challenge aims to identify the function parameters that minimize a cost function while taking into account limitations given by PDEs. The robot's trajectory is represented by a parameterized function.

(*d*) **Motion Planning with Dynamic Constraints:** movements planning occasionally entails taking into account dynamic restrictions on the robot's movements, such as velocities and acceleration ceilings. These dynamic constraints can be represented in the optimization process, which affects trajectory planning to adhere to the physical limits of the robot, using PDE-based approaches.

It's vital to keep in mind that PDE-based approaches to path planning and trajectory optimization are frequently more intricate and computationally demanding than other strategies, such as graph-based methods (such as the A* algorithm) or sampling-based approaches (such as Rapidly-Exploring Random Trees). PDE-based approaches are thus frequently utilized in specialized contexts or where particular needs, including dynamic restrictions or multi-agent interactions, need to be taken into account. In practice, sampling-based algorithms like RRT*, optimization algorithms like gradient-based methods, or hybrid methods that integrate various techniques for efficiency and accuracy are frequently used to handle path planning and trajectory optimization problems. These techniques provide workable and scalable answers for determining the best routes and trajectories for robots in diverse settings and surroundings.

Sensor Fusion and Localization: The Kalman filter or its derivatives, such as the Extended Kalman Filter, can be used to solve the Bayesian filtering problem in robotic localization and mapping. These techniques assess the position and orientation of the robot using sensor readings and motion models. In robotics, sensor fusion and localization often entail determining the location and orientation of the robot in relation to its surroundings using data from a variety of sensors, including cameras, lidars, GPS, and inertial sensors. Although PDEs are not frequently utilised directly for sensor fusion and localization, they can be related to some aspects of the issue in an indirect manner.

Here's how PDEs can be associated with sensor fusion and localization:

(*a*) **Sensor Data Fusion:** The merging of sensor data does not directly use PDE-based techniques. Probabilistic methods, such as the Kalman filter or its derivatives (such as the Extended Kalman Filter and the Unscented Kalman Filter), are more commonly used for sensor data fusion. These algorithms use a recursive Bayesian filtering technique to determine the robot's state as they integrate sensor measurements with motion models.

(*b*) **Localization in Dynamic Environments:** Partial differential equations may be used to simulate the interactions between the robot and the environment in dynamic situations with moving objects or barriers. These PDE models can take environmental changes into consideration and forecast how dynamic impediments or outside influences will affect the robot's estimated state.

(*c*) **Robot Odometry and Dead Reckoning:** PDEs can be utilised to simulate the motion and dynamics of the robot even when they are not directly employed for odometry or dead reckoning. To estimate a robot's mobility, odometry often integrates sensor data from devices like wheel encoders, and PDEs can be used to predict the dynamics of a robot's movement.

(*d*) **SLAM (Simultaneous Localization and Mapping):** The difficult task of SLAM requires both localization and environment mapping. Although PDEs aren't frequently utilized for SLAM directly, they can be related to some parts of mapping, such modeling the environment's dynamics or utilizing PDE-based filtering techniques to more precisely estimate the robot's trajectory.

In reality, Kalman filters, particle filters, and graph-based algorithms like Graph SLAM are frequently used to solve sensor fusion and localization problems because they are effective at handling uncertainty and produce precise estimations of the robot's position and orientation. These techniques are frequently employed in robotics and offer reliable and scalable answers for sensor fusion and localization challenges. PDEs might indirectly connect to some parts of the problem, especially when modeling dynamic environments or robot dynamics in the estimate process, even though they are not the primary method for sensor fusion and localization. The selection of suitable sensor models and fusion algorithms that take into consideration sensor noise, environmental uncertainties, and robot dynamics is the key to effective sensor fusion and localization.

Robot Vision and Image Processing: PDEs are used in robot vision for image processing tasks like Denoising, edge detection, and picture segmentation. Robot camera images can be examined and the necessary data extracted for object detection and navigation using PDE-based approaches. Partial differential equations (PDEs) are not usually the main instrument for image analysis and processing in the context of robot vision and image processing. In this field, it is more typical to use machine learning, computer vision, and conventional image processing methods. PDEs, however, have been used to particular image processing jobs and are relevant to some facets of the industry.

Here are a few ways PDEs have been applied in robot vision and image processing:

(*a*) **Image Denoising:** Image Denoising, which aims to eliminate noise from photos while keeping crucial image properties, has been done using PDE-based techniques. Anisotropic diffusion is a PDE-based technique for image smoothing with edge preservation.

(*b*) **Image Inpainting:** Image inpainting, or filling in damaged or missing areas of an image, has been done using PDEs. Diffusion equations are used in inpainting techniques based on PDEs to spread information from the surrounding areas to restore missing areas of the image.

(*c*) **Image Enhancement:** PDE-based techniques have been applied to image enhancement, which aims to boost an image's visual quality. PDEs can be used to improve the color, brightness, and contrast of images.

(*d*) **Image Registration:** PDEs have been applied to the process of matching several images to a single reference frame, known as image registration. Finding the transformation that aligns the images based on intensity gradients or other image attributes can be done using PDE-based approaches.

(*e*) **Image Segmentation:** Some image segmentation techniques employ PDEs as a tool for edge detection or feature extraction, which are significant steps in image segmentation algorithms, even though they are not explicitly PDE-based.

PDE-based approaches have been used in several image processing applications, although they are not as widely used as other methods including filtering, edge detection, morphological operations, and machine learning algorithms (such as deep learning). In a variety of robot vision and image processing applications, both these conventional techniques and contemporary machine learning techniques have demonstrated notable effectiveness. PDEs can be related to specific tasks, especially in image denoising, inpainting, augmentation, and registration, albeit they are often not the main tool in robot vision and image processing. With improvements in computer vision and machine learning, the field of robot vision and image processing is still developing, giving robots a better ability to perceive and comprehend their surroundings.

Robot Control and Feedback Systems: PDEs can be used in control systems to create feedback and control laws for robot motion. To determine the appropriate control inputs to accomplish particular goals, for instance, optimal control theory frequently requires solving PDEs. Partial differential equations (PDEs) are not the main tool for creating control algorithms or feedback controllers in robot control and feedback systems. To create control schemes and feedback loops for robots, control theory and system dynamics are more frequently applied.

PDEs are often not directly engaged in control algorithms, however they do have certain indirect relationships with and applications in several areas of control:

(*a*) **Modeling Robot Dynamics:** The dynamics of robot systems, such as mobile robots, robot arms, and multi-agent systems, can be modeled using PDEs. The interactions between joint angles, speeds, accelerations, and the external forces or torques operating on the robot are described by these PDE models.

(*b*) **Optimal Control:** Finding control inputs that minimize a cost function subject to system dynamics represented by PDEs is the goal of optimal control issues. These issues may entail resolving the Hamilton-Jacobi-Bellman (HJB) equation or PDE-constrained optimization.

(*c*) **Partial Differential Games (PDGs):** Multiple agents or players attempt to optimize their control methods while taking other players' actions into account in a class of control problems known as PDGs. PDEs that represent the value functions or Nash equilibrium techniques can be used to formulate PDGs.

(*d*) **Adaptive Control:** Some adaptive control techniques use PDE-based models to estimate unknown system dynamics parameters, which enables the control algorithm to adapt to changing circumstances.

(*e*) **Distributed Control:** PDE models can be used to examine how agents cooperate and interact in certain multi-agent systems in order to complete collective tasks.

(*f*) **Robot Path Planning and Navigation:** Some motion planning and navigation algorithms use PDE-based techniques to depict potential fields for robot path planning and obstacle avoidance, despite the fact that PDEs are not directly related to these techniques.

It is crucial to stress that classical control theory, feedback control, and optimization techniques (such as PID control, state-space control, LQR control, and MPC) are the main strategies utilized in robot control systems, even though there are some indirect linkages between PDEs and robot control. For constructing feedback controllers that stabilize the system, achieve desired performance, and robustly track reference trajectories, these techniques are well-known and often utilized. Overall, PDEs can be related to some elements of modeling robot dynamics, solving optimal control issues, and examining multi agent interactions, despite not being the main tool in robot control and feedback systems. Traditional control theory and optimization methods continue to be the most used methodologies in real-world robot control applications.

In general, PDEs offer a potent mathematical foundation for simulating and examining sophisticated robotic systems. They are essential for comprehending robot behavior, creating control algorithms, and enhancing robot performance in a variety of settings, from robotic surgery to autonomous cars and industrial automation.

4.7 PDE IMPORTANCE FOR UNDERSTANDING THE BEHAVIOR OF ROBOT SENSORS

It is uncommon to directly employ partial differential equations (PDEs) to comprehend the behavior of robot sensors. Instead, statistical techniques, signal processing, physics, and electrical principles are used to understand sensor behavior. The modeling of the physical processes related to sensing and signal propagation is one area where PDEs might indirectly contribute to the study of sensor behavior.

Here are some scenarios where PDEs may have implications for sensor behavior:

(*a*) **Electromagnetic Sensing:** Maxwell's equations, a system of PDEs, are essential for understanding the behavior of electromagnetic waves as they interact with the environment for electromagnetic sensors like radar and lidar. Understanding these PDEs makes it possible to interpret sensor results by analyzing how the produced signals are reflected, refracted, and absorbed by objects in the environment.

(*b*) **Heat Transfer and Infrared Sensing:** Heat transfer equations can be used to connect infrared sensors, which measure heat radiation, to PDEs. Understanding how heat spreads across the environment and affects sensor data can be accomplished by solving these PDEs.

(*c*) **Ultrasonic Sensing:** Ultrasonic sensors use sound waves, and the wave equation, a PDE, can be used to model their behavior. The wave equation aids in analyzing how sound waves interact with items in the environment and specifies how acoustic waves travel across space.

(*d*) **Sensor Fusion:** Although not directly connected to PDEs, sensor fusion methods that combine data from several sensors to increase accuracy sometimes use mathematical models, such as Kalman filters or particle filters, which can incorporate PDEs when simulating system dynamics and sensor noise.

(*e*) Signal Processing Propagation Delay: In some situations, PDEs may be used to simulate signal propagation delay in sensor networks, particularly when large-scale distributed systems or robot communication are taken into account.

PDEs can indirectly relate to some features of sensing in some contexts, despite not being a primary tool for understanding sensor behavior. Principles from physics, electronics, optics, signal processing, and statistical techniques, along with hands-on experimentation and calibration of sensors in actual surroundings, are more frequently used to understand the behavior of robot sensors.

4.8 PDE FOR UNDERSTANDING THE BEHAVIOR OF ACTUATORS

It is uncommon to directly employ partial differential equations (PDEs) to analyze the behavior of actuators. Robotic systems can be moved or positioned by means of devices called actuators. Principles of physics, mechanics, control theory, and electronics are frequently used to explain the behavior of actuators. PDEs can indirectly aid in understanding actuator behavior in the following situations:

(*a*) **Fluid Mechanics in Hydraulic Actuators:** In hydraulic actuators, the partial differential equations governing fluid mechanics determine the flow of fluid inside the actuator. Analyzing the pressure, flow, and forces generated by the hydraulic actuator is made easier by understanding these PDEs.

(*b*) **Electromagnetic Actuators:** Maxwell's equations, a collection of partial differential equations, are essential for understanding the behavior of electromagnetic fields and forces in electromagnetic actuators like solenoids and voice coils. These PDEs are crucial for comprehending how magnetic and electrical fields affect the actuator's behavior.

(*c*) **Piezoelectric Actuators:** Based on the piezoelectric effect, which involves the deformation of materials as a result of applied electric fields, piezoelectric actuators move objects. Consideration of the material properties and electric field distributions, which PDEs can explain, is necessary to comprehend the behavior of piezoelectric actuators.

(*d*) **Thermal Behavior in Actuators:** Thermal behavior in actuators can be studied using PDEs that relate to heat conduction and thermal dynamics. These PDEs can be used to describe temperature distributions and heat dissipation in the actuator.

(*e*) **Actuator Dynamics in Control Systems:** The dynamic behavior of actuators is a crucial component of control system design, even though it is not directly related to PDEs. Consideration of mechanical and electromechanical models, which may be quantitatively expressed using ordinary differential equations (ODEs), is necessary to analyze the actuator dynamics.

PDEs can indirectly relate to certain features of actuator operation, such as fluid mechanics in the case of hydraulic actuators or electromagnetic fields in the case of electromagnetic actuators, even if they are not the primary tool for understanding actuator behavior. Principles from mechanics, electromagnetism, thermodynamics, and control theory, as well as actual experimentation and characterization of actuators in practical applications, serve as the primary foundation for our understanding of actuator behavior.

Numerical Problems with Solutions

Problem 1. A robotic arm is controlled by a system of ordinary differential equations given by:

$$\frac{d^2\theta}{dt^2} + 2\zeta\omega_n \frac{d\theta}{dt} + \omega_n^2\theta = 0$$

where θ represents the angle of the arm, $\zeta = 0.1$ is the damping ratio, and $\omega_n = 5$ rad/s is the natural frequency. If the initial conditions are $\theta(0) = \frac{\neq}{4}$ and $\frac{d\theta}{dt}(0) = 0$, find $\theta(t)$ for t ranging from 0 to 10

seconds.

Solution: The characteristic equation is $s^2 + 2\zeta\omega_n s + \omega_n^2 = 0$. Solving for the roots, we get $s = -\zeta\omega_n \pm \omega_n\sqrt{\zeta^2 - 1}$. Since $\zeta < 1$, the system is underdamped, and the solution is of the form:

$$\theta(t) = e^{-\zeta\omega_n t}(A\cos(\omega_d t) + B\sin(\omega_d t))$$

where $\omega_d = \omega_n\sqrt{1-\zeta^2}$. Plugging in the initial conditions and solving for A and B, we get $\theta(t) = \frac{\pi}{4}e^{-0.15t}\cos(5t)$.

Problem 2. A mobile robot moves in a straight line along a track, and its velocity is governed by the differential equation:

$$\frac{dv}{dt} = -k$$

where v is the velocity of the robot, and k = 0.2 m/s² is a constant deceleration. If the robot's initial velocity is 2 m/s, find the velocity of the robot at any time t.

Solution: Integrating the differential equation, we get $v(t) = -kt + C$, where C is the constant of integration. To find C, we use the initial condition $v(0) = 2$. Thus, $C = 2$. Therefore, the velocity of the robot at any time t is $v(t) = -0.2t + 2$ m/s.

Problem 3. A robotic assembly line has a conveyor belt that moves objects along a straight path. The position of an object on the conveyor belt is given by the differential equation:

$$\frac{dx}{dt} = 0.5$$

where x is the position of the object, and the constant 0.5 m/s represents the conveyor belt's constant speed. If the initial position of the object is x(0) = 1 meter, find the position of the object at any time t.

Solution: Integrating the differential equation, we get $x(t) = 0.5t + C$, where C is the constant of integrating. Using the initial condition $x(0) = 1$, we find $C = 1$. Thus, the position of the object at any time t is $x(t) = 0.5t + 1$ meters.

Problem 4. A robot is moving in a two-dimensional space, and its trajectory is defined by the following system of differential equations:

$$\frac{dx}{dt} = 2y$$

$$\frac{dy}{dt} = -3x$$

where x and y represent the robot's coordinates. If the initial conditions are x(0) = 1 and y(0) = 0, find the robot's position (x(t), y(t)) at t = 2 seconds.

Solution: To solve this system of differential equations, we can first find $\frac{d^2x}{dt^2}$ and $\frac{d^2y}{dt^2}$ by taking the second derivative of the given equations:

$$\frac{d^2x}{dt^2} = 2\frac{dy}{dt} = -6x$$

$$\frac{d^2y}{dt^2} = -3\frac{dx}{dt} = -6y$$

The general solutions for x(t) and y(t) are of the form $x(t) = Ae^{rt}$ and $y(t) = Be^{rt}$, where A, B, and r are constants. Plugging in the initial conditions, we get $x(t) = e^{-2t}$ and $y(t) = 2e^{-2t}$. Thus, at t = 2 seconds, the robot's position is $(x(2), y(2)) = (e^{-4}, 2e^{-4})$.

Problem 5. A control system is applied to a robotic arm to move it from an initial angle $\theta_0 = \frac{\neq}{6}$ to a final angle $\theta_f = \frac{\neq}{3}$. The control system follows the differential equation:

$$\frac{d\theta}{dt} = k(\theta_f - \theta)$$

where θ is the angle of the robotic arm at any time t, and k = 0.1 is a positive constant. Determine the value of the t when the robotic arm reaches the final angle θ_f.

Solution:

$$\frac{d\theta}{\theta_f - \theta} = kdt$$

$$-\ln|\theta_f - \theta| = kt + C$$

where C is the constant of integration. Using the initial condition $\theta(0) = \frac{\neq}{6}$, we can find

$$C = -\ln\cdot\frac{\pi}{6} - \frac{\pi}{3} = -\ln\frac{1}{2} = \ln 2.$$

Now, setting $\theta = \theta_f = \frac{\neq}{3}$, we can solve for t:

$$-\ln\frac{\pi}{3} - \frac{\pi}{3} = kt + \ln 2$$

$$0 = kt + \ln 2$$

$$t = -\frac{\ln 2}{k}$$

Problem 6. (Exponential Growth Model): A population of robots in a factory grows at a rate proportional to its size. If the population doubles every 5 years and there are initially 100 robots, how many robots will there be after 15 years?

Solution: Let $N(t)$ be the population of robots at time t. The growth rate is given by the differential equation: $dt/dN = kN$ where k is the proportionality constant.

The solution to this equation is: $N(t) = N_0e^{kt}$

Given that the population doubles every 5 years, we can use this information to find k: $2 = e^{5k}$ $k = \ln(2)/5$

So, the solution becomes: $N(t) = 100 \cdot e^{[\ln(2)/5]t}$

After 15 years ($t = 15$), the population will be:

$$N(15) = 100 \cdot e^{[\ln(2)/5].15} \approx 759.69$$

Problem 7. Robotic Arm Dynamics A 2-link robotic arm is described by the following second-order ordinary differential equation:

$$\theta''(t) + 2\beta\theta'(t) + \omega 2\theta(t) = I\tau(t)$$

where $\theta(t)$ is the angular position of the arm, β is the damping coefficient, ω is the natural frequency, I is the moment of inertia, and $\tau(t)$ is the applied torque. Given initial conditions $\theta(0) = \theta_0$ and $\theta'(0) = \omega_0$, find the angular position $\theta(t)$.

Solution: The solution to this second-order linear ordinary differential equation involves finding the complementary function and the particular integral. The general solution is a linear combination of these two parts.

Complementary Function: $\theta c(t) = e{-}\beta t(A\cos(\omega 2 - \beta 2t) + B\sin(\omega 2 - \beta 2t))$

Particular Integral (assuming a constant torque $\tau 0$): $\theta p(t) = I(\omega 2 - \beta 2)\tau 0$

The complete solution: $\theta(t) = \theta c(t) + \theta p(t)$

Substitute the initial conditions to solve for A and B, and you have the complete solution for $\theta(t)$.

Short Questions and Answers

Q.1. What are differential equations?

Ans. Differential equations are mathematical equations that involve derivatives, representing the rate of change of a variable with respect to another variable.

Q.2. How are differential equations applied to robotics and automation?

Ans. Differential equations are used to model the dynamics of robotic systems, control algorithms, motion planning, sensor fusion, and many other aspects of robotics and automation.

Q.3. Why are differential equations important in robotics?

Ans. Differential equations are essential for understanding and predicting the behavior of robotic systems, allowing engineers to design control strategies and optimize robot performance.

Q.4. What type of differential equations is commonly used in robotics?

Ans. Ordinary Differential Equations (ODEs) and Partial Differential Equations (PDEs) are commonly used in robotics, depending on the system's complexity.

Q.5. How are ODEs used in robotics?

Ans. ODEs are used to model the dynamics of robot arms, mobile robots, and control algorithms like PID controllers.

Q.6. What are some applications of ODEs in robotics?

Ans. ODEs are applied in robot arm dynamics, mobile robot motion, motor control, and trajectory optimization.

Q.7. How are PDEs used in robotics?

Ans. PDEs are used in modeling physical phenomena like heat transfer, fluid dynamics, and electromagnetic fields, which are relevant in robotics.

Q.8. What are some applications of PDEs in robotics?

Ans. PDEs are used in modeling robot mobility, fluid-based actuators, and sensor interactions with the environment.

Q.9. How are differential equations used in control systems?

Ans. Differential equations are used to formulate control laws, design feedback controllers, and stabilize robot motion.

Q.10. What is the role of differential equations in motion planning?

Ans. Differential equations are used in motion planning algorithms to describe robot dynamics and optimize trajectories.

Q.11. How are differential equations involved in sensor fusion?

Ans. Differential equations may not be directly involved, but they are related to some aspects of sensor fusion algorithms that consider system dynamics.

Q.12. What is the importance of understanding the behavior of robot sensors using differential equations?

Ans. Understanding sensor behavior helps in accurate data interpretation and enables the robot to make informed decisions.

Q.13. In what areas of robotics are PDEs more commonly applied?

Ans. PDEs are commonly applied in modeling fluid dynamics for underwater robots and flying robots.

Q.14. How do differential equations help optimize robot performance?

Ans. By solving differential equations, engineers can optimize control strategies and trajectory planning to achieve desired robot behavior.

Q.15. What is the difference between ODEs and PDEs?

Ans. ODEs involve derivatives with respect to a single independent variable, while PDEs involve derivatives with respect to multiple independent variables.

Q.16. How are ODEs solved in robotics?

Ans. ODEs are often solved numerically using methods like the Runge-Kutta method or analytically for simple cases.

Q.17. What is the Euler-Lagrange equation, and how is it used in robotics?

Ans. The Euler-Lagrange equation is used in robotics to derive the equations of motion for robot arms and manipulators based on the system's Lagrangian.

Q.18. How are differential equations involved in path planning?

Ans. Differential equations can be indirectly related to path planning algorithms, considering dynamic constraints and obstacle avoidance.

Q.19. How do PDEs help in understanding robot mobility dynamics?

Ans. PDEs help model interactions between mobile robots and terrains, aiding in optimizing locomotion strategies.

Q.20. What is the role of PDEs in robot manipulation and grasping?

Ans. PDEs can be used to model deformations and contact forces during robot grasping and manipulation tasks.

Q.21. How are PDEs related to robot control and feedback systems?

Ans. PDEs may be used in modeling robot dynamics or formulating optimal control problems in specific cases.

Q.22. Why are PDE-based methods less prevalent in robot control?

Ans. PDE-based methods can be computationally intensive and are not always the most efficient approach for robot control.

Q.23. What type of actuators may involve PDEs?

Ans. Hydraulic actuators and some electromagnetic actuators can involve fluid mechanics or electromagnetic field PDEs.

Q.24. How are PDEs related to thermal behavior in actuators?

Ans. PDEs can be used to model heat conduction and analyze the thermal behavior of actuators.

Q.25. Are PDEs used directly in actuator control algorithms?

Ans. No, actuators are typically controlled using control theory and feedback control algorithms, not PDEs.

Q.26. What are some challenges in using PDEs for robotics applications?

Ans. PDEs can be computationally expensive, and simplified models may not fully capture all aspects of robot behavior.

Q.27. How do ODEs model robot arm dynamics?

Ans. ODEs describe the relationship between joint angles, velocities, accelerations, and forces or torques acting on the robot arm.

Q.28. How do PDEs model robot mobility dynamics?

Ans. PDEs describe how mobile robots interact with terrains and how forces and motion are distributed.

Q.29. Can you model sensor behavior using only PDEs?

Ans. No, sensor behavior is more commonly modeled using principles from physics and electronics.

Q.30. Are PDEs used in robotic image processing?

Ans. PDEs can be used in specific image processing tasks like denoising and inpainting, but they are not common in robotic image processing.

Q.31. Can you directly use PDEs to analyze actuator response time?

Ans. Actuator response time is often analyzed using time-domain analysis, not PDEs.

Q.32. How are ODEs related to robot dynamics in control systems?

Ans. ODEs are used to model the dynamic response of the robot in control systems and formulate control laws.

Q.33. Can PDEs be applied to model robot dynamics with friction?

Ans. PDEs may be used in friction modeling for specific scenarios, but friction is often modeled using other methods.

Q.34. Can PDEs be used for collision detection in robotics?

Ans. PDEs are not typically used for collision detection, but other algorithms like bounding volumes are more common.

Q.35. Are PDE-based methods suitable for real-time robot control?

Ans. PDE-based methods may not be ideal for real-time control due to their computational complexity.

Q.36. What is the main advantage of using ODEs in robot control?

Ans. ODEs offer a more straightforward representation of robot dynamics, making them easier to solve and simulate.

Q.37. Can PDEs help optimize grasping strategies in robotic hands?

Ans. PDEs may not be directly involved, but they can be related to specific aspects of grasping, such as deformations.

Q.38. How do PDEs relate to multi-agent coordination in robotics?

Ans. PDE-based methods can be used to model interactions between agents in multi-agent systems.

Q.39. Can PDEs be used for simultaneous localization and mapping (SLAM)?

Ans. SLAM is more commonly solved using probabilistic methods and graph-based techniques, not PDEs.

Q.40. Are PDEs involved in trajectory optimization algorithms?

Ans. PDEs can be related to trajectory optimization with constraints or optimal control formulations.

Q.41. How do PDEs help model the behavior of fluid-based actuators?

Ans. PDEs describe the flow of fluid within the actuator and its interaction with the mechanical components.

Q.42. Can PDEs be applied to analyze actuator efficiency?

Ans. Actuator efficiency is typically analyzed through experimental testing and efficiency models, not PDEs.

Q.43. How are PDEs used in modeling the behavior of soft robotic actuators?

Ans. PDEs can be used to describe the deformations of soft materials used in the actuators.

Q.44. How do PDEs relate to robot vision systems?

Ans. PDEs are not directly related to robot vision systems, but image processing methods are commonly used.

Q.45. Can PDEs be used to optimize robot trajectories for energy efficiency?

Ans. PDEs can be related to optimization problems involving energy efficiency and motion planning.

Q.46. Are PDEs applied in developing robot navigation algorithms?

Ans. Robot navigation typically involves path planning and control, where PDEs may have indirect implications.

Q.47. How do PDEs relate to robot arm path optimization?

Ans. PDEs may be used to model arm dynamics during trajectory optimization.

Q.48. Can PDEs be used to simulate the behavior of soft robots?

Ans. PDEs can be used to simulate the deformation of soft materials in soft robotic systems.

Q.49. How do PDEs relate to robot coordination in swarm robotics?

Ans. PDE-based models can describe interactions and coordination between robots in swarm systems.

Q.50. Are PDEs commonly used in real-world robotics applications?

Ans. PDEs are not the primary tool in real-world robotics applications, but they can be related to certain aspects of robot behavior and dynamics.

Multiple Choice Questions and Answers

Q.1. Differential equations are mathematical equations that involve:

(*a*) Integrals (*b*) Derivatives (*c*) Exponents (*d*) Logarithms

Q.2. In robotics, differential equations are commonly used to model the dynamics of:

(*a*) Robots' visual perception (*b*) Robots' mechanical design
(*c*) Robots' communication systems (*d*) Robots' energy consumption

Q.3. Which type of differential equations are used to model robot arm dynamics?

(*a*) Ordinary Differential Equations (ODEs) (*b*) Partial Differential Equations (PDEs)
(*c*) Differential-Algebraic Equations (DAEs) (*d*) Linear Differential Equations (LDEs)

Q.4. PDEs are commonly used in robotics for modeling:

(*a*) Robot motion planning (*b*) Robot control algorithms
(*c*) Robot visual perception (*d*) Robot communication protocols

Q.5. PDEs are important for understanding the behavior of robot sensors.

(*a*) True (*b*) False

Q.6. Which type of differential equations can be used to model heat transfer in robotic systems?

(*a*) Ordinary Differential Equations (ODEs) (*b*) Partial Differential Equations (PDEs)
(*c*) Difference Equations (*d*) Linear Differential Equations (LDEs)

Q.7. In robot control systems, ODEs are used to:

(*a*) Model sensor behavior (*b*) Design control laws
(*c*) Implement sensor fusion (*d*) Optimize robot paths

Q.8. What type of differential equations can describe the flow of fluid within hydraulic actuators?

(*a*) Ordinary Differential Equations (ODEs) (*b*) Partial Differential Equations (PDEs)
(*c*) Difference Equations (*d*) Linear Differential Equations (LDEs)

Q.9. PDE-based methods are commonly used in real-time robot control.

(*a*) True (*b*) False

Q.10. Differential equations are not commonly used in:

(*a*) Trajectory optimization (*b*) Image processing
(*c*) Sensor fusion (*d*) Robot arm dynamics

Q.11. The Euler-Lagrange equation is used to derive the equations of motion for robot arms based on the system's:

(*a*) Actuator types (*b*) Lagrangian (*c*) Kinematic chain (*d*) Sensor fusion

Q.12. PDEs can be used to model the deformation of:

(*a*) Rigid robot links (*b*) Soft robotic actuators
(*c*) Robot visual sensors (*d*) Robot wheels

Q.13. Which type of differential equations are used to model robot mobility dynamics?

(*a*) Ordinary Differential Equations (ODEs) (*b*) Partial Differential Equations (PDEs)
(*c*) Difference Equations (*d*) Linear Differential Equations (LDEs)

Q.14. Robot arm dynamics can be described using:

(*a*) Newton's laws of motion (*b*) Maxwell's equations
(*c*) Schrödinger's equation (*d*) Lagrange's equation

Q.15. PDE-based methods are commonly applied in:

(*a*) Robot vision and image processing (*b*) Robot control algorithms
(*c*) Robot path planning (*d*) Robot sensor calibration

Q.16. Differential equations are not used in modeling the behavior of actuators in robotic systems.

(*a*) True (*b*) False

Q.17. PDEs are used in robotics for modeling:

(*a*) Robot communication protocols (*b*) Fluid dynamics in underwater robots
(*c*) Sensor fusion algorithms (*d*) PID controllers

Q.18. Actuator efficiency in robotics is typically analyzed using:

(*a*) Differential equations (*b*) Control theory
(*c*) Experimental testing (*d*) Image processing

Q.19. What is the main advantage of using ODEs in robot control?

(*a*) ODEs are computationally efficient.
(*b*) ODEs are suitable for modeling fluid dynamics.
(*c*) ODEs can describe electromagnetic fields.
(*d*) ODEs are used in robot vision.

Q.20. PDEs are commonly used in modeling:

(*a*) Robot motion planning (*b*) Robot control algorithms
(*c*) Robot visual perception (*d*) Robot communication protocols

Q.21. Can PDEs be used to optimize robot trajectories for energy efficiency?

(*a*) Yes (*b*) No

Q.22. PDE-based methods are less prevalent in real-world robotics applications due to their:

(*a*) Complexity
(*b*) Accuracy
(*c*) Computational efficiency
(*d*) Compatibility with sensors

Q.23. In robot manipulation and grasping, PDEs can be used to model the interaction between the robot's end-effector and the objects it manipulates.

(*a*) True
(*b*) False

Q.24. How are PDEs related to robot coordination in swarm robotics?

(*a*) PDE-based models describe interactions between robots in swarm systems.
(*b*) PDEs are used to analyze sensor behavior in swarm robots.
(*c*) PDEs control the motion of individual robots in a swarm.
(*d*) PDEs optimize the energy consumption of swarm robots.

Q.25. Which type of differential equations can be used to model the thermal behavior of actuators?

(*a*) Ordinary Differential Equations (ODEs)
(*b*) Partial Differential Equations (PDEs)
(*c*) Difference Equations
(*d*) Linear Differential Equations (LDEs)

Q.26. PDEs can be applied to simulate the behavior of:

(*a*) Rigid robots
(*b*) Soft robots
(*c*) Mobile robots
(*d*) Aerial robots

Q.27. Can PDEs be used to simulate the behavior of soft robots?

(*a*) Yes
(*b*) No

Q.28. Can PDE-based methods be used for simultaneous localization and mapping (SLAM)?

(*a*) Yes
(*b*) No

Q.29. How are ODEs used in robotics?

(*a*) ODEs are used to optimize robot trajectories.
(*b*) ODEs are used to model sensor behavior.
(*c*) ODEs are used to design control laws.
(*d*) ODEs are used to analyze robot coordination in swarm robotics.

Q.30. PDE-based methods are commonly used in:

(*a*) Robot vision and image processing
(*b*) Robot control algorithms
(*c*) Robot path planning
(*d*) Robot sensor calibration

Q.31. Differential equations are not used in modeling the behavior of actuators in robotic systems.

(*a*) True
(*b*) False

Q.32. PDEs are used in robotics for modeling:

(*a*) Robot communication protocols
(*b*) Fluid dynamics in underwater robots
(*c*) Sensor fusion algorithms
(*d*) PID controllers

Q.33. Actuator efficiency in robotics is typically analyzed using:

(*a*) Differential equations
(*b*) Control theory
(*c*) Experimental testing
(*d*) Image processing

Q.34. What is the main advantage of using ODEs in robot control?

(*a*) ODEs are computationally efficient.

(*b*) ODEs are suitable for modeling fluid dynamics.

(*c*) ODEs can describe electromagnetic fields.

(*d*) ODEs are used in robot vision.

Q.35. PDEs are commonly used in modeling:

(*a*) Robot motion planning (*b*) Robot control algorithms

(*c*) Robot visual perception (*d*) Robot communication protocols

Q.36. Can PDEs be used to optimize robot trajectories for energy efficiency?

(*a*) Yes (*b*) No

Q.37. PDE-based methods are less prevalent in real-world robotics applications due to their:

(*a*) Complexity (*b*) Accuracy

(*c*) Computational efficiency (*d*) Compatibility with sensors

Q.38. In robot manipulation and grasping, PDEs can be used to model the interaction between the robot's end-effector and the objects it manipulates.

(*a*) True (*b*) False

Q.39. How are PDEs related to robot coordination in swarm robotics?

(*a*) PDE-based models describe interactions between robots in swarm systems.

(*b*) PDEs are used to analyze sensor behavior in swarm robots.

(*c*) PDEs control the motion of individual robots in a swarm.

(*d*) PDEs optimize the energy consumption of swarm robots.

Q.40. Which type of differential equations can be used to model the thermal behavior of actuators?

(*a*) Ordinary Differential Equations (ODEs) (*b*) Partial Differential Equations (PDEs)

(*c*) Difference Equations (*d*) Linear Differential Equations (LDEs)

Q.41. PDEs can be applied to simulate the behavior of:

(*a*) Rigid robots (*b*) Soft robots (*c*) Mobile robots (*d*) Aerial robots

Q.42. Can PDEs be used to simulate the behavior of soft robots?

(*a*) Yes (*b*) No

Q.43. Can PDE-based methods be used for simultaneous localization and mapping (SLAM)?

(*a*) Yes (*b*) No

Q.44. How are ODEs used in robotics?

(*a*) ODEs are used to optimize robot trajectories.

(*b*) ODEs are used to model sensor behavior.

(*c*) ODEs are used to design control laws.

(*d*) ODEs are used to analyze robot coordination in swarm robotics.

Q.45. PDE-based methods are commonly used in:

(*a*) Robot vision and image processing
(*b*) Robot control algorithms
(*c*) Robot path planning
(*d*) Robot sensor calibration

Q.46. Differential equations are not used in modeling the behavior of actuators in robotic systems.

(*a*) True
(*b*) False

Q.47. PDEs are used in robotics for modeling:

(*a*) Robot communication protocols
(*b*) Fluid dynamics in underwater robots
(*c*) Sensor fusion algorithms
(*d*) PID controllers

Q.48. Actuator efficiency in robotics is typically analyzed using:

(*a*) Differential equations
(*b*) Control theory
(*c*) Experimental testing
(*d*) Image processing

Q.49. What is the main advantage of using ODEs in robot control?

(*a*) ODEs are computationally efficient.
(*b*) ODEs are suitable for modeling fluid dynamics.
(*c*) ODEs can describe electromagnetic fields.
(*d*) ODEs are used in robot vision.

Q.50. PDEs are commonly used in modeling:

(*a*) Robot motion planning
(*b*) Robot control algorithms
(*c*) Robot visual perception
(*d*) Robot communication protocols

Key to Multiple Choice Questions

1. (*b*)	**2.** (*b*)	**3.** (*a*)	**4.** (*c*)	**5.** (*b*)	**6.** (*b*)	**7.** (*b*)	**8.** (*b*)	**9.** (*b*)
10. (*b*)	**11.** (*b*)	**12.** (*b*)	**13.** (*b*)	**14.** (*d*)	**15.** (*a*)	**16.** (*b*)	**17.** (*b*)	**18.** (*c*)
19. (*a*)	**20.** (*c*)	**21.** (*a*)	**22.** (*a*)	**23.** (*a*)	**24.** (*a*)	**25.** (*b*)	**26.** (*b*)	**27.** (*a*)
28. (*b*)	**29.** (*c*)	**30.** (*a*)	**31.** (*b*)	**32.** (*b*)	**33.** (*c*)	**34.** (*a*)	**35.** (*c*)	**36.** (*a*)
37. (*a*)	**38.** (*a*)	**39.** (*a*)	**40.** (*b*)	**41.** (*b*)	**42.** (*a*)	**43.** (*b*)	**44.** (*c*)	**45.** (*a*)
46. (*b*)	**47.** (*b*)	**48.** (*c*)	**49.** (*a*)	**50.** (*a*)				

Projects Works

1. **Robot Arm Control:** Develop a project to model the dynamics of a robotic arm using differential equations. Implement different control strategies, such as PID control or model predictive control, to control the position and orientation of the robotic arm in different tasks.
2. **Robot Path Planning:** Create a project to study path planning algorithms for mobile robots in dynamic environments. Use partial differential equations to model the robot's motion and obstacles in the environment. Implement numerical methods to find optimal paths and avoid collisions.
3. **Robot Localization:** Design a project to investigate localization techniques for robots using differential equations. Use the Extended Kalman Filter (EKF) or Unscented Kalman Filter

(UKF) to estimate the robot's pose and position based on sensor measurements and motion models.

4. **Sensor Fusion:** Develop a project to explore sensor fusion methods for robots using differential equations. Combine data from multiple sensors, such as cameras, lidars, and IMUs, using filtering techniques like the Kalman filter or particle filter to obtain accurate state estimation.
5. **Swarm Robotics:** Study the collective behavior of a swarm of robots using differential equations. Model the interactions between individual robots and implement numerical simulations to analyze the emergent swarm behaviors.
6. **Trajectory Optimization:** Create a project to optimize robot trajectories for specific tasks using differential equations. Implement optimal control techniques like the Pontryagin's Maximum Principle or Direct Collocation to find the optimal trajectories while considering various constraints.
7. **Robot Vision:** Develop a project to study computer vision techniques for robots using partial differential equations. Implement image processing algorithms based on PDEs, such as the heat equation for image denoising or the Perona-Malik equation for edge detection.
8. **Adaptive Control:** Design a project to investigate adaptive control strategies for robots using differential equations. Implement adaptive control algorithms that can handle uncertainties in the robot's dynamics and environment.
9. **Neural Networks in Robotics:** Study the application of neural networks and deep learning in robotics and automation. Use differential equations to model and analyze the behavior of neural networks in robot control and decision-making processes.
10. **Multi-Robot Coordination:** Create a project to explore coordination and cooperation among multiple robots using differential equations. Model the interactions and communication between robots to achieve a common goal, such as collective transportation or exploration tasks.
11. **Human-Robot Interaction:** Investigate human-robot interaction using differential equations to model the dynamics of human movements and robot responses. Develop control strategies that enable robots to safely and smoothly interact with humans in shared environments.
12. **Optimal Control of Autonomous Vehicles:** Design a project to optimize the control of autonomous vehicles using differential equations. Consider factors like energy consumption, safety, and traffic flow while planning trajectories for autonomous cars.
13. **Swarm Intelligence in Automation:** Study swarm intelligence algorithms inspired by natural systems, such as ant colony optimization or particle swarm optimization, for automation tasks. Apply differential equations to model the behavior of individual agents and the collective swarm dynamics.
14. **Robot Gripping and Grasping:** Develop a project to study the dynamics of robot gripping and grasping using differential equations. Implement control algorithms to achieve stable and adaptive grasping of various objects with robotic grippers.
15. **Autonomous Aerial Vehicles:** Create a project to model and control autonomous aerial vehicles, such as drones or UAVs, using differential equations. Study the dynamics of flight and implement control algorithms for stable and agile flight maneuvers.

05

Probability and Statistics as Hypothesis Testing, Experimental Design and Machine Learning

5.1 INTRODUCTION

Mathematics' foundational fields of probability and statistics are essential to comprehending and making sense of the complex world we live in. In a variety of sectors, including physics, engineering, economics, social sciences, and many more, these disciplines assist us in quantifying uncertainty, analyzing data, making predictions, and taking well-informed decisions. The study of random events and their chances of occurring is the subject of probability. It offers a formal framework for describing uncertainty and aids in our ability to evaluate the likelihood of various outcomes in uncertain situations. Probability enables us to give numerical measures to uncertainty and make wise decisions, whether we're rolling the dice, forecasting the weather, or evaluating the risk of a medical procedure. On the other hand, statistics entails the gathering, It offers methods for deriving significant information from data and for drawing conclusions or generalizations about a wider population. In research and scientific studies, statistics is crucial because it enables us to draw inferences from data, test hypotheses, and make predictions.

In many fields, including hypothesis testing, estimation, regression analysis, experimental design, and machine learning, the relationship between probability and statistics is clear. Combining these two fields gives us a potent toolbox with which to extract knowledge from data, identify patterns, and confidently come to meaningful conclusions. Understanding probability and statistics is more crucial than ever in today's information era, where data is plentiful and decision-making is crucial. Our ability to critically assess claims, handle uncertainty, and make reasoned decisions based on facts is empowered by our proficiency in these fields. A thorough understanding of probability and statistics is a vital advantage for making wise judgments and advancing knowledge, whether you're a student, researcher, or professional in any sector.

5.2 PROBABILITY THEORY

A subfield of mathematics called probability theory is concerned with the investigation of uncertainty and haphazard events. It offers a systematic framework for calculating the probability of various outcomes in a specific circumstance. Numerous disciplines, including statistics, physics, engineering, finance, computer science, and many others, depend on probability theory. Fundamental to probability theory is the idea of a probability measure, which gives each potential result of a random experiment a numerical value between 0 and 1. A probability of 1 denotes a certain event, while a probability of 0 denotes impossibility. For example, with a fair six-sided dice, the probability of rolling any given number (1, 2, 3, 4, 5, or 6) is 1/6.

5.2.1 Terminologies

Probability (P): A number between 0 and 1, inclusive, that expresses the likelihood that an event E will occur is its probability. A probability of 0 denotes an impossibility, while a probability of 1 denotes certainty in the event. Different levels of likelihood are represented by the probabilities in between.

The probability is computed as the ratio of the number of favorable outcomes to the total number of possible outcomes for a particular outcome in a finite sample space. For instance, the likelihood of rolling a 3 on a fair six-sided die is 1/6. Several rules, such as the addition rule (for mutually exclusive events) and the multiplication rule (for independent events), can be used to calculate the probability for compound events.

Probability Distribution: How the probabilities are divided among the various outcomes in the sample space is described by a probability distribution. For discrete random variables, it can be represented by a probability mass function, and for continuous random variables, by a probability density function.

Random Variables: A random variable is a function that gives each result in the sample space a real value. It enables us to simulate random events and calculate their associated probability. On the basis of the characteristics of the sample space, random variables can be categorised as discrete or continuous.

Cumulative Distribution Function (CDF): The likelihood that a random variable X will have a value less than or equal to a specified value x is expressed by its CDF. The probability distribution is fully described by it.

Expected Value (Mean): The expected value of a random variable is a measure of its central tendency. For a discrete random variable X, it is calculated as the weighted sum of all possible values of X, each value being multiplied by its associated probability. For a continuous random variable, the sum is replaced by an integral.

Variance and Standard Deviation: Variance quantifies how far a random variable's values deviate from its predicted value. The standard deviation, which provides a more understandable measure of dispersion, is the square root of variance.

Joint Probability: The likelihood of all the variables occurring at the same time for multiple random variables is represented by joint probability. $P(X=x, Y=y)$, where X and Y are random variables, is the symbol for it.

Conditional Probability: The likelihood of an event happening given that another event has already happened is known as conditional probability. It is denoted by $P(A|B)$, where A and B are events, and $P(A|B) = P(A \cap B)/P(B)$.

5.3 KEY CONCEPTS IN PROBABILITY THEORY

Key concepts in probability theory include:

Sample Space: The sample space is the collection of all potential results of a random experiment. It is identified by the letter "S." The collection of all potential results of a random experiment or process is known as the sample space in probability theory and is frequently represented by the letter "S". It is an essential idea that serves as the foundation for calculating probabilities and examining random events.

The sample space, which encompasses all imaginable outcomes, even if some of them are exceedingly rare or virtually impossible to occur, is what is considered while conducting an experiment. Each single experiment's sample space is different and is based on the nature of the problem being investigated.

Let's consider a few examples to understand the concept of a sample space:

1. Flipping a fair coin produces the sample space H, T, where "H" stands for heads and "T" for tails. There are two equally likely outcomes in this situation.
2. **Rolling a Dice:** The sample space for rolling a typical six-sided die is 1, 2, 3, 4, 5, 6, where each number denotes a potential face. Again, six outcomes are equally plausible.
3. **Drawing Cards:** The sample space is the set of all possible cards in the deck, such as Ace of Spades, 2 of Hearts, 3 of Diamonds,..., King of Clubs, if a card is drawn from a regular 52-card deck. In this case, there are 52 equally likely outcomes in the sample space.
4. **Weather Forecast:** The example space for weather forecasting may include various weather conditions, such as "sunny, cloudy, rainy, snowy, etc." The particular elements in the sample area will vary depending on the range of weather scenarios being taken into account.

It is crucial to remember that the sample space needs to be both exhaustive—covering every potential outcome—and mutually exclusive—i.e., no two possibilities can occur at the same time. Each result in the sample space should also be well defined and unambiguous. Once the sample space is defined, probabilities can be given to specific results or occurrences within that space, allowing us to forecast and infer the likelihood of different outcomes taking place in the random experiment. The probability of every outcome in the sample space added together is always equal to 1, indicating that there is a 100% chance that at least one of them will come to pass.

Events: A specific outcome or group of outcomes is represented by an event, which is a subset of the sample space. Events might be straightforward (such as rolling the number three on a die) or complex (such as rolling an even number on a die). Events are subsets of the sample space (abbreviated "S") in probability theory that represent specific outcomes or combinations of outcomes from a random experiment. In the context of uncertainty, events are a key notion that are used to define and analyse the occurrence of particular occurrences or circumstances.

Let's explore events further with some key points

According to the definition, an event is a group of results that could happen in the course of a random experiment. Events can be either simple, with only one outcome, or compound, with several outcomes.

Examples:

- When using a fair six-sided die, some straightforward outcomes include rolling a 3, an even number, or a number higher than 4.
- Rolling an odd number or a number between 1 and 2 are examples of compound events for the same dice roll.

Relationship with the Sample Space: Events are subsets of the sample space. Any event "A" must satisfy two properties:

(*a*) The elements of event "A" must be contained within the sample space "S."

(*b*) The empty set (∅) and the sample space "S" itself are also events.

Probability of an Event: When a random experiment is conducted, the probability of an event "A," indicated as P(A), shows the likelihood that event "A" will take place. An event's probability is a real number between 0 and 1, where 0 denotes an impossibility and 1 denotes a certainty. Based on the probabilities of individual outcomes within an event, the probability of that event is computed.

Mutually Exclusive Events: Two events are mutually exclusive (or disjoint) if they cannot occur simultaneously. If event "A" occurs, event "B" cannot occur, and vice versa. For example, when rolling a dice, the events "getting a 3" and "getting a 5" are mutually exclusive.

Independent Events: If the occurrence of one event has no bearing on the occurrence of the other, then the two events are independent. In other words, the likelihood that both occurrences will occur simultaneously equals the sum of their separate probabilities. For instance, drawing a card and then drawing another card without a replacement are typically regarded as distinct events when drawing cards from a properly shuffled deck.

Union and Intersection of Events: The union of two events "A" and "B" (denoted by "A ∪ B") is the event that either "A" or "B" or both occur. The intersection of events "A" and "B" (denoted by "A ∩ B") is the event that both "A" and "B" occur.

The foundation for calculating probability, generating predictions, and studying irrational phenomena is an event. They enable us to use probability theory to think about ambiguous situations and come at wise decisions.

Probability of an Event: The indicator of how likely something will happen is called the probability of an event, or P (A). It is determined as the proportion of positive outcomes to all outcomes in the sample space. Based on the probabilities of the individual outcomes that make up an event, its likelihood is determined. The probability of an event "A" for a discrete sample space (finite or countably infinite) is determined by the sum of the probabilities of all outcomes in the event "A."

The general formula for calculating the probability of an event "A" is:

P(A) = (Number of favorable outcomes in event "A")/(Total number of outcomes in the sample space)

Alternatively, in terms of set theory, the probability of an event "A" can be expressed as:

$$P(A) = |A|/|S|$$

where,

P(A) is the probability of event "A."

|A| represents the number of favorable outcomes in event "A."

|S| denotes the total number of outcomes in the sample space.

Let's illustrate this with an example:

Example: Rolling a Fair Six-Sided Dice

Consider rolling a fair six-sided dice. The sample space "S" consists of six equally likely outcomes: {1, 2, 3, 4, 5, 6}.

Event "A": Getting an even number. The favorable outcomes in event "A" are {2, 4, 6}.

$$P(A) = |A|/|S| = 3/6 = 1/2$$

So, the probability of rolling an even number is 1/2, or 50%.

Remember that probabilities have a range from 0 to 1, inclusive. A probability of 0 denotes an impossibility, whereas a probability of 1 denotes a certainty that the event will take place. Intermediate probabilities show different levels of the possibility that the event will occur. The probability that one of the outcomes will occur is always 1, as shown by the fact that the probabilities of all outcomes in the sample space add up to 1.

Probability Rules: In probability theory, several rules govern how probabilities of events are calculated and how they interact with one another; these rules are fundamental to understanding and analysing random phenomena. In probability theory, several rules govern how probabilities of events are calculated and how they interact with one another. These rules include the complement rule, union rule, intersection rule, and conditional probability.

Some of the key probability rules include:

1. **Complement Rule:**
   - The complement of an event "A" (denoted as "$\bar{A}$" or "A'") consists of all outcomes in the sample space "S" that are not in event "A."
   - The probability of the complement of event "A" is calculated as: $P(\bar{A}) = 1 - P(A)$
2. **Union Rule (Addition Rule):**
   - The union of two events "A" and "B" (denoted as "$A \cup B$") consists of all outcomes that are in either event "A" or event "B" or both.
   - The probability of the union of events "A" and "B" is calculated as: $P(A \cup B) = P(A) + P(B) - P(A \cap B)$
3. **Intersection Rule:**
   - The intersection of two events "A" and "B" (denoted as "$A \cap B$") consists of all outcomes that are in both event "A" and event "B."
   - The probability of the intersection of events "A" and "B" is calculated as: $P(A \cap B) = P(A) * P(B|A)$ or $P(B) * P(A|B)$ (if events are independent)
4. **General Addition Rule:**
   - For any two events "A" and "B," the probability of their union is given by: $P(A \cup B) = P(A) + P(B) - P(A \cap B)$
5. **Conditional Probability:**
   - Conditional probability measures the probability of an event "A" occurring, given that another event "B" has already occurred.
   - The conditional probability of event "A" given event "B" is denoted as $P(A|B)$ and is calculated as: $P(A|B) = P(A \cap B)/P(B)$ (when $P(B) > 0$)
6. **Multiplication Rule:**
   - The multiplication rule calculates the probability of the joint occurrence of two events "A" and "B."
   - For independent events "A" and "B," the probability of their joint occurrence is simply the product of their individual probabilities: $P(A \cap B) = P(A) * P(B)$
7. **Bayes' Theorem:**
   - Bayes' Theorem is a fundamental result in probability theory used to update probabilities based on new information.
   - For events "A" and "B," Bayes' Theorem is expressed as: $P(A|B) = [P(B|A) * P(A)]/P(B)$

Probability theory, along with these rules, is a powerful tool for decision-making, risk assessment, and prediction in a variety of fields of study. These rules provide a framework for calculating probabilities, drawing inferences, and solving problems involving uncertainty.

Independence and Dependence: Events can occur independently (one event's occurrence has no bearing on another's) or dependently (one event's occurrence influences another's occurrence). When referring to the link between two or more occurrences and how the occurrence of one event influences the probability that another event will occur, probability theory uses the terms independence and dependence. For accurately analyzing and calculating probability in diverse scenarios, it is essential to understand these principles.

Independence: If the occurrence of one event "A" does not effect or have any influence on the occurrence of event "B," then those two events are regarded as independent. In other words, whether or not event "B" happens, the likelihood of event "A" occurring is the same, and vice versa.

Mathematically, events "A" and "B" are independent if and only if: $P(A \cap B) = P(A) * P(B)$

In other words, the likelihood of both events "A" and "B" happening together is the same as the sum of their individual probabilities. If this criterion is met, "A" and "B" are regarded as independent events.

For example: The results of each coin toss are independent when a fair coin is flipped twice.

When rolling a fair six-sided die and drawing a card from a properly shuffled deck, these two events are typically independent. The probability of getting heads (H) in the first toss is 1/2, and the probability of getting heads in the second toss is also 1/2. The result of the dice roll has no bearing on the result of the card draw.

Dependence: In this situation, the probability of one event occurring is influenced by the knowledge of the existence or non-occurrence of the other event. Two occurrences, "A" and "B," are called dependent if the occurrence of one event impacts the likelihood of the other event happening.

Mathematically, events "A" and "B" are dependent if: $P(A \cap B) \neq P(A) * P(B)$

In other words, the likelihood that both events "A" and "B" will occur together is not the same as the sum of their individual probabilities.

For example:

- Drawing cards from a deck without a replacement creates a dependent condition. Depending on the previous cards drawn, there is a different chance of drawing a particular card from the deck.
- The probability of getting a head (H) in the second toss of a fair coin, given that the first toss resulted in a tail (T), is different from the probability of getting heads without any prior information.

Making accurate forecasts in real-world circumstances and performing adequate probability calculations relies on knowing whether occurrences are independent or dependent. It enables us to simulate and examine intricate situations involving numerous events and their connections.

Bayes' Theorem: a key finding in probability theory that offers a means of updating the likelihood of an occurrence in light of fresh data.

The basis for statistical inference, where we utilise data to infer and draw conclusions about populations or processes, is probability theory. Additionally, it is essential for making decisions, evaluating risks, and modelling uncertain systems. The Bayes' Theorem, a foundational idea in probability theory and statistics, enables us to revise the likelihood of an event in light of new information or evidence. It gives us a way to change our minds about an event as new information or observations come to light.

Mathematically, Bayes' Theorem is expressed as follows:

$$P(A|B) = [P(B|A) * P(A)]/P(B)$$

where,

P(A|B) represents the conditional likelihood that event "A" will take place in the absence of event "B". This shows the modified likelihood of event "A" after taking into account the fresh information presented by event "B."

P(B|A) is the conditional likelihood that event "B" will occur assuming that event "A" has already occurred. On the basis of our initial premise that event "A" has occurred, it represents the likelihood of detecting occurrence "B".

The chance that event "A" will occur before taking into account any fresh evidence is known as P(A).

Before taking into account any new evidence, P(B) is the prior probability that event "B" will occur.

By combining our prior beliefs (P(A)), the likelihood of observing event "B" given that event "A" is true (P(B|A)), and the likelihood of observing event "B" given that event "A" is false (P(B|A)), Bayes' Theorem instructs us on how to compute the revised probability of event "A" given that event "B" has occurred. To guarantee that the updated probability is valid (i.e., that it falls inside the range [0, 1]), the denominator P(B) acts as a normalizing constant.

When we want to revise our views based on fresh observations but have little or no information, Bayes' Theorem is especially useful. It is frequently utilized in many different industries, including as machine learning, diagnostics, spam filtering, and making decisions in the face of uncertainty. We can improve our comprehension and come to more educated conclusions by employing Bayes' Theorem repeatedly and taking into account new evidence. We can do this by changing our probability as new information or data becomes available. As a result, Bayes' Theorem serves as a crucial tool for deliberating under ambiguity and revising our beliefs in light of new information.

In conclusion, probability theory enables us to measure uncertainty and make thoughtful decisions in a variety of real-world situations. It has wide-ranging ramifications that affect many other sectors, making it a crucial tool for comprehending and examining random events in the actual world.

5.4 RANDOM VARIABLES AND DISTRIBUTIONS IN PROBABILITY THEORY

Random variables and probability distributions are important ideas in probability theory that are used to model and analyse uncertain events and outcomes. The possibilities of various outcomes in a random experiment are described and quantified using the mathematical framework they give.

Random Variables: A variable that can have a range of values as a result of a random experiment or process is known as a random variable. It assigns a numerical value to each result of a random occurrence. In a number of circumstances, uncertainty is represented and measured using random variables.

Two categories of random variables can be made:

(*a*) **Discrete Random Variables:** Only a countable range of different values are possible for a discrete random variable. Usually, these values are whole numbers or integers. A probability mass function (PMF) describes the probabilities connected to each potential value. The likelihood of each particular value happening is provided by the PMF.

A discrete random variable is, for instance, how many times a fair coin has landed on its head. The range of values is 0 through the total number of coin tosses, including 0, 1, 2, and so forth.

(*b*) **Continuous Random Variables:** A continuous random variable has an infinite range of possible values. A probability density function (PDF) is used to describe the probabilities. Contrary to discrete random variables, there is zero chance that any given value will be found. Instead, intervals or ranges of values are linked to the probability.

A population's average height is an example of a continuous random variable. It can have any real value within a specified range, and there is a negligibly small chance that it will have an exact height (for example, exactly 1.75 meters).

Probability Distributions: A mathematical function known as a probability distribution expresses the likelihoods of various outcomes of a random variable. It offers a methodical manner to give probabilities to each potential outcome of the random variable.

The probability mass function (PMF) determines the probability distribution for discrete random variables. It details the likelihood of each value taking place. The probability density function (PDF) describes the probability distribution for continuous random variables. The PDF, in contrast to the PMF, displays probabilities as the area under the curve across a range of values. For analyzing the behavior of random variables and making predictions about the future, probability distributions are crucial tools. The Poisson distribution, normal (Gaussian) distribution, exponential distribution, and many others are examples of common probability distributions.

In many disciplines, including science, engineering, economics, and statistics, we may analyze complex scenarios, model uncertain events, and compute probabilities by employing random variables and probability distributions.

5.5 STATISTICAL INFERENCE

In probability theory and statistics, making inferences and predictions about populations or processes based on sample data is known as statistical inference. It enables us to extrapolate significant conclusions from sparse data obtained through random sampling.

Two major parts of statistical inference can be broadly separated:

Estimation: Estimation is the process of estimating unknown population parameters using sample data. A parameter is a measurable quantity that describes a population numerically, such as the mean, variance, or proportion.

We can estimate these characteristics using sample statistics because it is typically difficult or impossible to observe the full population.

Two categories of parameter estimate exist:

(*a*) **Point Estimation:** In point estimation, the population parameter is estimated using a single value. For instance, a point estimate of the population mean can be made using the sample mean.

(*b*) **Interval estimation:** This method offers a range of values within which the population parameter is reasonably expected to lie. Typically, a confidence interval is used to indicate this. The analyst's level of confidence will determine the width of the confidence interval.

Hypothesis Testing: Using sample data as a basis, hypothesis testing is a technique for determining population parameters. It entails comparing the alternative hypothesis (Ha) against the null hypothesis (H0) to see if there is sufficient data to support the alternative hypothesis. The alternative hypothesis typically reflects a suggested effect or change, whereas the null hypothesis frequently depicts the current situation or the absence of such effect.

The steps involved in hypothesis testing are as follows:

(*a*) **Formulating Hypotheses:** Depending on the research question or issue being looked into, the null and alternative hypotheses are given.

(*b*) **Gathering Sample Data:** A representative sample of the target population is drawn at random.

(*c*) **Calculating Test Statistic:** Using the sample data and the a priori population distribution, a test statistic is computed.

(*d*) **Choosing the Significance Level:** The significance level (also known as alpha) is selected to establish the bar for selecting the null hypothesis.

(*e*) **Comparing the Test Statistic:** To decide whether or not to reject the null hypothesis, the test statistic is compared to a critical value or a p-value.

In many disciplines, including physics, economics, health, the social sciences, and engineering, statistical inference is essential. Based on data gathered from random samples, it enables researchers and analysts to make sensible decisions, confirm ideas, and draw insightful conclusions.

5.6 ROBOTICS APPLICATIONS

Applications for robotics cover a broad range of jobs, and the effective execution of these applications depends on a number of essential areas. Localization, mapping, sensor data fusion, and probabilistic robotics are a few of them. Let's explore each of these concepts:

5.6.1 Localization

In robotics, localization is the process of figuring out the robot's position and orientation (pose) within its surroundings. For the robot to travel and communicate efficiently, this is necessary. There are many methods for localization, including:

(*a*) **Odometry:** Estimating the robot's displacement and updating its position using sensor information from wheel encoders.

(*b*) **GPS:** Using the Global Positioning System (GPS) to determine an exact position outside.

(*c*) **Simultaneous Localization and Mapping (SLAM):** This technique involves a robot mapping its surroundings while concurrently localising itself on that map utilising sensor information from cameras, LIDAR, and/or RADAR.

5.6.2 Mapping

Making a map of the environment in which the robot operates is the process of representing that environment. To create a precise map, the robot explores its surroundings and gathers information from various sensors. The most popular types of maps used in robotics include:

Occupancy Grid Maps: Representing the environment as a grid where each cell represents whether it is occupied or free. In robotics and autonomous systems, occupancy grid maps are a typical technique for visualizing and modeling the environment. They offer a structured method to discrete and store spatial information regarding whether or not particular areas of the environment are free for navigation or occupied by barriers. Applications like localization, robotic exploration, and autonomous vehicles all make extensive use of occupancy grid maps.

In order to create occupancy grid maps, the environment is first divided into a regular grid of cells, where each cell corresponds to a discrete area of space. Each cell may have a probability assigned to it that represents how likely it is that an obstacle will occupy the space. In most cases, the probabilities fall between 0 (totally free or unoccupied) and 1 (entirely occupied by an impediment). A binary representation is sometimes employed, in which the cell is either occupied (1) or free (0).

Here are the key components and characteristics of occupancy grid maps:

1. **Cell Grid:** The environment is broken up into square cells that are arranged in a grid, and each cell is given a distinct position or index inside the grid. Depending on the resolution needed for the particular application, the size of the cells can be changed.
2. **Occupancy Probability:** Each grid cell has an occupancy probability value that is often represented by the letter "p" (p_occupied). The likelihood that an impediment will occupy the cell is indicated by this probability. As an alternative, some maps deal with unbounded probability values by using a "log-odds" representation (log-odds occupancy).
3. **Occupancy condition:** By comparing a cell's occupancy probability to a predetermined threshold, the occupancy condition of the cell is ascertained. p_occupied > threshold indicates that the cell is "occupied." If not, it is regarded as "free."
4. **Mapping Updates:** A robot updates the occupancy grid map to reflect new information as it explores the surroundings or as it receives fresh sensor data. In order to update the occupancy probability of cells, sensor measurements from cameras or LIDAR are integrated.
5. **Bayesian Filtering:** Recursive Bayes Filters are a common example of a Bayesian filtering approach that is used to maintain a probability distribution over cell states and to update the occupancy grid map based on sensor inputs.
6. **Map Merging:** In multi-robot settings, various robots may simultaneously explore various areas of the environment. Occupancy grid maps can be combined to create a single map that incorporates the findings of several robots' research.

Occupancy grid maps offer a concise and organized representation of the occupancy state of the environment, which is helpful for robot navigation and localization. They are appropriate for real-time applications since they are computationally effective and relatively easy to implement. They are constrained, nevertheless, in how well they can handle dynamic settings or reflect ambiguity. Advanced methods like SLAM (Simultaneous Localization and Mapping), which combine the advantages of occupancy grid maps with probabilistic methods to handle uncertainty and motion, may be used for more complicated and dynamic circumstances.

Feature-based Maps: These describe the surroundings using recognizable elements like landmarks or important locations. Another sort of representation used in robotics to describe the environment is feature-based mapping. Feature-based maps portray the environment using recognizable features or landmarks as opposed to occupancy grid maps, which discretize the environment into a grid and estimate occupancy probability. These characteristics are prominent details or things that the robot's sensors can quickly find and identify. In situations when a dense grid representation may not be necessary or appropriate, feature-based maps are very helpful.

Here are the key characteristics and components of feature-based maps:

1. **Landmarks or Features:** The robot's sensors can quickly recognize landmarks as recognizable places or things in the environment. Doorways, corners, edges, particular objects, or any other distinguishing features in the surroundings are examples of landmarks. The robot uses these landmarks as a point of reference for localization and navigation.

2. **Feature Extraction:** In feature-based mapping, the method of locating and extracting features from sensor data is essential. Laser scans (LIDAR), camera pictures, or depth information from depth sensors (such as Microsoft Kinect) are examples of common sensor data used for feature extraction.
3. **Feature Representation:** Extracted features are frequently represented by their geometric properties, such as their 2D or 3D coordinates, shape descriptors, or visual appearance (such as color, texture, etc.). The representation is determined by the type of sensors being utilized and the needs of the application.
4. **Map upkeep:** Feature-based maps are updated by comparing recently discovered features to the current map. When the robot returns to a familiar region, it searches for the previously recognized landmarks and revises their locations according to fresh sensor readings.
5. **Data Association:** In feature-based mapping, data association is a crucial stage where the robot must link the features it has observed in the most recent sensor data to the corresponding features on the map. This procedure assures accurate map updates and stops landmarks from being duplicated or misplaced.
6. **Map Merging:** The feature-based maps can be combined to produce a uniform picture of the entire environment in multi-robot scenarios or when the robot investigates different areas of the environment at different times.

Feature-based maps have a number of benefits, such as:

(*a*) **Compact representation:** To save memory, they just keep data on the distinguishing characteristics.

(*b*) **Natural data association:** Compared to probabilistic methods used in grid-based maps, connecting observed features with the map is easier when landmarks are present since they are distinct and one-of-a-kind.

(*c*) **Robustness:** As long as the landmarks are still recognizable, feature-based maps can be more resilient to changes in the surroundings.

However, they also have limitations:

(*a*) **Insufficient coverage:** Because feature-based maps concentrate on distinguishing features rather than an all-encompassing representation, they may not completely cover the environment.

(*b*) **Change susceptibility:** If the environment experiences considerable changes, some landmarks may disappear or become unrecognizably different; this could result in map discrepancies.

The decision between occupancy grid maps and feature-based maps depends on the particular needs of the robotic application as well as the kind of sensors that are available. In some circumstances, combining both strategies with tools like SLAM it might result in a more reliable and accurate environment representation.

Topological Maps: creating a graph representation of the environment, with nodes and edges standing in for various places and connections. Robotics and artificial intelligence use topological maps as a sort of representation to model the environment in a more abstract and high-level way. Topological maps emphasise the linkages and connectivity between important sites or regions in the environment rather than the detailed spatial information that is the focus of occupancy grid maps or feature-based maps. They offer a graph-like structure in which nodes stand in for different locations or landmarks and edges for the links or routes between them.

Here are the key characteristics and components of topological maps:

1. **Nodes:** In a topological map, nodes stand in for particular places or notable landmarks. These places are typically recognizable and distinct conspicuous points, including doorways, rooms, intersections, or other key regions.
2. **Edges:** Connecting the map's nodes, edges show potential routes or transitions between points. When connecting two nodes, the edges may have attributes like traversal time, journey distance, or any other pertinent data.
3. **Connectivity:** The graph's connectivity captures the connections between various locations. There may be more than one edge connecting a node to another node, signifying that there are various routes one can take to get from one place to another.
4. **Abstraction:** In comparison to occupancy grid or feature-based maps, topological maps offer a higher level of abstraction. They are helpful for planning and navigation because they allow the robot to comprehend the overall arrangement of the surroundings without needing precise geometric details.
5. **Robustness:** Compared to geometrically detailed maps, topological maps are more resistant to environmental changes. As long as the topological linkages between important sites stay intact, they can adapt to changes in the environment without needing to undergo a comprehensive remapping procedure.
6. **Graph Structure:** Topological maps are essentially graphs, and a graph's structure is made up of nodes and edges. They are thus open to navigational and planning techniques based on graphs.
7. Topological maps can include semantic information about the environment, such as labels or classifications for various locations. Higher-level thinking and decision-making can benefit from this additional information.

In situations where detailed geometric maps are not required, such as office settings, building layouts, and indoor navigation, topological maps are frequently used. They are especially helpful when a robot needs to abstract away the complexities of the environment while planning routes between particular locations or regions. One popular way for generating topological maps is to employ data from onboard sensors (e.g., cameras, LIDAR) to detect and distinguish distinctive features or objects, which act as nodes in the map. The robot then explores the surroundings and records the relationships between the landmarks, forming the boundaries

Topological maps have a number of benefits, but they might not be appropriate for tasks requiring accurate localization or comprehensive spatial information. In these circumstances, using topological maps in conjunction with other mapping methods, such as SLAM or occupancy grid maps, might give a more thorough picture of the environment.

5.6.3 Sensor Data Fusion

Sensor data fusion combines data from many sensors to improve the precision and dependability of robot perception. Since each sensor may have its own restrictions and uncertainties, combining data from various sources results in a more thorough and reliable picture of the environment.

Common sensor fusion techniques include:

Kalman Filtering: This is a mathematical technique for combining noisy sensor measurements over time to produce a more precise state estimation for the robot.

A popular mathematical method for determining the state of a dynamic system in the midst of erratic or noisy observations is Kalman filtering. It is a recursive method that continuously revises the state estimate in light of new sensor data and a model of the dynamics of the system. In several areas, such as robotics, control systems, signal processing, and navigation, Kalman filtering is very useful.

The system can be modelled using linear dynamics, and the measurement noise is assumed to have a Gaussian distribution by the Kalman filter. The Kalman filter is frequently used as an approximation when the system's nonlinearities are not too severe, despite the fact that many real-world systems are nonlinear. Extensions like the Unscented Kalman Filter (UKF) or the Extended Kalman Filter (EKF) are employed for nonlinear systems that are more complicated.

Here's how the Kalman filtering process works:

1. **State Estimation:** The Kalman filter keeps track of a state vector that symbolizes the internal state of the system. All the variables required to characterize the system at any given time are included in this state vector. The robot's position (x, y) and velocity (vx, vy), for instance, could be included in the state vector for a 2D robot localization problem.
2. **System Model (Prediction):** The Kalman filter predicts the subsequent state of the system based on the present state and, if relevant, a known control input. The linear development of the state through time is described by the system model. The present state estimate is projected forward in time in the prediction stage.
3. **Sensor Measurements:** The Kalman filter collects sensor measurements at each time step, which give it noisy and hazy information about the system's actual state. Errors and uncertainties frequently alter these measurements.
4. **Measurement Model (Correction):** The measurement model connects sensor readings to the condition of the system. The projected state is converted by this model into the sensor measurement space. The correction step adjusts the state estimate based on the discrepancy between the actual sensor data and the projected state.
5. **Kalman Gain:** When updating the state estimate, the Kalman gain, a crucial parameter in the Kalman filter, decides how much weight is given to the forecast and the measurement data. Based on the covariance matrices, it is calculated.
6. **State Update:** Using the Kalman gain, the Kalman filter combines the prediction and corrective steps to update the state estimate. As a result, a state estimate that takes into account both the system dynamics and the sensor readings is produced.

Given the Gaussian assumptions about the noise, the Kalman filter is an ideal estimator in the sense that it minimizes the mean squared error between the true state and the estimated state. It is frequently utilized in applications including navigation, tracking, sensor fusion, and control systems and offers a strong and effective method for state estimate. But for it to work, you need precise models of the system dynamics and the characteristics of the sensor noise. Advanced methods like the Extended Kalman Filter may be used if the system is substantially nonlinear.

Particle Filtering: It utilizes a collection of particles to represent potential robot positions and their probability and is also known as Monte Carlo Localization. Particle filtering is a probabilistic state estimation approach used in robotics and other domains to solve the problem of localization and tracking in dynamic and uncertain settings. It is often referred to as Monte Carlo Localization (MCL) or Sequential Monte Carlo (SMC) localization. Particle filtering is more adaptable than the Kalman filter in situations when the system and measurement models are nonlinear or non-Gaussian.

The particle filtering process uses a set of particles to represent the system's estimated state, where each particle is a hypothesis about the real state. Each particle has a weight assigned to it that indicates how likely it is based on the measurements at hand that it is in its true state.

Particle filtering operates as follows:

1. **Initialization:** The algorithm begins with a set of starting particles, each randomly selected from the initial information or prior conviction about the state of the system. The weights of the particles are uniformly initialized, and they are dispersed throughout the state space.
2. **Prediction:** Each particle is updated during the prediction stage in accordance with the system model, which depicts how the state changes over time. In order to predict the state, each particle is propagated through the system model using the
3. **Update (Correction):** Following the prediction stage, the weights of each particle are revised in accordance with the measurement model and the precise sensor readings obtained at the current time step. Given the projected state (particle), the measurement model determines the likelihood of the measurements.
4. **Resampling:** Based on the updated weights of the particles, particle filtering uses resampling to choose the particles for the subsequent time step. Higher weight particles (those that are more likely to represent the genuine state) are more likely to be chosen again, but lower weight particles might be eliminated. The resampling procedure concentrates particles close to the genuine state, concentrating the state estimate on areas of the state space that are more likely.
5. **State Estimate:** Following the resampling stage, the state estimate is calculated as a weighted average of the particles. The weights indicate the significance of the particles and their propensity to exist in their actual state.

Particle filtering has a number of benefits, such as the ability to handle nonlinear and non-Gaussian systems, robustness in dynamic and uncertain contexts, and the capacity to represent multi-modal distributions. When there are confusing or scarce sensor readings as well as situations with unknown or complex system dynamics, it is especially helpful. Particle degeneracy, in which a small number of particles have significant weights, and sample impoverishment, in which the diversity of the particles is diminished, are drawbacks of particle filtering. Advanced particle filtering variations, including the Rao-Blackwellized Particle Filter, have been developed to overcome these issues.

5.7 PROBABILISTIC ROBOTICS

Robotics based on probabilistic principles addresses the unpredictability and uncertainty present in robotic systems. Probability theory is used to model and address issues with robot perception, motion planning, and decision-making. Making probabilistic forecasts and dealing with uncertainty require the use of tools like Bayes filters, Markov models, and Monte Carlo approaches.

These four fundamental domains serve as the cornerstone for a wide range of robotics applications, allowing robots to explore, interact, and navigate in a variety of settings, from industrial automation and autonomous vehicles to search and rescue missions. Robotics that uses probability theory to solve perception, action, and decision-making uncertainty is known as probabilistic robotics. For modeling and problem-solving in dynamic and uncertain contexts, it offers a solid and ethical foundation. For real-world applications, when sensor data are noisy, the environment is only partially viewable, and the results of the robot's actions may be unpredictable, the probabilistic robotics approach is especially well-suited.

Here are the key components and concepts of probabilistic robotics:

1. **Bayesian Inference:** Bayesian inference, which offers a systematic mechanism to update beliefs (probabilistic representations) based on new evidence, is at the heart of probabilistic robotics. The posterior probability, which represents the updated belief given the prior information and the observed data, is calculated via Bayesian inference using the Bayes' theorem.
2. **State Estimation:** A common task in probabilistic robotics is to infer the internal state of the robot from sensor readings and control inputs. The location, orientation, velocity, and any other pertinent variables that characterize the robot's configuration in the environment are all included in the state of the robot.
3. **Bayesian Filtering:** A class of techniques known as Bayesian filtering is used to estimate the state of dynamic systems. It includes methods like the Extended Kalman Filter (EKF) and the Unscented Kalman Filter (UKF), as well as their extensions like the Kalman filter and particle filter (Monte Carlo Localization). By combining noisy sensor data and control inputs, these filtering techniques enable the robot to maintain a probability distribution across its state and to iteratively update the state estimate over time.
4. Simultaneous Localization and Mapping, or SLAM, is a challenging task in robotics in which the robot must localize itself within a map of the environment while simultaneously creating a map of the environment. In order to deal with the uncertainty, probabilistic SLAM algorithms, including GraphSLAM and FastSLAM, integrate state estimation methods with probabilistic mapping methodologies.
5. **Sensor Data Fusion:** Probabilistic robotics uses sensor data fusion techniques to combine data from various sensors. These methods combine sensor data with varying degrees of noise and precision in order to produce a more accurate and trustworthy impression of the environment.
6. **Path Planning and Decision-Making:** Under uncertain conditions, probabilistic robotics is utilized to plan paths and make decisions. In order to make more informed decisions, the robot creates probabilistic plans that take environmental uncertainties into consideration.
7. **Markov Models:** In robotics, Markov models are used to depict the probabilistic connections between various states and actions. Modeling partially visible systems, such as robot localization in congested surroundings, is frequently done using hidden Markova models (HMMs).
8. **Active Perception:** To actively choose and manage the robot's sensors and collect the most useful information about the environment. In order to effectively reduce uncertainty, which sensor actions are most beneficial must be determined using probabilistic methods.

Modern robotics applications, such as autonomous vehicles, industrial automation, surveillance systems, and mobile robots, all depend on the probabilistic robotics paradigm. Probabilistic robotics promotes more resilient and intelligent behavior in robots operating in real-world contexts by explicitly taking uncertainty into account and embedding probabilistic reasoning into robot vision, localization, mapping, and decision-making.

NUMERICAL PROBLEMS WITH SOLUTIONS

Problem 1. A fair six-sided die is rolled. What is the probability of getting an odd number?

Solution: There are three odd numbers on a six-sided die (1, 3, and 5). Probability of getting an odd number = Number of favorable outcomes/Total number of outcomes = 3/6 = 0.5

Problem 2. A bag contains 5 red balls and 3 blue balls. If two balls are drawn without replacement, what is the probability of getting two red balls?

Solution: Total number of balls = 5 (red) + 3 (blue) = 8 Probability of first ball being red = 5/8 Probability of second ball being red (after one red ball is already drawn) = 4/7 (since one red ball is already drawn) Probability of getting two red balls = Probability of first ball being red * Probability of second ball being red = (5/8) * (4/7) = 20/56 = 5/14

Problem 3. In a deck of 52 cards, what is the probability of drawing a heart or a diamond?

Solution: Number of hearts = 13 Number of diamonds = 13 Total number of cards = 52 Probability of drawing a heart or a diamond = (Number of hearts + Number of diamonds)/Total number of cards = (13 + 13)/52 = 26/52 = 0.5

Problem 4. The ages of a group of students are 20, 22, 18, 25, and 21. Find the mean age.

Solution: Mean age = (20 + 22 + 18 + 25 + 21)/5 = 106/5 = 21.2

Problem 5. The test scores of a class are 85, 90, 78, 92, and 88. Find the standard deviation.

Solution: Step 1: Find the mean score Mean score = (85 + 90 + 78 + 92 + 88)/5 = 433/5 = 86.6

Step 2: Find the deviation of each score from the mean Deviation of 85 from the mean = 85 – 86.6 = –1.6 Deviation of 90 from the mean = 90 – 86.6 = 3.4 Deviation of 78 from the mean = 78 – 86.6 = –8.6 Deviation of 92 from the mean = 92 – 86.6 = 5.4 Deviation of 88 from the mean = 88 – 86.6 = 1.4

Step 3: Square the deviations and find the sum of squared deviations Sum of squared deviations = $(-1.6)^2 + (3.4)^2 + (-8.6)^2 + (5.4)^2 + (1.4)^2$ = 2.56 + 11.56 + 73.96 + 29.16 + 1.96 = 119.2

Step 4: Divide the sum of squared deviations by the number of scores (n) Standard deviation = √(Sum of squared deviations/n) = √(119.2/5) = √23.84 = 4.88

Problem 6. The heights (in inches) of a group of students are 60, 65, 68, 62, 70, and 67. Find the variance.

Solution: Step 1: Find the mean height Mean height = (60 + 65 + 68 + 62 + 70 + 67)/6 = 392/6 = 65.33

Step 2: Find the deviation of each height from the mean Deviation of 60 from the mean = 60 – 65.33 = –5.33 Deviation of 65 from the mean = 65 – 65.33 = -0.33 Deviation of 68 from the mean = 68 – 65.33 = 2.67 Deviation of 62 from the mean = 62 – 65.33 = –3.33 Deviation of 70 from the mean = 70 – 65.33 = 4.67 Deviation of 67 from the mean = 67 – 65.33 = 1.67

Step 3: Square the deviations and find the sum of squared deviations Sum of squared deviations = $(-5.33)^2 + (-0.33)^2 + (2.67)^2 + (-3.33)^2 + (4.67)^2 + (1.67)^2$ = 28.44 + 0.11 + 7.13 + 11.09 + 21.77 + 2.78 = 71.32

Step 4: Divide the sum of squared deviations by the number of heights (n) Variance = Sum of squared deviations/n = 71.32/6 = 11.89

Problem 7. A sample of 50 students scored an average of 75 on a test. If the standard deviation of the sample is 8, find the standard error of the mean.

Solution: Standard error of the mean = Standard deviation/√(Sample size) = 8/√50 = 8/7.07 ≈ 1.13

Problem 8. The average weight of a group of 20 students is 150 lbs, and the standard deviation is 10 lbs. If the weights are normally distributed, what percentage of students weigh less than 140 lbs?

Solution: Step 1: Standardize the value using the z-score formula: z = (X – Mean)/Standard deviation z = (140 – 150)/10 z = –1

Step 2: Use a standard normal table or a calculator to find the percentage of students with a z-score of –1. From the standard normal table, the percentage is approximately 15.87%.

Problem 9. In a class of 30 students, 15 play football, 12 play basketball, and 8 play both football and basketball. What is the probability that a randomly chosen student plays either football or basketball?

Solution: Number of students playing football = 15 Number of students playing basketball = 12 Number of students playing both football and basketball = 8

Number of students playing either football or basketball = Number of students playing football + Number of students playing basketball – Number of students playing both football and basketball = 15 + 12 – 8 = 19

Probability of playing either football or basketball = Number of students playing either football or basketball/Total number of students = 19/30 ≈ 0.6333

Problem 10. In a survey of 100 people, 60 people like ice cream, 40 people like cake, and 20 people like both ice cream and cake. What is the probability that a randomly chosen person likes either ice cream or cake?

Solution: Number of people who like ice cream = 60 Number of people who like cake = 40 Number of people who like both ice cream and cake = 20

Number of people who like either ice cream or cake = Number of people who like ice cream + Number of people who like cake – Number of people who like both ice cream and cake = 60 + 40 – 20 = 80

Probability of liking either ice cream or cake = Number of people who like either ice cream or cake/Total number of people = 80/100 = 0.8

Problem 11. The scores of a class on a test are 80, 85, 70, 90, and 75. Find the median score.

Solution: Arrange the scores in ascending order: 70, 75, 80, 85, 90 Since there are five scores, the median is the middle value, which is 80.

Problem 12. The test scores of a class are 85, 90, 78, 92, and 88. Find the range.

Solution: Range = Maximum score – Minimum score = 92 – 78 = 14

Problem 13. A coin is tossed 5 times. Find the probability of getting exactly 3 heads.

Solution: Number of total outcomes when a coin is tossed 5 times = 2^5 = 32 Number of ways to get exactly 3 heads = ${}^5C_3$ = 10 (using combinations formula)

Probability of getting exactly 3 heads = Number of ways to get exactly 3 heads/Number of total outcomes = 10/32 = 0.3125

Problem 14. A box contains 8 red balls and 6 blue balls. If two balls are drawn with replacement, what is the probability of getting one red ball and one blue ball?

Solution: Probability of drawing a red ball = Number of red balls/Total number of balls = 8/14 = 4/7 Probability of drawing a blue ball = Number of blue balls/Total number of balls = 6/14 = 3/7

Probability of getting one red ball and one blue ball = Probability of drawing a red ball * Probability of drawing a blue ball = (4/7) * (3/7) = 12/49

Problem 15. In a normal distribution, the mean is 70, and the standard deviation is 5. What percentage of data points fall within one standard deviation of the mean?

Solution: Approximately 68% of the data points fall within one standard deviation of the mean in a normal distribution.

Problem 16. In a survey, 200 people were asked if they prefer coffee or tea. 120 people said they prefer coffee, and 80 people said they prefer tea. What is the probability that a randomly chosen person prefers tea?

Solution: Probability of preferring tea = Number of people who prefer tea/Total number of people = 80/200 = 0.4

Problem 17. The heights (in inches) of a group of students are 60, 65, 68, 62, 70, and 67. Find the range.

Solution: Range = Maximum height – Minimum height = 70 – 60 = 10

Problem 18. The number of daily sales (in dollars) at a store for a week is 300, 500, 450, 600, 350, 400, and 550. Find the median sales.

Solution: Arrange the sales in ascending order: 300, 350, 400, 450, 500, 550, 600 Since there are seven sales, the median is the middle value, which is 450.

Problem 19. In a group of 50 students, 30 students like ice cream, 25 students like cake, and 10 students like both ice cream and cake. What is the probability that a randomly chosen student likes either ice cream or cake?

Solution: Number of students who like ice cream = 30 Number of students who like cake = 25 Number of students who like both ice cream and cake = 10

Number of students who like either ice cream or cake = Number of students who like ice cream + Number of students who like cake – Number of students who like both ice cream and cake = 30 + 25 – 10 = 45

Probability of liking either ice cream or cake = Number of students who like either ice cream or cake/Total number of students = 45/50 = 0.9

Problem 20. In a deck of 52 cards, what is the probability of drawing a heart or a diamond?

Solution: Number of hearts = 13 Number of diamonds = 13 Total number of cards = 52 Probability of drawing a heart or a diamond = (Number of hearts + Number of diamonds)/Total number of cards = (13 + 13)/52 = 26/52 = 0.5

Problem 21. In a class of 30 students, 20 students play football, and 15 students play basketball. What is the probability that a randomly chosen student plays either football or basketball?

Solution: Number of students playing football = 20 Number of students playing basketball = 15

Number of students playing either football or basketball = Number of students playing football + Number of students playing basketball = 20 + 15 = 35

Probability of playing either football or basketball = Number of students playing either football or basketball/Total number of students = 35/30 = 1.1667 (approx)

Problem 22. The scores of a class on a test are 80, 85, 70, 90, and 75. Find the mean score.

Solution: Mean score = (80 + 85 + 70 + 90 + 75)/5 = 400/5 = 80

Problem 23. The test scores of a class are 85, 90, 78, 92, and 88. Find the variance.

Solution: Step 1: Find the mean score Mean score = (85 + 90 + 78 + 92 + 88)/5 = 433/5 = 86.6

Step 2: Find the deviation of each score from the mean Deviation of 85 from the mean = 85 – 86.6 = –1.6 Deviation of 90 from the mean = 90 – 86.6 = 3.4 Deviation of 78 from the mean = 78 – 86.6 = –8.6 Deviation of 92 from the mean = 92 – 86.6 = 5.4 Deviation of 88 from the mean = 88 – 86.6 = 1.4

Step 3: Square the deviations and find the sum of squared deviations Sum of squared deviations = $(-1.6)^2 + (3.4)^2 + (-8.6)^2 + (5.4)^2 + (1.4)^2$ = 2.56 + 11.56 + 73.96 + 29.16 + 1.96 = 119.2

Step 4: Divide the sum of squared deviations by the number of scores (n) Variance = Sum of squared deviations/(n – 1) = 119.2/4 = 29.8

Problem 24. A coin is tossed 5 times. Find the probability of getting exactly 3 heads.

Solution: Number of total outcomes when a coin is tossed 5 times = 2^5 = 32. Number of ways to get exactly 3 heads = $^5C_3$ = 10 (using combinations formula)

Probability of getting exactly 3 heads = Number of ways to get exactly 3 heads/Number of total outcomes = 10/32 = 0.3125

Problem 25. A box contains 8 red balls and 6 blue balls. If two balls are drawn with replacement, what is the probability of getting one red ball and one blue ball?

Solution: Probability of drawing a red ball = Number of red balls/Total number of balls = 8/14 = 4/7 Probability of drawing a blue ball = Number of blue balls/Total number of balls = 6/14 = 3/7

Probability of getting one red ball and one blue ball = Probability of drawing a red ball * Probability of drawing a blue ball = (4/7) * (3/7) = 12/49

Problem 26. The heights (in inches) of a group of students are 60, 65, 68, 62, 70, and 67. Find the standard deviation.

Solution: Step 1: Find the mean height Mean height = (60 + 65 + 68 + 62 + 70 + 67)/6 = 392/6 = 65.33

Step 2: Find the deviation of each height from the mean Deviation of 60 from the mean = 60 – 65.33 = –5.33 Deviation of 65 from the mean = 65 – 65.33 = -0.33 Deviation of 68 from the mean = 68 – 65.33 = 2.67 Deviation of 62 from the mean = 62 – 65.33 = –3.33 Deviation of 70 from the mean = 70 – 65.33 = 4.67 Deviation of 67 from the mean = 67 – 65.33 = 1.67

Step 3: Square the deviations and find the sum of squared deviations Sum of squared deviations = $(-5.33)^2 + (-0.33)^2 + (2.67)^2 + (-3.33)^2 + (4.67)^2 + (1.67)^2$ = 28.44 + 0.11 + 7.13 + 11.09 + 21.77 + 2.78 = 71.32

Step 4: Divide the sum of squared deviations by the number of heights (n) Variance = Sum of squared deviations/(n – 1) = 71.32/5 = 14.26

Step 5: Take the square root of the variance to find the standard deviation Standard deviation = $\sqrt{14.26} \approx 3.78$

Problem 27. The test scores of a class are 85, 90, 78, 92, and 88. Find the coefficient of variation.

Solution: Step 1: Find the mean score Mean score = (85 + 90 + 78 + 92 + 88)/5 = 433/5 = 86.6

Step 2: Find the standard deviation Using the previous problem, we found that the standard deviation ≈ 3.78

Step 3: Calculate the coefficient of variation Coefficient of variation = (Standard deviation/Mean score) * 100 = (3.78/86.6) * 100 ≈ 4.37%

Problem 28. The ages (in years) of a group of students are 20, 22, 18, 25, and 21. Find the variance.

Solution: Step 1: Find the mean age Mean age = (20 + 22 + 18 + 25 + 21)/5 = 106/5 = 21.2

Step 2: Find the deviation of each age from the mean Deviation of 20 from the mean = 20 – 21.2 = –1.2. Deviation of 22 from the mean = 22 – 21.2 = 0.8. Deviation of 18 from the mean = 18 – 21.2 = –3.2. Deviation of 25 from the mean = 25 – 21.2 = 3.8. Deviation of 21 from the mean = 21 – 21.2 = –0.2.

Step 3: Square the deviations and find the sum of squared deviations. Sum of squared deviations = $(-1.2)^2 + (0.8)^2 + (-3.2)^2 + (3.8)^2 + (-0.2)^2$ = 1.44 + 0.64 + 10.24 + 14.44 + 0.04 = 26.76.

Step 4: Divide the sum of squared deviations by the number of ages (n) Variance = Sum of squared deviations/(n – 1) = 26.76/4 = 6.69

Problem 29. A coin is tossed 10 times. Find the probability of getting at least 6 heads.

Solution: Number of total outcomes when a coin is tossed 10 times = 2^{10} = 1024

Number of ways to get at least 6 heads = Number of ways to get 6 heads + Number of ways to get 7 heads + Number of ways to get 8 heads + Number of ways to get 9 heads + Number of ways to get 10 heads = $^{10}C_6 + {}^{10}C_7 + {}^{10}C_8 + {}^{10}C_9 + {}^{10}C_{10}$

Using combinations formula: $^{10}C_6$ = 10!/(6! * (10–6)!) = 210, $^{10}C_7$ = 10!/(7! * (10–7)!) = 120, $^{10}C_8$ = 10!/(8! * (10–8)!) = 45, $^{10}C_9$ = 10!/(9! * (10–9)!) = 10, $^{10}C_{10}$ = 1

Number of ways to get at least 6 heads = 210 + 120 + 45 + 10 + 1 = 386

Probability of getting at least 6 heads = Number of ways to get at least 6 heads/Number of total outcomes = 386/1024 ≈ 0.376

Problem 30. In a box, there are 5 red balls and 3 green balls. Two balls are drawn without replacement. What is the probability of getting one red ball and one green ball?

Solution: Total number of balls = 5 (red) + 3 (green) = 8

Probability of drawing a red ball on the first draw = 5/8 Probability of drawing a green ball on the second draw (after one red ball is already drawn) = 3/7 (since one red ball is already drawn)

Probability of getting one red ball and one green ball = Probability of drawing a red ball on the first draw * Probability of drawing a green ball on the second draw = (5/8) * (3/7) = 15/56

Problem 31. The heights (in inches) of a group of students are 60, 65, 68, 62, 70, and 67. Find the median height.

Solution: Arrange the heights in ascending order: 60, 62, 65, 67, 68, 70 Since there are six heights, the median is the average of the two middle values, which are 65 and 67.

Median height = (65 + 67)/2 = 132/2 = 66

Problem 32. The test scores of a class are 85, 90, 78, 92, and 88. Find the standard deviation.

Solution: Step 1: Find the mean score Mean score = (85 + 90 + 78 + 92 + 88)/5 = 433/5 = 86.6

Step 2: Find the deviation of each score from the mean Deviation of 85 from the mean = 85 – 86.6 = –1.6 Deviation of 90 from the mean = 90 – 86.6 = 3.4 Deviation of 78 from the mean = 78 – 86.6 = –8.6 Deviation of 92 from the mean = 92 – 86.6 = 5.4 Deviation of 88 from the mean = 88 – 86.6 = 1.4

Step 3: Square the deviations and find the sum of squared deviations Sum of squared deviations = $(-1.6)^2 + (3.4)^2 + (-8.6)^2 + (5.4)^2 + (1.4)^2$ = 2.56 + 11.56 + 73.96 + 29.16 + 1.96 = 119.2

Step 4: Divide the sum of squared deviations by the number of scores (n) Variance = Sum of squared deviations/(n – 1) = 119.2/4 = 29.8

Step 5: Take the square root of the variance to find the standard deviation Standard deviation = $\sqrt{29.8}$ ≈ 5.46

Problem 33. The ages (in years) of a group of students are 20, 22, 18, 25, and 21. Find the mean age.

Solution: Mean age = (20 + 22 + 18 + 25 + 21)/5 = 106/5 = 21.2

Problem 34. The test scores of a class are 85, 90, 78, 92, and 88. Find the range.

Solution: Range = Maximum score – Minimum score = 92 – 78 = 14

Problem 35. A bag contains 8 red balls and 6 blue balls. If two balls are drawn without replacement, what is the probability of getting one red ball and one blue ball?

Solution: Total number of balls = 8 (red) + 6 (blue) = 14

Probability of drawing a red ball on the first draw = 8/14 Probability of drawing a blue ball on the second draw (after one red ball is already drawn) = 6/13 (since one red ball is already drawn)

Probability of getting one red ball and one blue ball = Probability of drawing a red ball on the first draw * Probability of drawing a blue ball on the second draw = (8/14) * (6/13) ≈ 0.2857

Problem 36. The heights (in inches) of a group of students are 60, 65, 68, 62, 70, and 67. Find the variance.

Solution: Step 1: Find the mean height Mean height = (60 + 65 + 68 + 62 + 70 + 67)/6 = 392/6 = 65.33

Step 2: Find the deviation of each height from the mean Deviation of 60 from the mean = 60 – 65.33 = –5.33 Deviation of 65 from the mean = 65 – 65.33 = -0.33 Deviation of 68 from the mean = 68 – 65.33 = 2.67 Deviation of 62 from the mean = 62 – 65.33 = –3.33 Deviation of 70 from the mean = 70 – 65.33 = 4.67 Deviation of 67 from the mean = 67 – 65.33 = 1.67

Step 3: Square the deviations and find the sum of squared deviations Sum of squared deviations = $(-5.33)^2 + (-0.33)^2 + (2.67)^2 + (-3.33)^2 + (4.67)^2 + (1.67)^2$ = 28.44 + 0.11 + 7.13 + 11.09 + 21.77 + 2.78 = 71.32

Step 4: Divide the sum of squared deviations by the number of heights (n) Variance = Sum of squared deviations/(n – 1) = 71.32/5 = 14.26

Problem 37. A coin is tossed 5 times. Find the probability of getting exactly 3 heads.

Solution: Number of total outcomes when a coin is tossed 5 times = 2^5 = 32 Number of ways to get exactly 3 heads = ${}^5C_3$ = 10 (using combinations formula)

Probability of getting exactly 3 heads = Number of ways to get exactly 3 heads/Number of total outcomes = 10/32 ≈ 0.3125

Problem 38. A box contains 8 red balls and 6 blue balls. If two balls are drawn with replacement, what is the probability of getting one red ball and one blue ball?

Solution: Probability of drawing a red ball = Number of red balls/Total number of balls = 8/14 = 4/7 Probability of drawing a blue ball = Number of blue balls/Total number of balls = 6/14 = 3/7

Probability of getting one red ball and one blue ball = Probability of drawing a red ball * Probability of drawing a blue ball = (4/7) * (3/7) = 12/49

Problem 39. The heights (in inches) of a group of students are 60, 65, 68, 62, 70, and 67. Find the median height.

Solution: Arrange the heights in ascending order: 60, 62, 65, 67, 68, 70 Since there are six heights, the median is the average of the two middle values, which are 65 and 67.

Median height = (65 + 67)/2 = 132/2 = 66

Problem 40. The test scores of a class are 85, 90, 78, 92, and 88. Find the standard deviation.

Solution: Step 1: Find the mean score Mean score = (85 + 90 + 78 + 92 + 88)/5 = 433/5 = 86.6

Step 2: Find the deviation of each score from the mean Deviation of 85 from the mean = 85 – 86.6 = –1.6 Deviation of 90 from the mean = 90 – 86.6 = 3.4 Deviation of 78 from the mean = 78 – 86.6 = –8.6 Deviation of 92 from the mean = 92 – 86.6 = 5.4 Deviation of 88 from the mean = 88 – 86.6 = 1.4

Step 3: Square the deviations and find the sum of squared deviations Sum of squared deviations = $(-1.6)^2 + (3.4)^2 + (-8.6)^2 + (5.4)^2 + (1.4)^2$ = 2.56 + 11.56 + 73.96 + 29.16 + 1.96 = 119.2

Step 4: Divide the sum of squared deviations by the number of scores (n) Variance = Sum of squared deviations/(n – 1) = 119.2/4 = 29.8

Step 5: Take the square root of the variance to find the standard deviation Standard deviation = $\sqrt{29.8} \approx 5.46$

Problem 41. The ages (in years) of a group of students are 20, 22, 18, 25, and 21. Find the mean age.

Solution: Mean age = (20 + 22 + 18 + 25 + 21)/5 = 106/5 = 21.2

Problem 42. The test scores of a class are 85, 90, 78, 92, and 88. Find the range.

Solution: Range = Maximum score – Minimum score = 92 – 78 = 14

Problem 43. A bag contains 8 red balls and 6 blue balls. If two balls are drawn without replacement, what is the probability of getting one red ball and one blue ball?

Solution: Total number of balls = 8 (red) + 6 (blue) = 14

Probability of drawing a red ball on the first draw = 8/14 Probability of drawing a blue ball on the second draw (after one red ball is already drawn) = 6/13 (since one red ball is already drawn)

Probability of getting one red ball and one blue ball = Probability of drawing a red ball on the first draw * Probability of drawing a blue ball on the second draw = (8/14) * (6/13) ≈ 0.2857

Problem 44. The heights (in inches) of a group of students are 60, 65, 68, 62, 70, and 67. Find the variance.

Solution: Step 1: Find the mean height Mean height = (60 + 65 + 68 + 62 + 70 + 67)/6 = 392/6 = 65.33

Step 2: Find the deviation of each height from the mean Deviation of 60 from the mean = 60 – 65.33 = –5.33. Deviation of 65 from the mean = 65 – 65.33 = –0.33. Deviation of 68 from the mean = 68 – 65.33 = 2.67. Deviation of 62 from the mean = 62 – 65.33 = –3.33. Deviation of 70 from the mean = 70 – 65.33 = 4.67. Deviation of 67 from the mean = 67 – 65.33 = 1.67

Step 3: Square the deviations and find the sum of squared deviations. Sum of squared deviations = $(-5.33)^2 + (-0.33)^2 + (2.67)^2 + (-3.33)^2 + (4.67)^2 + (1.67)^2$ = 28.44 + 0.11 + 7.13 + 11.09 + 21.77 + 2.78 = 71.32

Step 4: Divide the sum of squared deviations by the number of heights (n) Variance = Sum of squared deviations/(n – 1) = 71.32/5 = 14.26

Problem 45. A coin is tossed 5 times. Find the probability of getting exactly 3 heads.

Solution: Number of total outcomes when a coin is tossed 5 times = 2^5 = 32 Number of ways to get exactly 3 heads = ${}^5C_3$ = 10 (using combinations formula)

Probability of getting exactly 3 heads = Number of ways to get exactly 3 heads/Number of total outcomes = 10/32 ≈ 0.3125

Problem 46. A box contains 8 red balls and 6 blue balls. If two balls are drawn with replacement, what is the probability of getting one red ball and one blue ball?

Solution: Probability of drawing a red ball = Number of red balls/Total number of balls = 8/14 = 4/7 Probability of drawing a blue ball = Number of blue balls/Total number of balls = 6/14 = 3/7

Probability of getting one red ball and one blue ball = Probability of drawing a red ball on the first draw * Probability of drawing a blue ball on the second draw = (4/7) * (3/7) = 12/49

Problem 47. The heights (in inches) of a group of students are 60, 65, 68, 62, 70, and 67. Find the median height.

Solution: Arrange the heights in ascending order: 60, 62, 65, 67, 68, 70 Since there are six heights, the median is the average of the two middle values, which are 65 and 67.

Median height = (65 + 67)/2 = 132/2 = 66

Problem 48. The test scores of a class are 85, 90, 78, 92, and 88. Find the standard deviation.

Solution: Step 1: Find the mean score Mean score = (85 + 90 + 78 + 92 + 88)/5 = 433/5 = 86.6

Step 2: Find the deviation of each score from the mean Deviation of 85 from the mean = 85 – 86.6 = –1.6 Deviation of 90 from the mean = 90 – 86.6 = 3.4 Deviation of 78 from the mean = 78 – 86.6 = –8.6 Deviation of 92 from the mean = 92 – 86.6 = 5.4 Deviation of 88 from the mean = 88 – 86.6 = 1.4

Step 3: Square the deviations and find the sum of squared deviations Sum of squared deviations = $(-1.6)^2 + (3.4)^2 + (-8.6)^2 + (5.4)^2 + (1.4)^2 = 2.56 + 11.56 + 73.96 + 29.16 + 1.96 = 119.2$

Step 4: Divide the sum of squared deviations by the number of scores (n) Variance = Sum of squared deviations/(n – 1) = 119.2/4 = 29.8

Step 5: Take the square root of the variance to find the standard deviation Standard deviation = $\sqrt{29.8} \approx 5.46$

Problem 49. The ages (in years) of a group of students are 20, 22, 18, 25, and 21. Find the mean age.

Solution: Mean age = (20 + 22 + 18 + 25 + 21)/5 = 106/5 = 21.2

Problem 50. The test scores of a class are 85, 90, 78, 92, and 88. Find the range.

Solution: Range = Maximum score – Minimum score = 92 – 78 = 14

Short Questions and Answers

Q.1. What is probability?

Ans. Probability is a measure of the likelihood that an event will occur. It is expressed as a value between 0 and 1, where 0 indicates impossibility, and 1 indicates certainty.

Q.2. What is a random variable?

Ans. A random variable is a variable whose value is determined by the outcome of a random event. It can take on different values with associated probabilities.

Q.3. What is the difference between discrete and continuous random variables?

Ans. Discrete random variables have a countable number of possible outcomes, while continuous random variables have an uncountable and infinite number of possible outcomes within a range.

Q.4. What is the sample space?

Ans. The sample space is the set of all possible outcomes of a random experiment.

Q.5. What is a probability distribution?

Ans. A probability distribution describes the probabilities of all possible outcomes of a random variable. It can be represented as a table, formula, or graph.

Q.6. What is the mean (expected value) of a random variable?

Ans. The mean (expected value) of a random variable is the average value it takes over a large number of trials.

Q.7. What is variance?

Ans. Variance is a measure of the spread or dispersion of the values of a random variable around its mean.

Q.8. What is standard deviation?

Ans. Standard deviation is the square root of the variance and represents the typical amount of deviation from the mean.

Q.9. What is the Central Limit Theorem?

Ans. The Central Limit Theorem states that the distribution of the sample means of a large number of independent, identically distributed random variables approximates a normal distribution, regardless of the original distribution.

Q.10. What is the difference between a population and a sample?

Ans. A population includes all members of a group being studied, while a sample is a subset of the population used to make inferences about the entire population.

Q.11. What is a confidence interval?

Ans. A confidence interval is a range of values within which the true population parameter is likely to fall with a specified level of confidence.

Q.12. What is hypothesis testing?

Ans. Hypothesis testing is a statistical method used to determine whether a claim about a population is supported by sample data.

Q.13. What is a p-value?

Ans. The p-value is the probability of obtaining a test statistic as extreme as the one observed or even more extreme, assuming that the null hypothesis is true.

Q.14. What is a type I error?

Ans. A type I error occurs when a true null hypothesis is rejected.

Q.15. What is a type II error?

Ans. A type II error occurs when a false null hypothesis is not rejected.

Q.16. What is a one-tailed (one-sided) test?

Ans. A one-tailed test is a hypothesis test that focuses on one direction of difference from the hypothesized value.

Q.17. What is a two-tailed (two-sided) test?

Ans. A two-tailed test is a hypothesis test that tests for a difference in either direction from the hypothesized value.

Q.18. What is correlation?

Ans. Correlation measures the strength and direction of the linear relationship between two variables.

Q.19. What is the correlation coefficient?

Ans. The correlation coefficient is a statistic that quantifies the degree of correlation between two variables, ranging from –1 (perfect negative correlation) to +1 (perfect positive correlation).

Q.20. What is a scatter plot?

Ans. A scatter plot is a graph that shows the relationship between two variables, with one variable represented on the x-axis and the other on the y-axis.

Q.21. What is simple linear regression?

Ans. Simple linear regression is a statistical method used to model the relationship between two variables with a straight line equation.

Q.22. What is multiple linear regressions?

Ans. Multiple linear regressions are a statistical method used to model the relationship between multiple independent variables and a single dependent variable.

Q.23. What is the coefficient of determination (R-squared)?

Ans. R-squared is a statistic that represents the proportion of the variance in the dependent variable that can be explained by the independent variables in a regression model.

Q.24. What is the difference between correlation and causation?

Ans. Correlation indicates a statistical relationship between two variables, but it does not imply a cause-and-effect relationship. Causation requires additional evidence to establish a cause-and-effect relationship.

Q.25. What is the law of large numbers?

Ans. The law of large numbers states that as the number of independent trials increases, the sample mean will converge to the true population mean.

Q.26. What is a histogram?

Ans. A histogram is a graphical representation of the distribution of a continuous random variable. It consists of bars that represent the frequency of values within specific intervals.

Q.27. What is a normal distribution?

Ans. A normal distribution, also known as a Gaussian distribution, is a bell-shaped probability distribution characterized by its mean and standard deviation.

Q.28. What is the empirical rule (68–95–99.7 rule)?

Ans. The empirical rule states that for a normal distribution, approximately 68% of data falls within one standard deviation of the mean, 95% within two standard deviations, and 99.7% within three standard deviations.

Q.29. What is the difference between the population mean and sample mean?

Ans. The population mean represents the average of all values in the entire population, while the sample mean represents the average of values in a sample drawn from the population.

Q.30. What is a box plot?

Ans. A box plot, also known as a box-and-whisker plot, is a graphical representation of the distribution of a dataset, showing the median, quartiles, and outliers.

Q.31. What is the mode?

Ans. The mode is the value that occurs most frequently in a dataset.

Q.32. What is skewness?

Ans. Skewness measures the asymmetry of the probability distribution of a random variable.

Q.33. What is kurtosis?

Ans. Kurtosis measures the degree of peakedness or flatness of a probability distribution compared to the normal distribution.

Q.34. What is the difference between a parameter and a statistic?

Ans. A parameter is a characteristic of a population, while a statistic is a characteristic of a sample.

Q.35. What is a sampling distribution?

Ans. A sampling distribution is the probability distribution of a sample statistic, such as the mean or standard deviation, obtained from multiple random samples of the same size from a population.

Q.36. What is random sampling?

Ans. Random sampling is a sampling method where each member of the population has an equal chance of being selected for the sample.

Q.37. What is stratified sampling?

Ans. Stratified sampling is a sampling method where the population is divided into subgroups (strata), and random samples are taken from each stratum.

Q.38. What is systematic sampling?

Ans. Systematic sampling is a sampling method where samples are selected at fixed intervals from the population.

Q.39. What is convenience sampling?

Ans. Convenience sampling is a non-probabilistic sampling method where samples are selected based on what is readily available and accessible.

Q.40. What is the difference between an outlier and an influential point?

Ans. An outlier is a data point that is significantly different from the other data points in a dataset. An influential point is a data point that has a strong influence on the regression model's parameters.

Q.41. What is the law of total probability?

Ans. The Law of Total Probability states that for any event A and partition of the sample space S into mutually exclusive events $B_1, B_2,..., B_n$, the probability of A can be calculated as the sum of the probabilities of A given each partition multiplied by the probability of each partition.

Q.42. What is conditional probability?

Ans. Conditional probability is the probability of an event occurring given that another event has already occurred. It is denoted as $P(A|B)$, where A and B are events.

Q.43. What is Bayes' theorem?

Ans. Bayes' theorem is a fundamental result in probability theory that allows the revision of probabilities based on new evidence. It is given by:

$$P(A|B) = (P(B|A) * P(A))/P(B)$$

Q.44. What is a probability mass function (PMF)?

Ans. A probability mass function (PMF) is a function that gives the probability of each discrete value of a discrete random variable.

Q.45. What is a probability density function (PDF)?

Ans. A probability density function (PDF) is a function that gives the probability of a continuous random variable taking on a particular value within a given interval.

Q.46. What is the law of large numbers?

Ans. The law of large numbers states that as the number of trials or samples increases, the observed frequency of an event will converge to its true probability.

Q.47. What is statistical significance?

Ans. Statistical significance refers to the likelihood that the observed results in an experiment or study are not due to random chance but are instead a result of the variables being studied.

Q.48. What is a null hypothesis?

Ans. A null hypothesis is a statement that assumes there is no significant difference or effect in the variables being studied.

Q.49. What is a alternative hypothesis?

Ans. An alternative hypothesis is a statement that contradicts the null hypothesis and suggests there is a significant difference or effect in the variables being studied.

Q.50. What is thc chi-square test?

Ans. The chi-square test is a statistical test used to determine if there is a significant association between two categorical variables.

Multiple Choice Questions and Answers

Q.1. What is the definition of probability?

(*a*) A measure of the likelihood of an event occurring
(*b*) A measure of the size of a sample space
(*c*) The number of favorable outcomes divided by the total number of outcomes
(*d*) The sum of all possible outcomes

Q.2. Which of the following is an example of a discrete random variable?

(*a*) Temperature in degrees Celsius
(*b*) Height in centimeters
(*c*) Number of heads obtained in tossing a coin
(*d*) Weight in kilograms

Q.3. What is the range of a continuous random variable?

(*a*) All real numbers
(*b*) Positive integers
(*c*) Natural numbers
(*d*) Whole numbers

Q.4. The mean of a data set is equal to:

(*a*) The sum of all data values
(*b*) The middle value in the data set
(*c*) The average of all data values
(*d*) The most frequently occurring value

Q.5. What is the variance of a data set?

(*a*) A measure of the spread of data around the mean
(*b*) The average of all data values
(*c*) The sum of all data values
(*d*) The middle value in the data set

Q.6. The standard deviation is the:

(*a*) Range of data values
(*b*) Variance multiplied by the sample size
(*c*) Square root of the variance
(*d*) Average of all data values

Q.7. The Central Limit Theorem states that:

(*a*) The mean of the sample will be equal to the population mean

(*b*) The sample size should be larger than the population size for accurate results

(*c*) The distribution of the sample means approaches a normal distribution as the sample size increases

(*d*) The sample variance will be equal to the population variance

Q.8. What is a confidence interval?

(*a*) A range of values within which the true population parameter is likely to fall with a specified level of confidence

(*b*) The range of data values in a dataset

(*c*) The standard error of the sample mean

(*d*) The difference between the sample mean and the population mean

Q.9. A p-value less than the significance level (alpha) indicates:

(*a*) Rejection of the null hypothesis (*b*) Failure to reject the null hypothesis

(*c*) Acceptance of the null hypothesis (*d*) The test is inconclusive

Q.10. Which type of error occurs when a true null hypothesis is rejected?

(*a*) Type I error (*b*) Type II error (*c*) Type III error (*d*) Type IV error

Q.11. Which type of error occurs when a false null hypothesis is not rejected?

(*a*) Type I error (*b*) Type II error (*c*) Type III error (*d*) Type IV error

Q.12. A one-tailed test is used when:

(*a*) The researcher is interested in both directions of difference from the hypothesized value

(*b*) The researcher is only interested in one direction of difference from the hypothesized value

(*c*) The sample size is large (*d*) The data is continuous

Q.13. A two-tailed test is used when:

(*a*) The researcher is interested in both directions of difference from the hypothesized value

(*b*) The researcher is only interested in one direction of difference from the hypothesized value

(*c*) The sample size is large (*d*) The data is continuous

Q.14. What is correlation?

(*a*) A measure of the spread of data values

(*b*) A measure of the strength and direction of the linear relationship between two variables

(*c*) The sum of all data values

(*d*) The average of all data values

Q.15. The correlation coefficient ranges from:

(*a*) –1 to 0 (*b*) 0 to 1 (*c*) –1 to 1 (*d*) 0 to infinity

Q.16. What type of plot is used to represent the relationship between two variables in a scatterplot?

(*a*) Bar plot (*b*) Line plot (*c*) Histogram (*d*) Scatter plot

Q.17. Simple linear regression is used to model the relationship between:

(*a*) Two or more independent variables and one dependent variable

(*b*) One independent variable and two or more dependent variables

(*c*) One independent variable and one dependent variable

(*d*) Multiple independent variables and multiple dependent variables

Q.18. What does R-squared represent in a regression model?

(*a*) The strength of the relationship between the independent and dependent variables

(*b*) The sum of the squared residuals

(*c*) The mean of the dependent variable

(*d*) The proportion of the variance in the dependent variable explained by the independent variable

Q.19. What is the difference between correlation and causation?

(*a*) Correlation implies causation, while causation implies correlation.

(*b*) Correlation measures the strength of the relationship between two variables, while causation indicates a cause-and-effect relationship.

(*c*) Correlation and causation are interchangeable terms.

(*d*) Correlation and causation are unrelated concepts.

Q.20. What is the law of large numbers?

(*a*) A statistical principle that states the sample mean will approach the population mean as the sample size increases.

(*b*) A mathematical theorem that describes the probability of an event.

(*c*) A rule stating that the variance of a dataset decreases with increasing sample size.

(*d*) A principle that explains the relationship between the median and mean of a dataset.

Q.21. What is a histogram used for?

(*a*) Representing the relationship between two variables in a scatterplot.

(*b*) Visualizing the distribution of a continuous random variable.

(*c*) Comparing the means of two groups in a statistical test.

(*d*) Estimating the correlation coefficient between two variables.

Q.22. What is a normal distribution?

(*a*) A distribution where all data values are equal.

(*b*) A distribution with a perfect linear relationship between two variables.

(*c*) A distribution with a bell-shaped curve, characterized by its mean and standard deviation.

(*d*) A distribution with a uniform spread of data values.

Q.23. What does the empirical rule (68–95–99.7 rule) state?

(*a*) The probability that a data point falls within one standard deviation of the mean is 68%.

(*b*) The probability that a data point falls within two standard deviations of the mean is 95%.

(*c*) The probability that a data point falls within three standard deviations of the mean is 99.7%.

(*d*) All of the above.

Q.24. The sample mean is calculated as:

(*a*) The sum of all data values divided by the sample size.

(*b*) The median of all data values.

(*c*) The mode of all data values.

(*d*) The difference between the largest and smallest data values.

Q.25. What is a box plot used for?

(*a*) Comparing two or more groups in a statistical test.

(*b*) Visualizing the spread and distribution of data values, including outliers.

(*c*) Estimating the correlation coefficient between two variables.

(*d*) Representing the relationship between two variables in a scatter plot.

Q.26. What is the mode of a dataset?

(*a*) The middle value in the data set.

(*b*) The average of all data values.

(*c*) The value that occurs most frequently in the dataset.

(*d*) The range of data values.

Q.27. Skewness measures:

(*a*) The spread of data values.

(*b*) The degree of peakedness or flatness of a probability distribution.

(*c*) The asymmetry of the probability distribution.

(*d*) The strength and direction of the linear relationship between two variables.

Q.28. Kurtosis measures:

(*a*) The spread of data values.

(*b*) The degree of peakedness or flatness of a probability distribution.

(*c*) The asymmetry of the probability distribution.

(*d*) The strength and direction of the linear relationship between two variables.

Q.29. What is the difference between a parameter and a statistic?

(*a*) A parameter is a characteristic of a sample, while a statistic is a characteristic of a population.

(*b*) A parameter is a characteristic of a population, while a statistic is a characteristic of a sample.

(*c*) Parameters and statistics are interchangeable terms.

(*d*) Parameters and statistics are unrelated concepts.

Q.30. What is a sampling distribution?

(*a*) The distribution of a dataset.

(*b*) The distribution of a sample mean.

(*c*) The distribution of a population.

(*d*) The distribution of a sample median.

Q.31. What is random sampling?

(*a*) A sampling method where samples are selected at fixed intervals from the population.

(*b*) A sampling method where samples are selected based on what is readily available and accessible.

(c) A sampling method where each member of the population has an equal chance of being selected for the sample.

(d) A sampling method where the population is divided into subgroups, and random samples are taken from each stratum.

Q.32. What is stratified sampling?

(a) A sampling method where samples are selected at fixed intervals from the population.

(b) A sampling method where samples are selected based on what is readily available and accessible.

(c) A sampling method where each member of the population has an equal chance of being selected for the sample.

(d) A sampling method where the population is divided into subgroups, and random samples are taken from each stratum.

Q.33. What is systematic sampling?

(a) A sampling method where samples are selected at fixed intervals from the population.

(b) A sampling method where samples are selected based on what is readily available and accessible.

(c) A sampling method where each member of the population has an equal chance of being selected for the sample.

(d) A sampling method where the population is divided into subgroups, and random samples are taken from each stratum.

Q.34. What is convenience sampling?

(a) A sampling method where samples are selected at fixed intervals from the population.

(b) A sampling method where samples are selected based on what is readily available and accessible.

(c) A sampling method where each member of the population has an equal chance of being selected for the sample.

(d) A sampling method where the population is divided into subgroups, and random samples are taken from each stratum.

Q.35. What is the difference between an outlier and an influential point?

(a) An outlier is a data point that is significantly different from the other data points in a dataset, while an influential point has a strong influence on the regression model's parameters.

(b) An outlier has a strong influence on the regression model's parameters, while an influential point is a data point that is significantly different from the other data points in a dataset.

(c) An outlier and an influential point are synonymous terms.

(d) An outlier and an influential point are unrelated concepts.

Q.36. What is the Law of Total Probability?

(a) A statistical principle that states the sample mean will approach the population mean as the sample size increases.

(b) A mathematical theorem that describes the probability of an event.

(*c*) A rule stating that the variance of a dataset decreases with increasing sample size.
(*d*) A principle that explains the relationship between the median and mean of a dataset.

Q.37. What is conditional probability?
(*a*) The probability of an event occurring given that another event has already occurred.
(*b*) The probability that two events occur simultaneously.
(*c*) The probability that one event occurs, and the other does not.
(*d*) The probability that at least one of two events occurs.

Q.38. What is Bayes' theorem used for?
(*a*) Calculating the probability of an event occurring given the probability of another event.
(*b*) Calculating the average of all data values.
(*c*) Calculating the standard deviation of a dataset.
(*d*) Calculating the variance of a dataset.

Q.39. What is a probability mass function (PMF)?
(*a*) A function that gives the probability of each discrete value of a discrete random variable.
(*b*) A function that gives the probability of a continuous random variable taking on a particular value within a given interval.
(*c*) A function used to calculate the mean of a dataset.
(*d*) A function used to calculate the standard deviation of a dataset.

Q.40. What is a probability density function (PDF)?
(*a*) A function that gives the probability of each discrete value of a discrete random variable.
(*b*) A function that gives the probability of a continuous random variable taking on a particular value within a given interval.
(*c*) A function used to calculate the mean of a dataset.
(*d*) A function used to calculate the standard deviation of a dataset.

Q.41. What is the difference between a population mean and a sample mean?
(*a*) The population mean is a measure of the spread of data values, while the sample mean is the average of all data values.
(*b*) The population mean represents the average of all data values, while the sample mean represents the average of values in a sample drawn from the population.
(*c*) The population mean is calculated using a sample, while the sample mean is calculated using the entire population.
(*d*) The population mean and the sample mean are unrelated concepts.

Q.42. What is a box plot used for?
(*a*) Representing the relationship between two variables in a scatterplot.
(*b*) Visualizing the spread and distribution of data values, including outliers.
(*c*) Comparing the means of two groups in a statistical test.
(*d*) Estimating the correlation coefficient between two variables.

Q.43. What is the mode of a dataset?

(*a*) The middle value in the data set.

(*b*) The average of all data values.

(*c*) The value that occurs most frequently in the dataset.

(*d*) The range of data values.

Q.44. Skewness measures:

(*a*) The spread of data values.

(*b*) The degree of peakedness or flatness of a probability distribution.

(*c*) The asymmetry of the probability distribution.

(*d*) The strength and direction of the linear relationship between two variables.

Q.45. Kurtosis measures:

(*a*) The spread of data values.

(*b*) The degree of peakedness or flatness of a probability distribution.

(*c*) The asymmetry of the probability distribution.

(*d*) The strength and direction of the linear relationship between two variables.

Q.46. What is the difference between a parameter and a statistic?

(*a*) A parameter is a characteristic of a sample, while a statistic is a characteristic of a population.

(*b*) A parameter is a characteristic of a population, while a statistic is a characteristic of a sample.

(*c*) Parameters and statistics are interchangeable terms.

(*d*) Parameters and statistics are unrelated concepts.

Q.47. What is a sampling distribution?

(*a*) The distribution of a dataset.

(*b*) The distribution of a sample mean.

(*c*) The distribution of a population.

(*d*) The distribution of a sample median.

Q.48. What is random sampling?

(*a*) A sampling method where samples are selected at fixed intervals from the population.

(*b*) A sampling method where samples are selected based on what is readily available and accessible.

(*c*) A sampling method where each member of the population has an equal chance of being selected for the sample.

(*d*) A sampling method where the population is divided into subgroups, and random samples are taken from each stratum.

Q.49. What is stratified sampling?

(*a*) A sampling method where samples are selected at fixed intervals from the population.

(*b*) A sampling method where samples are selected based on what is readily available and accessible.

(*c*) A sampling method where each member of the population has an equal chance of being selected for the sample.

(*d*) A sampling method where the population is divided into subgroups, and random samples are taken from each stratum.

Q.50. What is systematic sampling?

(*a*) A sampling method where samples are selected at fixed intervals from the population.

(*b*) A sampling method where samples are selected based on what is readily available and accessible.

(*c*) A sampling method where each member of the population has an equal chance of being selected for the sample.

(*d*) A sampling method where the population is divided into subgroups, and random samples are taken from each stratum.

Key to Multiple Choice Questions

1. (*a*)	**2.** (*c*)	**3.** (*a*)	**4.** (*c*)	**5.** (*a*)	**6.** (*c*)	**7.** (*c*)	**8.** (*a*)	**9.** (*a*)
10. (*a*)	**11.** (*b*)	**12.** (*b*)	**13.** (*a*)	**14.** (*b*)	**15.** (*c*)	**16.** (*d*)	**17.** (*c*)	**18.** (*d*)
19. (*b*)	**20.** (*a*)	**21.** (*b*)	**22.** (*c*)	**23.** (*d*)	**24.** (*a*)	**25.** (*b*)	**26.** (*c*)	**27.** (*c*)
28. (*b*)	**29.** (*b*)	**30.** (*b*)	**31.** (*c*)	**32.** (*d*)	**33.** (*a*)	**34.** (*b*)	**35.** (*a*)	**36.** (*b*)
37. (*a*)	**38.** (*a*)	**39.** (*a*)	**40.** (*b*)	**41.** (*b*)	**42.** (*b*)	**43.** (*c*)	**44.** (*c*)	**45.** (*b*)
46. (*b*)	**47.** (*b*)	**48.** (*c*)	**49.** (*d*)	**50.** (*a*)				

Projects Works

1. **Sensor Fusion for Robot Localization:** Develop a probabilistic algorithm that fuses data from multiple sensors (e.g., LiDAR, cameras, IMU) to accurately estimate the robot's position and orientation in an environment. Use statistical methods to handle sensor uncertainties and noise.
2. **Path Planning with Uncertainty:** Create a path planning algorithm for a robot that takes into account the uncertainties in the environment, such as uncertain obstacle positions and sensor measurements. Utilize probabilistic methods like Monte Carlo Localization and Bayesian optimization for efficient and safe path planning.
3. **Robot Swarm Behavior Analysis:** Investigate the statistical properties of robot swarm behavior in different scenarios. Analyze metrics like swarm density, cohesion, and dispersion to understand emergent behaviors and optimize swarm control strategies.
4. **Object Recognition and Classification:** Develop a probabilistic approach for robot vision to recognize and classify objects in the environment. Implement algorithms like Gaussian Mixture Models (GMM) or Hidden Markov Models (HMM) to handle object uncertainties and variations.
5. **Fault Detection and Diagnosis:** Design a statistical method for fault detection and diagnosis in robot systems. Use data-driven approaches like Bayesian networks or Support Vector Machines (SVM) to identify and isolate faulty components in the robot.
6. **Adaptive Control for Robot Manipulators:** Create an adaptive control system for robot manipulators that can adapt to uncertainties in the robot's dynamics and environment. Utilize statistical learning techniques like reinforcement learning or Gaussian Processes to optimize the control policy.
7. **Human-Robot Interaction Analysis:** Analyze human-robot interaction data to understand user behavior and preferences. Use statistical methods like hypothesis testing and regression analysis to evaluate the effectiveness of different robot interaction strategies.

8. **Predictive Maintenance for Robots:** Develop a predictive maintenance system for robots based on historical sensor data. Utilize time-series analysis and probabilistic models to predict the remaining useful life of robot components and schedule maintenance accordingly.
9. **Optimal Sensor Placement:** Use statistical optimization techniques to determine the optimal placement of sensors on a robot for efficient data gathering and localization accuracy. Consider factors like sensor range, coverage, and cost.
10. **Robot Learning from Demonstration:** Implement a probabilistic framework for robot learning from demonstration. Use techniques like Gaussian Process Regression or Bayesian Nonparametric models to learn the underlying task and uncertainty from human demonstrations.
11. **Collaborative Robot Task Allocation:** Develop a statistical approach for task allocation in a team of collaborative robots. Use methods like Multi-Armed Bandits or Markov Decision Processes to optimize task assignment based on team performance and uncertainties.
12. **Robot Anomaly Detection:** Build a system that uses statistical anomaly detection methods to identify unusual behaviors or events in a robot's operation. Use techniques like One-Class SVM or Auto encoders for unsupervised anomaly detection.
13. **Multi-Robot Localization and Mapping:** Develop a probabilistic algorithm that enables a team of robots to collaboratively build a map of their environment while localizing themselves. Use data fusion techniques like Extended Kalman Filters or Particle Filters for robust localization and mapping.
14. **Reinforcement Learning with Uncertainty:** Investigate how to incorporate uncertainty modeling into reinforcement learning algorithms for robotics. Explore approaches like Bayesian Reinforcement Learning or Distributional RL to handle uncertainties in reward signals and environment dynamics.
15. **Probabilistic Path Following:** Implement a path-following algorithm for a robot that accounts for uncertainties in the control inputs and the environment. Utilize probabilistic control methods like Stochastic Optimal Control or Probabilistic Model Predictive Control.

06

Control Theory with the Study of controlling the behavior of Dynamic Systems

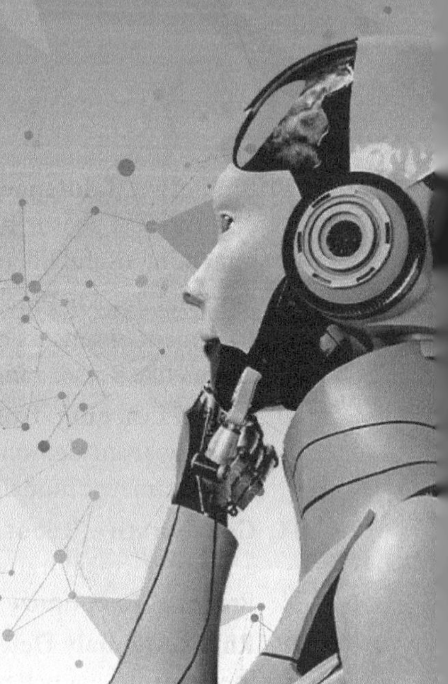

6.1 INTRODUCTION

Engineering and mathematics' area of control theory is concerned with the study of regulating the behavior of dynamic systems. It offers methods and tools for planning and analyzing systems to produce desired behavior and output. Making systems behave predictably and steadily while keeping specific desired qualities is the basic objective of control theory.

6.2 KEY CONCEPTS

Key Concepts in Control Theory are as follows:

Dynamic Systems: A dynamic system is one that can be modeled mathematically and evolves over time. These systems can be physical (such as robots, automobiles, and airplanes) or non-physical (such as financial and biological systems).

Dynamic systems are defined as systems in control theory that vary over time and have their behavior predicted by mathematical models. Differential equations are frequently used in these models to link the system's state variables to its inputs and outputs. Control theory relies heavily on the study of dynamic systems because it explains how systems change in response to various inputs and control actions.

In control theory, dynamic systems primarily fall into two categories:

1. **Systems with Continuous-Time Dynamics:**
   - Dynamic systems that change continuously over time are known as continuous-time dynamic systems. Ordinary differential equations (ODEs) or partial differential equations (PDEs) are commonly used to describe the behaviour of these systems.
   - Pendulum motion, cooling system temperature dynamics, and robotic arm position control are a few examples of continuous-time dynamic systems.
2. **Discrete-Time Dynamic Systems:** These systems experience state variable changes at discrete time steps rather than continuously. Difference equations are often used to describe the behavior of these systems.
   - Digital control systems, sampled data systems, and systems that get regular updates, like some digital circuits, are examples of discrete-time dynamic systems.

Designing efficient control strategies requires a thorough understanding of a system's dynamics. Control engineers examine the stability, controllability, and observability of dynamic systems using mathematical tools and techniques. These systems exhibit a wide range of behaviors, from straightforward first-order systems with linear dynamics to intricate higher-order systems with nonlinear dynamics. The construction of controllers that may modify the behavior of dynamic systems to achieve desired performance and stability is made possible by the framework provided by control theory.

Control: Control is the capacity to influence a dynamic system's behavior in order to fulfill predetermined goals. Different control actions, such as modifying inputs, parameters, or feedback, can be used. In control theory, the term "control" refers to the process of influencing or controlling a dynamic system's behavior in order to accomplish specified goals or performance standards. The control action is used to direct the system towards the desired behavior by responding to the system's present state and the target state or reference.

The process of control involves several key elements:

1. **System:** The dynamic system under management is referred to as the system. It could be an abstract system like an economic model or a population dynamics model, or it could be a physical system like a robot, an airplane, or a chemical factory.
2. **Input:** The control input, frequently represented by the letter "u," is the signal or action applied to the system to affect its behavior. Any factor that has the ability to change the state of the system qualifies, including force, voltage, set points, and more.
3. **Output:** The system's output, frequently indicated by the letter "y," stands for the quantifiable or observable quality that best describes its reaction. It is frequently necessary to manage the output in order to attain the desired state or performance.
4. **State:** The system's state, represented by the letter "x," refers to the internal variables that characterize its current state. The system's current state contains all the data required to forecast its future behavior.
5. **Reference or Set point:** The reference or set point, frequently represented by the letter "r," is the ideal value or course that the system ought to take. The goal of the control system is to have the system's output follow or match this desired reference.
6. **Controller:** The key element of the control system is the controller. It is in charge of figuring out the control input (u) based on the intended reference (r), the desired reference (r), and the present state (x). Depending on the type of control strategy being employed, such as proportional control, integral control, or derivative control, the controller's architecture will vary.
7. **Feedback:** One common type of control is feedback control. To determine the error ($e = r - y$), one must measure the system's output (y) and compare it to the desired reference (r). The controller then receives the error signal and generates the control input (u), which directs the system towards the intended behavior.
8. **Closed-Loop Control vs. Open-Loop Control:** With open-loop control, the control input is chosen without taking the system output into account. Closed-loop control, also known as feedback control, improves accuracy and disturbance robustness by computing the control input from the system output.

In control theory, achieving stability, accuracy, and optimal performance of the system is the main goal of control. To make sure the controlled system complies with particular requirements and restrictions, control engineers employ a variety of analytical and design methodologies. Stability, precision in steady state, transient responsiveness, disturbance rejection, and resilience against uncertainties are a few examples of these needs. In numerous industries, such as robotics, aircraft, process control, automation, and others, control theory is essential for enhancing the effectiveness and performance of complex systems.

Feedback Control: A typical technique in control theory is feedback control, which involves measuring the system's output, comparing it to the desired output (the reference), and computing the control action based on the error (the difference between the two). This enables the system to continuously correct itself. A key idea in control theory is feedback control, commonly referred to as closed-loop control. In order to produce the control signal, which is subsequently sent back to the system to govern its behaviour, it is necessary to use knowledge about the system's output. In many engineering applications, this control method is employed to achieve stability, accuracy, and robustness.

The feedback control loop consists of the following components:

1. **Plant (System):** The plant, or the system, is the dynamic system being regulated. It could be a physical system (e.g., a mechanical system, an electrical circuit) or an abstract system (e.g., an economic model, a population dynamics model).
2. **Sensor:** The sensor is in charge of gauging the system's output. It offers details about the system's current condition, frequently represented as "y" (the measured output).
3. **Reference (Set point):** The reference (also known as a set point), represented by the letter "r," is the ideal value or trajectory that the system ought to follow. Making the system's output follow or match this intended reference is the control goal.
4. **Controller:** The feedback control system's central element is the controller. It computes the control signal or control input "u" using the measured output "y" and the reference "r" as inputs. The type of control strategy being utilized, such as proportional control, integral control, derivative control, or a mix of these (PID control), determines the design of the controller.
5. **Actuator:** The actuator applies the corresponding action to the plant after receiving the control signal "u" from the controller. A physical device (such as a motor or valve) or an algorithm that modifies the system's inputs can both function as the actuator.
6. **Feedback Loop:** The controller receives the system's output, "y," and compares it to the reference value, "r," to determine the error signal, "e" (error = r – y). The difference between the system's intended output and its actual output is represented by the error signal.

The feedback control loop operates as follows:

1. The controller computes the control input "u" based on the error "e" (the difference between the desired output "r" and the measured output "y") and compares it to the error.
2. The actuator receives the control input "u" and changes the inputs to the system.
3. The system reacts to changes in its inputs, and its output "y" adjusts as a result.
4. The sensor continuously measures the output "y" and feeds that information back to the controller, which uses it to create the new control signal.
5. The controller keeps modifying the control input based on feedback data during this loop-like procedure until the system's output closely resembles the desired reference "r."

The feedback control strategy offers several advantages:

- Because it continuously modifies the control input in response to the output being monitored and ensures that the system's response is rectified in real-time, it enables the control system to handle uncertainties and disruptions.

- It enables precise and accurate control of the system's behavior because the feedback data enables the controller to adjust the control input in accordance with the system's current state, thereby minimizing or eliminating steady-state errors and preventing erratic behavior or instability.
- As many real-world systems can be successfully controlled using fundamental control techniques including proportional, integral, and derivative control, it enables simple and effective control design.

In numerous engineering and control applications, such as robotics, process control, automotive control, aerospace engineering, and many more, feedback control is a frequently utilized and effective control method. It has evolved into a pillar of contemporary control systems, enabling the creation of intricate, effective, and dependable control solutions for a variety of dynamic systems.

Contrasting open-loop and closed-loop control, open-loop control chooses its course of action without taking the system's output into account. Closed-loop control, also known as feedback control, improves accuracy and disturbance resistance by computing the control action from the system output. Two main control techniques used in control theory to manage the behavior of dynamic systems are open-loop control and closed-loop control. Their approaches involve utilizing feedback data to make adjustments to the

1. **Open-Loop Control:**
    - In open-loop control, the control action is chosen without taking into account the system's performance or output. The application of the control input is based on an open-loop control law or preset control sequence that is independent of the system's present state or output.
    - With open-loop control, there is no feedback mechanism to modify the control action in response to the behavior of the system. It makes the unwavering assumption that the system will act as expected and exclusively depends on the accuracy of the control model.
    - Open-loop control may be less precise and robust in real-world circumstances since it ignores any interruptions or modifications to the system's behavior.
    - Some examples of open-loop control include programming a timer to turn on a light after a predetermined amount of time, giving a motor a constant voltage input for a predetermined period of time, or directing a robotic arm along a predetermined trajectory without monitoring its real position.
2. **Closed-Loop Control (Feedback Control):**
    - In closed-loop control, often referred to as feedback control, the control action is modified depending on feedback from the output or performance of the system. The behavior of the system is continuously observed and compared to the desired reference point or set point by the control system.
    - A sensor measures the output of the system, a controller calculates the control input based on the output and a reference, and an actuator applies the control signal to the system. This is known as closed-loop control.
    - The closed-loop control is more accurate, stable, and resistant to disturbances and uncertainties thanks to the feedback mechanism, which enables the control system to continuously correct mistakes between the desired output and the actual output.
    - By using feedback information, the control system is able to adjust to shifting circumstances and disturbances in real-time, ensuring that the system continues to behave as intended even in the midst of outside disturbances or parameter changes.

- Using a thermostat to keep a consistent room temperature, a cruise control system in a car to maintain a set speed, or a PID controller to stabilize the position of a robotic arm are all examples of closed-loop control.

In conclusion, closed-loop control uses feedback information to continuously alter the control action based on the behavior of the system, whereas open-loop control exclusively depends on predefined control actions without any input. Due to its capacity to manage uncertainties, disruptions, and shifting conditions, closed-loop control is typically favored in many real-world applications, resulting in more precise and steady system performance. When the system's behavior is predictable and feedback is either unnecessary or impractical to implement, open-loop control can nevertheless be helpful.

Controllers: The things or algorithms known as controllers are in charge of creating the control signals that move the system. Proportional-Integral-Derivative (PID) controllers, state-space controllers, and model-based controllers are examples of common controller types. The devices or methods that produce the control signals that steer a dynamic system in the direction of a desired state or trajectory are known as controllers in control theory. In order to control system behavior and meet certain performance goals, controllers are essential. They come in different shapes, and each has a unique control scheme and design philosophy.

Here are some common types of controllers used in control theory:

1. **Proportional (P) Controller:** A control signal is produced by the proportional controller that is proportionate to the difference between the desired set point and the system's present state.
2. **Integral (I) Controller:** The control signal is produced by the integral controller using the cumulative error over time. By guiding the system to the intended reference, it aids in the elimination of steady-state errors.
3. **Derivative (D) Controller:** The derivative controller takes into account the error's rate of change to predict how the system will behave and to perform corrective action to lessen overshooting and enhance stability.
4. **Proportional-Integral (PI) Controller:** The proportional and integral actions are combined by the PI controller. Comparatively speaking to a pure proportional controller, it can solve steady-state defects and offer a quicker reaction.
5. **Proportional-Integral-Derivative (PID) Controller:** The proportional, integral, and derivative actions are all combined by the PID controller. It is one of the most often used controllers because it can deliver precise and reliable control in a variety of applications.
6. **State-Space Controller:** Systems expressed in state-space form are controlled by state-space controllers. These controllers compute the control input by using the desired state and the system's state variables. Based on a state-feedback law, which commonly combines proportional, integral, and derivative actions on the state variables, the control input (u) is established.

The particular control requirements and characteristics of the system being controlled determine the controller to be used. To achieve stability, desirable performance, and robustness in the face of shocks and uncertainties, controllers are built and tuned utilizing a variety of control theory approaches and tools.

Stability: The ability of a system to resume its target state or trajectory following a disruption is known as stability, and it is a key notion in control theory. Unpredictable behavior and uncontrollable oscillations are not present in stable systems. The behavior of a dynamic system over time in reaction to disturbances or changes in the system's initial circumstances is referred to as stability, which is a basic term in control theory. An unstable system will display unbounded or oscillatory behavior, whereas a stable system will, after suffering disturbances, return to its desired or equilibrium state. Stability is important in control theory because an unstable system can produce unintended and unforeseen results, making the control system ineffectual and even damaging or failing.

There are two main types of stability commonly studied in control theory:

1. **Stability in the Sense of Bounded Inputs, Bounded Outputs (BIBO Stability):** Concerned with the system's output behavior when subjected to bounded input signals is BIBO stability. A system is BIBO stable if the output remains bounded for any bounded input signal.

   If the transfer function of the system contains all poles with negative real portions, then mathematically speaking, BIBO stability is attained for a continuous-time system. The transfer function of a discrete-time system must have magnitudes less than one at each of its poles in order for BIBO stability to exist.

2. **Stability in the Sense of Asymptotic Stability:** A more powerful type of stability known as asymptotic stability ensures that the system's output converges to a certain equilibrium or desirable state as time approaches infinity in addition to guaranteeing bounded output for bounded inputs.

   When all of the system's transfer function's poles have negative real portions, continuous-time systems are said to be asymptotically stable. Asymptotic stability for discrete-time systems necessitates that all of the system's transfer function poles have magnitudes that are strictly less than one. It is crucial to remember that stability is intimately tied to where the system's transfer function's poles are situated. The characteristic roots of the differential or difference equations for the system are represented by the poles, which are vital in establishing the stability characteristics of the system.

In addition to these two main types of stability, there are other stability concepts studied in control theory, such as:

- **Borderline Stability:** In a system with marginal stability, the poles are located on the dividing line between stable and unstable zones. The system exhibits persistent oscillations or other distinct behaviors since it neither degrades nor increases.
- **Lyapunov Stability:** Lyapunov stability is a technique for demonstrating a system's stability by building a Lyapunov function that gauges the system's energy or another characteristic. If the Lyapunov function declines or stays constant over time, a system is stable.
- **Exponential Stability:** If a system's response to perturbations eventually decays exponentially to zero, it is said to be exponentially stable.
- **Uniform Stability:** Within a specific circumscribed region, a system is considered uniformly stable if it maintains stability for all beginning circumstances.

In the control system engineering process, stability analysis and control design are crucial processes. To analyze the stability of systems and create controllers that guarantee stable behavior, desirable performance, and robustness against uncertainties and shocks, engineers employ a variety of mathematical tools and methodologies.

***Performance Metrics*:** Performance criteria for control systems, such as response time, precision, settling time, overshoot, and steady-state error, are frequently used in evaluations. These indicators aid in determining how well a control design is working. A control system's performance is evaluated and assessed quantitatively using performance metrics, which are utilized in control theory. These metrics aid researchers and control engineers in quantifying system behavior and evaluating various control schemes or system architectures. The precise control requirements and objectives of the system under examination determine the performance measures that should be used.

Some common performance metrics in control theory include:

1. **Steady-State Error:** Steady-state error calculates the discrepancy between the system output at equilibrium or in a steady-state condition and the desired setpoint or reference. It shows how successfully the system has produced the desired output as well as how precisely it can hold the set point.
2. **Rise Time:** From the initial set point or reference, the rise time is the amount of time required for the output of the system to reach and settle within a specific percentage (e.g., 90%, 95%) of the final target value.
3. **Settling Time:** Following the rise time, the settling time is the amount of time required for the system's output to remain within a predetermined range (for example, 5% of the final value) of the target value.
4. **Overshoot:** During a transient response, overshoot is the greatest percentage by which a system's output exceeds the desired value. The amount of oscillations or departure from the setpoint is indicated by overshoot.
5. **Peak Time:** During the transient reaction, the peak time is the amount of time it takes for the system's output to reach its maximum value.
6. **Stability Margin:** Stability margins, like phase margin and gain margin, offer information about how resilient and stable the system is in the face of uncertainties and shocks. Greater stability and robustness are indicated by a higher stability margin.
7. **Integral of Time-weighted Absolute Error (ITAE):** This performance metric measures the total absolute error over time. It is frequently applied to challenges involving system optimization and optimal control.
8. **Integral of Squared Error (ISE):** This performance metric calculates the total squared error over a period of time. It is frequently applied to tuning and optimizing control systems.
9. **Integral of Time-weighted Squared Error (ITSE):** The weighted sum of squared errors and time are combined to create the ITSE performance index. To strike a compromise between tracking accuracy and control effort in optimization issues, it is utilized.
10. **Control Effort:** The control system's necessary energy or effort to produce the desired performance is measured by control effort. Since it uses less energy and requires less actuation, lower control effort is often preferred.

With the help of these performance measures, control engineers can fine-tune the controllers, change the parameters, and optimize the control system to achieve particular design objectives. Depending on the needs and limitations of the application, a trade-off between several performance measures is frequently required in the design of realistic control systems.

6.2.1 Applications of Control Theory

Control theory finds application in various fields, including:

(*a*) **Industrial automation:** Regulating production procedures, robotics, and equipment for maximum effectiveness and efficiency.

(*b*) **Aerospace engineering:** Managing an aircraft's or spacecraft's flight dynamics to maintain stability and security.

(*c*) **Electrical engineering:** Controlling frequency and voltage in power systems.

(*d*) **Automotive Engineering:** Creating control systems for automobiles to increase their safety and fuel economy.

(*e*) **Biomedical engineering:** Creating equipment for treating patients and systems to regulate bodily functions.

(*f*) **Economics:** Researching economic systems and formulating economic policies.

(*g*) **Environmental engineering:** Controlling pollutants and managing environmental systems.

A wide range of applications can be found for the powerful and adaptable field of control theory. In many branches of science and engineering, its concepts are applied to enhance the performance, security, and dependability of complex systems.

6.3 TRANSFER FUNCTIONS AND BLOCK DIAGRAMS

The basic tools in control theory for analyzing and creating control systems are block diagrams and transfer functions. They aid in understanding system dynamics and control methods by providing a graphical and algebraic depiction of the behavior of the system.

Let's explore each of these concepts:

1. **Transfer Function:**
   - A transfer function, in control theory, is a mathematical illustration of the relationship between the Laplace transforms of the system's input and output, presuming all initial conditions are zero.
   - Systems with constant coefficients and time-invariant behavior are known as linear time-invariant (LTI) systems, and transfer functions are used to characterize their dynamics.
   - The common notation for the transfer function of a linear time-invariant system is "G(s)," where "s" stands for the complex variable that symbolizes the Laplace transform.
   - Control engineers may examine stability, frequency response, and other performance traits thanks to the transfer function's simple and frequency-domain description of the system.
2. **Block Diagrams:**
   - Block diagrams are graphical representations of control systems that show various system components and their interactions as interconnected blocks.
   - Each block in the diagram represents a mathematical operation that is carried out on the signals going through it, such as a transfer function.
   - The arrows linking the blocks illustrate the input-output relationships between the system's components by showing the signal flow between the blocks.
   - Control system structure and operation are frequently depicted using block diagrams, which facilitates the analysis and design of complex systems.
   - Block diagrams can be used to illustrate single-input, single-output (SISO) or multi-input, multi-output (MIMO) systems in both continuous-time and discrete-time systems.

The following are some common elements in a control system block diagram:

(*a*) The block known as the Summing Point (Summing Junction) symbolizes the algebraic sum of the input signals. To compute the error signal, reference signals and feedback signals are frequently combined.

(*b*) **Transfer Function Block:** This block depicts how a component or subsystem in the control system transfers data. It may stand in for a sensor, actuator, controller, or any other system component.

(*c*) **Output:** The output block displays the system's measured or controlled output.

(*d*) **Controller:** The control technique or strategy utilized to produce the control signal based on the error signal is represented by the controller block.

(*e*) **Actuator:** The actuator block designates the element in charge of transforming the control signal into a physical action, such as starting a motor or operating a device.

Control engineers can model, analyze, and design control systems, comprehend their behavior, and optimize their performance for numerous applications across engineering disciplines by using transfer functions and block diagrams. These tools are crucial for providing dynamic system control that is reliable, accurate, and stable.

6.4 PID CONTROLLERS IN CONTROL THEORY

Proportional-Integral-Derivative (PID) controllers, often known as feedback control systems, are frequently employed in control theory and industrial automation. They are very good at controlling the behavior of dynamic systems so that desired set points are reached and stability is maintained. The proportional, integral, and derivative control actions—each of which contributes differently to the control signal—are combined by the PID controller.

The three components of the PID controller:

1. **Proportional (P) Control:**
   - The error signal (current departure from the set point) and the proportional control action are inversely correlated. In an effort to swiftly reduce the error, it makes the control output proportional to the error.
   - The transient response can be enhanced and the steady-state error can be decreased with a large proportional gain.
2. **Integral (I) Control:**
   - The accumulated inaccuracy over time is accounted for by the integral control action. Since the error signal is continually integrated, the controller is able to get rid of steady-state errors that could arise from disturbances or inaccurate models.
   - Although a greater integral gain makes the controller better at catching long-term mistakes, it can also cause a slower response time and possible overshoot.
3. **Derivative (D) Control:**
   - By taking into account the error signal's rate of change, the derivative control action predicts the error signal's future behavior. It aids in dampening the reaction and minimizing oscillations. A greater derivative gain can increase system noise while simultaneously enhancing stability and lowering overshoot.

PID controllers are widely utilized in numerous fields, including process control, temperature control, motor control, and robotics. Since choosing and fine-tuning PID gains entails balancing the trade-offs between transient response, stability, and robustness against disturbances and uncertainties, doing so is essential for obtaining optimal performance. PID gains for certain control systems are chosen using a variety of tuning techniques, including Ziegler-Nichols, Cohen-Coon, and trial-and-error. The performance of control systems can also be improved by using sophisticated PID variations, such as PID with anti-windup methods, cascade control, and gain scheduling.

6.5 STATE-SPACE REPRESENTATION IN CONTROL THEORY

A mathematical modeling method used in control theory to describe the behavior of a dynamic system is state-space representation. Using matrices and differential equations, it offers a clear and complete way to express the system's state variables, inputs, outputs, and dynamics. For the analysis and design of MIMO and nonlinear dynamics-based control systems, state-space representation is very helpful.

The state-space model of a dynamic system consists of the following elements:

1. **State Variables (x):** A group of variables known as state variables are used to characterize the internal state of a system at any one time. They stand for the bare minimum of variables required to forecast the system's future behavior. The order of the differential equations for the system corresponds normally to the number of state variables.
2. **Input Variables (u):** The input variables stand for the system's control inputs or outside impacts. The system's behavior can be managed by changing these inputs.
3. **Output Variables (y):** The quantifiable elements that define the system's performance and behavior are known as the output variables. Typically, a set of output equations connects them to the state variables.
4. **State Equations:** Using the time derivatives of the state variables (dx/dt), state equations describe the dynamic behaviour of the system. Usually, they are shown as a collection of first-order ordinary differential equations.
5. **Output Equations:** The output equations specify how the state variables and the output variables relate to one another. As linear combinations of the state variables and, presumably, the input variables, they express the outputs.

The state-space representation allows for more flexibility in modeling complex systems and provides several advantages, including:

(*a*) It supports numerous input/output systems, which are typical in real-world applications.

(*b*) It is appropriate for a variety of control issues since it can explain both linear and nonlinear systems.

(*c*) It can be examined using a variety of control design techniques, including observer design, state feedback, and pole positioning.

(*d*) It offers a precise physical understanding of the dynamics and internal states of the system.

A key tool in control system engineering, state-space representation forms the foundation of numerous contemporary control design techniques and algorithms.

6.6 STABILITY AND CONTROL DESIGN IN CONTROL THEORY

In control theory, stability and control design are ideas that are closely related. The main goal of control design is to create a controller that can control a dynamic system's behavior in a desirable way while maintaining stability. The design of control systems must take stability into account since an unstable system can exhibit unintended and unanticipated behavior, making the control system ineffective or even dangerous.

Here's how stability and control design are interconnected in control theory:

1. **Stability Analysis:**
   - It is crucial to assess the stability of the open-loop system, or the system with no control action, before building a control system. When a system is subjected to disturbances or changes in initial circumstances, stability analysis can be used to assess if the system will continue to be constrained over time.
   - The eigenvalues of the system's state-space representation or the system's poles in the transfer function are examined during the stability study. The system is stable if all of the poles or eigenvalues have negative real components. On the other hand, the system is unstable if any pole or eigenvalue has a positive real portion.
   - In general, unstable systems are unsuitable for practical control because they may exhibit unbounded or oscillatory behavior, which makes control challenging or impossible.
2. **Stability in Closed-Loop Systems:**
   - Control design focuses on adding a controller to create a closed-loop system once the stability of the open-loop system has been proven. The closed-loop system generates the control input by using feedback from the system's output, which aids in regulating the system's behavior.
   - The stability of the closed-loop system depends critically on the controller's design. An initially unsteady system should be stabilized by a well-designed controller, and stability should be maintained in the face of disturbances and uncertainty.
   - The closed-loop transfer function or state-space representation's poles or eigenvalues are examined as part of the closed-loop system's stability study. For the closed-loop system to continue to function properly, the poles must still have negative real parts.
3. **Control Design for Stability:**
   - Stability is the goal of control design techniques, which also attempt to meet other performance objectives including rapid reaction, precise tracking, and resilience against disturbances and uncertainty.
   - PID control, state-feedback control, pole placement, and Lyapunov-based control design are typical techniques to control design for stability.
   - To achieve stability and the desired performance, PID control combines proportional, integral, and derivative actions.
   - State-feedback control creates a feedback control law that stabilizes the system using a state-space model of the system.
   - To accomplish stability and performance goals, pole placement entails selecting the controller's parameters and positioning the closed-loop poles in the desired positions.
   - Lyapunov-based control design creates control rules that ensure stability and convergence to a chosen equilibrium point using the Lyapunov stability theory.

Control engineers analyze a system's stability, design a controller, and then reevaluate the stability of the closed-loop system. This process is known as iterative stability and control design. In order to ensure steady and effective performance of the controlled system, the design procedure entails adjusting controller parameters and selecting suitable control techniques. A well-designed control system demonstrates robustness in the presence of uncertainties and disruptions, maintains stability, and precisely tracks desired set points.

6.7 ROBOTIC CONTROL SYSTEMS

Robotic control systems are essential to many robotics applications because they allow robots to interact with their surroundings, execute difficult tasks, and accomplish desired goals. These control systems are in charge of directing the robot's behaviors, movements, and interactions with its environment.

Some common robotics applications where control systems are used include:

1. **Industrial Automation:** Robots are frequently employed in manufacturing and industrial environments to do operations including welding, assembling, pick-and-place, and material handling. Control systems make ensuring that these duties are carried out precisely and effectively, increasing output and quality.
2. **Mobile Robotics and Autonomous Vehicles:** For autonomous mobile robots, such as self-driving automobiles, delivery drones, and warehouse robots, control systems are crucial. To navigate, avoid obstructions, and create the best courses, these systems employ sensors and algorithms.
3. **Robotic Manipulation and Grasping:** Robotic grippers and arms use control systems to manipulate and grasp things with precision. They enhance robot dexterity and efficiency by assisting with object form and size adaptation.
4. **Medical Robotics:** Robotics help in surgical procedures, physical therapy, and diagnostics in medical applications. Control systems improve the precision of medical devices and provide stability and accuracy during surgical procedures.
5. **Service Robotics:** Robots that are used in service industries like home assistance, elder care, and hospitality need on control systems to interact with people securely and move around enclosed spaces.
6. **Agricultural Robotics:** Robots used in agriculture can execute duties including planting, harvesting, and crop monitoring thanks to control systems. These robots can increase agricultural productivity and decrease the need for labor.
7. **Underwater and Aerial Robotics:** Robots in the air and under the ocean are employed for environmental monitoring, surveillance, and exploration. In difficult situations, control systems ensure steady flying and navigation.
8. **Humanoid Robotics:** Robots designed to resemble humans are called humanoids. Control systems are essential for producing humanoid motion, which enables them to walk, dance, and communicate with others.
9. **Space Robotics:** Robots are used in space missions to help with activities including servicing satellites, planetary exploration, and helping astronauts. In space conditions, control systems are essential for upholding stability and carrying out precise maneuvers.
10. **Entertainment and Gaming:** Control systems are utilized in animatronics and robotic characters for interactive experiences, theme park rides, and virtual reality applications in the entertainment and gaming sectors.

These are a few illustrations of the various uses for robotic control systems. More powerful, adaptable, and intelligent robotic systems are being created for a variety of sectors and disciplines thanks to the continued development of improved control algorithms, artificial intelligence, and sensor technologies.

6.8 ROBOTIC STABILITY ANALYSIS IN CONTROL THEORY

To ensure secure and dependable operation, robotic stability analysis in control theory entails assessing the stability of robotic systems. Because an unstable robot might behave in unforeseen ways, which can result in instability, oscillations, or even damage, stability analysis is crucial. Engineers can identify possible problems and build control solutions to achieve stable and robust performance by evaluating the stability of a robot's control system.

Various techniques can be used to analyze the stability of robotic systems, depending on the control architecture and mathematical model that are applied to the robot. Typical methods for stability analysis include:

1. **Linear Stability Analysis:**
   - For linear time-invariant (LTI) robotic systems, where the dynamic behavior can be modeled by linear differential equations, linear stability analysis is applied.
   - The eigenvalues of the transfer function or state-space representation of the system are used to assess the stability of the system.
   - Stability for continuous-time systems is guaranteed if all state-space matrix eigenvalues have negative real portions. The state-transition matrix of a discrete-time system must have all of its eigenvalues with magnitudes smaller than one in order for the system to be stable.
2. **Lyapunov Stability Analysis:**
   - A more versatile technique that can be used with nonlinear and time-varying robotic systems is Lyapunov stability analysis.
   - It entails building a Lyapunov function that acts as an energy-like function and captures the system's stability characteristics.
   - For all states besides the equilibrium point, a system is regarded as stable if the Lyapunov function is positive definite and its time derivative is negative definite.
3. **Input-Output Stability Analysis:**
   - Input-output stability analysis focuses on the connection between inputs and outputs as it investigates the stability of the system in the frequency domain.
   - Transfer functions and frequency response are ideas from control theory that are used to evaluate the system's robustness and stability to outside shocks.
4. **Passivity and Incremental Stability Analysis:**
   - Techniques for passive and incremental stability analysis are utilized for robotic systems that display passive or incremental behavior.
   - Passivity analysis ensures energy stability by determining if the system acts as a passive energy-storing mechanism.
   - Incremental stability analysis looks at whether minor system perturbations result in minute changes in the output, which would indicate local stability.

Numerical simulation and experimental testing are also frequently employed to evaluate the stability of robotic systems in real-world settings, in addition to these analytical techniques. Engineers can mimic robot behavior and assess stability under various settings using simulation tools like MATLAB/Simulink and ROS (Robot Operating System). Stability analysis is essential to the design and operation of robots because it helps to assure predictable and safe performance in a variety of operational settings. Engineers can choose control strategies, parameter tuning, and system architecture to provide stable and dependable robotic behavior by knowing the stability attributes of a robotic system.

6.9 ROBOTIC MOTION CONTROL IN CONTROL THEORY

In the field of control theory, the regulation of the motion and trajectory of robotic systems is referred to as robotic motion control. Motion control's main objective is to move robots precisely, smoothly, and effectively in order to carry out various duties and accomplish specified goals. In order to move the robot's joints or end-effector in a controlled manner, control signals must be generated to direct the actuators of the robot.

Here are some key aspects of robotic motion control in control theory:

1. **Kinematic and Dynamic Models:**
    - Engineers must create precise kinematic and dynamic models of the robot before building a motion control system. These models serve as the basis for motion planning and control, providing the necessary mathematical descriptions of the robot's motion behavior.
    - The kinematic model relates the joint motions to the end-effector position and orientation, while the dynamic model describes how the robot responds to external forces and torques.
2. **Trajectory Planning:**
    - Creating a smooth and practical path for the robot's end-effector or joints to follow from their current location to the intended position at the end is known as trajectory planning.
    - Planning trajectories that adhere to kinematic and dynamic requirements while minimizing jerk and accelerations can be done using a variety of techniques, including polynomial interpolation, and motion profiles.
3. **Feedback Control:**
    - A key strategy in robotic motion control is feedback control. It entails tracking the robot's position, speed, or other pertinent parameters continuously and modifying the control inputs in response to deviations from the intended trajectory.
    - Proportional derivative (PD) control, proportional-integral-derivative (PID) control, and model-based control are examples of common feedback control systems.
4. **Path Following:**
    - Path following techniques are used for robots having specified pathways, such as mobile robots or robotic arms operating along fixed trajectories, to ensure reliable tracking of the required path.
    - To accomplish precise path following, feedback controllers, such as PID controllers or model predictive controllers, are frequently utilized.
5. **Inverse Kinematics:**
    - Inverse kinematics is a crucial component of manipulator and robotic arm motion control. It entails figuring out the joint angles or speeds necessary to obtain the end-effector position and orientation that are desired.
    - Planning and managing the motion of robotic arms in a Cartesian space requires the solution of inverse kinematics.

6. **Motion Control of Redundant Robots:**
   - Robots that are redundant have more degrees of freedom than are required to do a certain task. Choosing the best configuration for redundant robot motion control requires optimizing performance criteria like manipulability, obstacle avoidance, or energy consumption.
7. **Adaptive Control:**
   - Robotic motion control employs adaptive control techniques to deal with system uncertainties, such as parameter changes or disturbances.

Robotic motion control is essential in a variety of applications, including mobile robotics, robotic surgery, industrial automation, and robotic manufacturing. To achieve precise and effective robotic motion that increases the capabilities of robotic systems in diverse real-world circumstances, effective motion control procedures are crucial.

NUMERICAL PROBLEMS WITH SOLUTIONS

Problem 1. Given the transfer function of a control system as $G(s) = (s + 2)/(s^2 + 3s + 2)$, determine the poles and zeros of the system and check its stability.

Solution: To find the poles and zeros, set the denominator $(s^2 + 3s + 2)$ equal to zero: $s^2 + 3s + 2 = 0$ Solving the quadratic equation, we get: $s = -1$, $s = -2$

The poles of the system are –1 and –2, and the zero is –2.

Since all the poles have negative real parts, the system is stable.

Problem 2. Consider a proportional-derivative (PD) control system with the transfer function C(s) = Kp + Kd*s, where Kp = 2 and Kd = 1. Find the closed-loop transfer function T(s) = Y(s)/R(s), where Y(s) is the output and R(s) is the reference input.

Solution: The closed-loop transfer function is given by: T(s) = C(s)/(1 + C(s)*G(s))

where G(s) is the transfer function of the plant. In this case, G(s) = 1 (assuming unity gain).

$$T(s) = (Kp + Kds)/(1 + (Kp + Kds)*1)\ T(s)$$
$$= (2 + s)/(1 + 2 + s)\ T(s) = (s + 2)/(s + 3)$$

Problem 3. A control system has the open-loop transfer function $G(s) = 1/(s^2 + 4s + 3)$. Use the Nyquist criterion to determine the system's stability.

Solution: The Nyquist criterion states that the system is stable if the Nyquist plot encircles the critical point –1 + j0 in a counterclockwise direction as many times as the number of poles of the transfer function in the right-half plane.

The transfer function has two poles: s = –1 and s = –3. Both poles have negative real parts, so there are no poles in the right-half plane.

To draw the Nyquist plot, evaluate G(jω) for ω ranging from 0 to infinity. The Nyquist plot is a curve in the complex plane. If the plot encircles –1 + j0 zero times, the system is stable.

Problem 4. A system with the transfer function G(s) = 1/(s + 2) is controlled using a PID controller with the transfer function C(s) = Kp + Ki/s + Kd*s, where Kp = 2, Ki = 1, and Kd = 0.5. Determine the closed-loop transfer function T(s) = Y(s)/R(s), where Y(s) is the output and R(s) is the reference input.

Solution: The closed-loop transfer function is given by: T(s) = C(s) * G(s)/(1 + C(s) * G(s))

Substitute the given values into C(s) and G(s): T(s) = (Kp + Ki/s + Kds) * (1/(s + 2))/(1 + (Kp + Ki/s + Kds) * (1/(s + 2)))

$$T(s) = (2 + 1/s + 0.5s)/(s + 2 + 2 + s + 0.5s(s + 2))\ T(s)$$
$$= (2 + 1/s + 0.5s)/(s^2 + 4s + 5)$$

Problem 5. A system with the transfer function G(s) = 10/(s^2 + 5s + 10) is controlled using a lead compensator with the transfer function C(s) = (s + 2)/(s + 5). Determine the closed-loop transfer function T(s) = Y(s)/R(s), where Y(s) is the output and R(s) is the reference input.

Solution: The closed-loop transfer function is given by: T(s) = C(s) * G(s)/(1 + C(s) * G(s))

Substitute the given values into C(s) and G(s): T(s) = ((s + 2)/(s + 5)) * (10/(s^2 + 5s + 10))/(1 + ((s + 2)/(s + 5)) * (10/(s^2 + 5s + 10)))

$$T(s) = (10(s + 2))/((s + 5)(s^2 + 5s + 10 + 10(s + 2)))\ T(s)$$
$$= (10(s + 2))/(s^3 + 7s^2 + 15s + 20)$$

Problem 6. Given a control system with the transfer function G(s) = (s + 2)/(s^2 + 3s + 2), determine the stability of the system using the Routh-Hurwitz criterion.

Solution: The characteristic equation of the system is $s^2 + 3s + 2 = 0$.

The Routh-Hurwitz table for this characteristic equation is as follows:

| s^2 | 1 || s^1 | 3 || s^0 | 2 |

Since there are no sign changes in the first column, the system is stable.

Problem 7. Consider a unity feedback control system with the open-loop transfer function G(s) = 10/(s^2 + 2s + 10). Find the closed-loop transfer function T(s) = Y(s)/R(s), where Y(s) is the output and R(s) is the reference input.

Solution: The closed-loop transfer function is given by: T(s) = G(s)/(1 + G(s))

Substitute the given transfer function: T(s) = (10/(s^2 + 2s + 10))/(1 + 10/(s^2 + 2s + 10))

$$T(s) = 10/((s^2 + 2s + 10) + 10)$$
$$T(s) = 10/(s^2 + 2s + 20)$$

Problem 8. A control system has a proportional controller with a gain of Kp – 5. Determine the value of Kp that results in a closed-loop system with a damping ratio of 0.7.

Solution: The damping ratio (ζ) is related to the gain (Kp) of the proportional controller by the formula: ζ = Kp/(2 * sqrt(Kp))

Substitute ζ = 0.7 into the equation and solve for Kp: 0.7 = Kp/(2 * sqrt(Kp)) 0.7 * (2 * sqrt(Kp)) = Kp 1.4 * sqrt(Kp) = Kp 1.4 = sqrt(Kp) Kp = 1.4^2 Kp = 1.96

So, the value of Kp that results in a closed-loop system with a damping ratio of 0.7 is approximately 1.96.

Problem 9. A control system has a transfer function G(s) = 1/(s^2 + 4s + 5). Find the natural frequency (ωn) and damping ratio (ζ) of the system.

Solution: The transfer function can be written as G(s) = 1/($(s + 2)^2$ + 1)

Comparing it with the standard form of a second-order system: G(s) = $\omega n^2/(s^2 + 2\zeta\omega n\ s + \omega n^2)$

We get $\omega n^2 = 1$ and $2\zeta\omega n = 4$

From $\omega n^2 = 1$, we get ωn = 1.

From 2ζωn = 4, we get ζ = 2.

So, the natural frequency (ωn) is 1 rad/s, and the damping ratio (ζ) is 2.

Problem 10. A control system has a transfer function $G(s) = 10/(s^2 + 6s + 25)$. Find the resonant peak (M_p) and resonant frequency (ω_r) of the system.

Solution: The resonant peak (M_p) and resonant frequency (ω_r) can be calculated using the following formulas: $M_p = 1/(2\zeta * \text{sqrt}(1 - \zeta^2))$ and $\omega_r = \omega n * \text{sqrt}(1 - \zeta^2)$

From Problem 9, we know that $\omega n = 1$ and $\zeta = 2$.

$$M_p = 1/(2 * 2 * \text{sqrt}(1 - 2^2)) = 1/(4 * \text{sqrt}(-3)) = 1/(4i\sqrt{3})$$

$$\omega_r = 1 * \text{sqrt}(1 - 2^2) = 1 * \text{sqrt}(1 - 4) = 1 * \text{sqrt}(-3) = i\sqrt{3}$$

So, the resonant peak (M_p) is $1/(4i\sqrt{3})$, and the resonant frequency (ω_r) is $i\sqrt{3}$ rad/s.

Note: In control theory, the resonant peak and resonant frequency are typically expressed in polar form as $M_p \angle \theta$ and $\omega_r \angle \phi$, respectively, where θ and φ are angles in degrees or radians.

Short Questions and Answers

Q.1. What is control theory?

Ans. Control theory is a branch of engineering and mathematics that deals with the analysis and design of systems to maintain desired behavior or performance over time.

Q.2. What is a control system?

Ans. A control system is a collection of components and processes that work together to regulate the behavior of a dynamic system.

Q.3. What is feedback in control theory?

Ans. Feedback is the process of using the system's output information to adjust the control input, allowing the control system to correct errors and achieve the desired behavior.

Q.4. What is an open-loop control system?

Ans. In an open-loop control system, the control action is predetermined and not adjusted based on the system's output.

Q.5. What is a closed-loop control system?

Ans. In a closed-loop control system, feedback is used to adjust the control action based on the system's output, improving accuracy and stability.

Q.6. What is a transfer function?

Ans. The transfer function is a mathematical representation of the relationship between the input and output of a linear time-invariant system.

Q.7. What is a state-space representation?

Ans. State-space representation is a mathematical modeling technique that describes the dynamics of a system using differential equations and state variables.

Q.8. What are the three components of a PID controller?

Ans. The three components of a PID controller are proportional (P), integral (I), and derivative (D) actions.

Q.9. What is the purpose of the proportional control action in a PID controller?

Ans. The proportional control action provides a control signal proportional to the error, aiming to reduce the error quickly.

Q.10. What is the purpose of the integral control action in a PID controller?

Ans. The integral control action accumulates the error over time to eliminate steady-state errors.

Q.11. What is the purpose of the derivative control action in a PID controller?

Ans. The derivative control action anticipates the future behavior of the error to reduce overshooting and dampen the response.

Q.12. What is stability in control theory?

Ans. Stability refers to the ability of a system to maintain a bounded output in response to disturbances or changes in initial conditions.

Q.13. What is the Nyquist criterion used for?

Ans. The Nyquist criterion is used to assess the stability of a control system by examining the system's frequency response.

Q.14. What is the Bode plot used for?

Ans. The Bode plot is a graph that shows the frequency response of a control system, providing insights into its stability and performance.

Q.15. What is the steady-state error in control systems?

Ans. Steady-state error is the difference between the desired setpoint and the output of the control system in a steady-state condition.

Q.16. What is the purpose of pole placement in control design?

Ans. Pole placement involves choosing the controller's parameters to place the closed-loop poles in desired locations, achieving stability and performance objectives.

Q.17. What is the role of a Kalman filter in control systems?

Ans. The Kalman filter is used to estimate the state of a system from noisy measurements, improving the accuracy of state feedback control.

Q.18. What is the difference between LQR and LQG control?

Ans. LQR (Linear Quadratic Regulator) is a control strategy for systems with perfect state feedback, while LQG (Linear Quadratic Gaussian) control incorporates state estimation using Kalman filtering.

Q.19. What is controllability in control theory?

Ans. Controllability refers to the property of a system where it can be driven from any initial state to any desired final state using control inputs.

Q.20. What is observability in control theory?

Ans. Observability refers to the property of a system where its internal state can be determined from its outputs using measurements.

Q.21. What is adaptive control?

Ans. Adaptive control is a control strategy that adjusts the controller's parameters based on the system's behavior to handle uncertainties and variations.

Q.22. What is the role of a compensator in control systems?

Ans. A compensator is a controller used to modify the system's transfer function to achieve desired performance and stability.

Q.23. What is gain margin in control systems?

Ans. Gain margin is a measure of how much the system's gain can be increased before it becomes unstable.

Q.24. What is phase margin in control systems?

Ans. Phase margin is a measure of how much the phase shift of the system can be increased before it becomes unstable.

Q.25. What is the settling time of a control system?

Ans. Settling time is the time it takes for the system's output to stay within a specified range of the desired value after the initial transient response.

Q.26. What is the rise time of a control system?

Ans. Rise time is the time it takes for the system's output to reach and settle within a certain percentage of the final desired value from the initial setpoint.

Q.27. What is the bandwidth of a control system?

Ans. The bandwidth is the frequency range over which the control system can effectively respond to changes in the input.

Q.28. What is the role of a lead compensator in control systems?

Ans. A lead compensator is used to increase the phase margin and improve the stability of the control system.

Q.29. What is the role of a lag compensator in control systems?

Ans. A lag compensator is used to reduce the bandwidth and improve the stability of the control system.

Q.30. What is the root locus method used for in control theory?

Ans. The root locus method is used to analyze the poles of a control system and study how they change with variations in the controller's parameters.

Q.31. What is the stability margin of a control system?

Ans. Stability margin is a measure of how far the system is from becoming unstable, quantifying its robustness against uncertainties.

Q.32. What is the difference between continuous-time and discrete-time control systems?

Ans. Continuous-time control systems deal with continuous signals and time, while discrete-time control systems operate with discrete signals and time steps.

Q.33. What is a zero in control systems?

Ans. A zero is a value of the transfer function's numerator that makes the system's output zero when the input is equal to that value.

Q.34. What is the role of pole-zero cancellation in control systems?

Ans. Pole-zero cancellation occurs when a pole and zero of the transfer function cancel each other, affecting the system's dynamics.

Q.35. What is robust control?

Ans. Robust control involves designing control systems that are insensitive to variations and uncertainties in the system's parameters.

Q.36. What is the role of feedforward control in control systems?

Ans. Feedforward control compensates for known disturbances in advance, improving the control system's performance.

Q.37. What is the difference between lead-lag compensators and PID controllers?

Ans. Lead-lag compensators are used for specific frequency shaping, while PID controllers are general-purpose controllers that can handle a wide range of control tasks.

Q.38. What is the role of a sensor in a control system?

Ans. Sensors provide measurements of the system's output or state variables, allowing feedback control to be implemented.

Q.39. What is the role of an actuator in a control system?

Ans. Actuators convert control signals into physical action, influencing the system's behavior.

Q.40. What is the role of hysteresis in control systems?

Ans. Hysteresis is a nonlinear phenomenon where the output depends not only on the current input but also on the past inputs.

Q.41. What is system identification in control theory?

Ans. System identification is the process of experimentally determining the mathematical model of a system from input-output data.

Q.42. What is a state observer in control systems?

Ans. A state observer is used to estimate the unmeasured state variables of a system, enabling state feedback control without direct state measurements.

Q.43. What is the role of deadband in control systems?

Ans. Deadband is a range of input values where the control system does not respond, reducing the control action's sensitivity to small changes.

Q.44. What is the role of a relay in control systems?

Ans. A relay is used to switch the control system between different modes of operation.

Q.45. What is the minimum-phase property in control systems?

Ans. The minimum-phase property indicates that the system has no zeros with positive real parts, ensuring stability.

Q.46. What is robust stability analysis?

Ans. Robust stability analysis evaluates the stability of a control system under uncertainties and parameter variations.

Q.47. What is the role of a control algorithm in control systems?

Ans. A control algorithm computes the control action based on the system's input and feedback information.

Q.48. What is the Nyquist stability criterion?

Ans. The Nyquist stability criterion determines the stability of a control system by examining the encirclement of the critical point in the Nyquist diagram.

Q.49. What is the role of integral windup prevention in control systems?

Ans. Integral windup prevention techniques are used to prevent the integral action from causing overshoot and instability.

Q.50. What is the concept of equilibrium point in control theory?

Ans. An equilibrium point is a state where the system's output and state variables remain constant with no input or disturbance.

Multiple Choice Questions and Answers

Q.1. What is control theory?

(*a*) A branch of biology (*b*) A branch of mathematics
(*c*) A branch of engineering (*d*) A branch of chemistry

Q.2. What is the primary goal of control theory?

(*a*) To design rockets (*b*) To control robots
(*c*) To regulate system behavior (*d*) To study animal behavior

Q.3. What does feedback control in a control system mean?

(*a*) Adjusting the control input based on the system output
(*b*) Keeping the control input fixed
(*c*) Controlling multiple systems simultaneously
(*d*) Randomly changing the control input

Q.4. What is an open-loop control system?

(*a*) A system with feedback (*b*) A system with sensors
(*c*) A system without feedback (*d*) A system without a controller

Q.5. What is a closed-loop control system?

(*a*) A system with sensors (*b*) A system with feedback
(*c*) A system without a controller (*d*) A system without feedback

Q.6. What is the transfer function in control theory?

(*a*) A function that transfers control to another system
(*b*) A function that relates input and output in a control system
(*c*) A function that transfers energy between components
(*d*) A function used to transfer data between computers

Q.7. What is the PID controller's full form?

(*a*) Proportional-Integral-Derivative (*b*) Proportional-Integral-Differentiate
(*c*) Proportional-Integral-Decrease (*d*) Proportional-Integral-Damping

Q.8. What does the integral action in a PID controller do?

(*a*) Provides a control signal proportional to the error
(*b*) Accumulates the error over time to eliminate steady-state errors
(*c*) Anticipates the future behavior of the error
(*d*) Improves stability by reducing overshoot

Q.9. What is stability in control theory?

(*a*) The ability to perform a task without errors

(*b*) The ability to control multiple systems at once

(*c*) The ability to maintain a bounded output in response to disturbances

(*d*) The ability to adapt to changing environments

Q.10. What is a Bode plot used for?

(*a*) Analyzing the poles of a control system

(*b*) Assessing the stability of a control system

(*c*) Evaluating system performance in the frequency domain

(*d*) Determining system gain and phase margins

Q.11. What does steady-state error in control systems represent?

(*a*) The error that occurs during system startup

(*b*) The error that occurs when the control system is not stable

(*c*) The difference between the desired setpoint and the output in steady-state conditions

(*d*) The error due to system disturbances

Q.12. What is the role of pole placement in control design?

(*a*) To calculate the stability margin of the control system

(*b*) To optimize system performance by placing poles in desired locations

(*c*) To minimize the control effort required to achieve stability

(*d*) To reduce the bandwidth of the control system

Q.13. What is the purpose of a Kalman filter in control systems?

(*a*) To remove noise from sensor measurements

(*b*) To estimate the system's state from noisy measurements

(*c*) To design optimal control laws

(*d*) To prevent integral windup in PID controllers

Q.14. What is observability in control theory?

(*a*) The ability to control a system effectively

(*b*) The ability to determine the system's internal state from its outputs

(*c*) The ability to measure the system's output accurately

(*d*) The ability to stabilize a system

Q.15. What is the role of a compensator in control systems?

(*a*) To provide feedback to the system

(*b*) To modify the system's transfer function to achieve desired performance

(*c*) To prevent integral windup in PID controllers

(*d*) To measure the system's output accurately A:

Q.16. What is the gain margin in control systems?

(*a*) A measure of the system's sensitivity to disturbances

(*b*) A measure of how much the system's gain can be increased before it becomes unstable
(*c*) A measure of the system's ability to achieve desired performance
(*d*) A measure of the system's output accuracy

Q.17. What is the role of a sensor in a control system?
(*a*) To convert the control signal into physical action
(*b*) To provide feedback to the controller
(*c*) To regulate the system's output
(*d*) To measure the system's input

Q.18. What is the role of an actuator in a control system?
(*a*) To provide feedback to the controller (*b*) To measure the system's output
(*c*) To convert the control signal into physical action
(*d*) To regulate the system's output

Q.19. What is system identification in control theory?
(*a*) The process of estimating the system's state from noisy measurements
(*b*) The process of determining the mathematical model of a system from input-output data
(*c*) The process of placing poles and zeros in the transfer function
(*d*) The process of selecting the controller's parameters

Q.20. What is a state observer in control systems?
(*a*) A device used to measure the system's output
(*b*) A controller that adjusts its parameters based on the system's behavior
(*c*) A mechanism used to estimate the unmeasured state variables of a system
(*d*) A compensator used to modify the system's transfer function

Q.21. What is the role of dead band in control systems?
(*a*) To prevent overshooting in the control action
(*b*) To prevent the system from responding to small changes in the input
(*c*) To improve the system's stability
(*d*) To compensate for disturbances in the system

Q.22. What is robust control?
(*a*) A control strategy that adjusts its parameters based on the system's behavior
(*b*) A control strategy that is insensitive to uncertainties and variations
(*c*) A control strategy that uses multiple actuators to control the system
(*d*) A control strategy that adapts to changing environments

Q.23. What does the Nyquist stability criterion determine?
(*a*) The stability margin of the control system
(*b*) The stability of a control system by examining the frequency response
(*c*) The time response of the control system
(*d*) The accuracy of the control system's output A:b)

Q.24. What does the integral windup prevention in a PID controller aim to prevent?

(*a*) Overshoot
(*b*) Instability
(*c*) Integral windup
(*d*) Steady-state error

Q.25. What is the minimum-phase property in control systems?

(*a*) The system's output is minimized for a given input
(*b*) The system has no zeros with positive real parts, ensuring stability
(*c*) The system has no poles with negative real parts, ensuring stability
(*d*) The system's output is maximized for a given input

Q.26. What is the Nyquist stability criterion?

(*a*) A method for analyzing poles of a control system
(*b*) A method for determining the gain and phase margins of a control system
(*c*) A graphical method for analyzing the stability of a control system
(*d*) A method for analyzing the bandwidth of a control system

Q.27. What is adaptive control?

(*a*) A control strategy that adjusts its parameters based on the system's behavior
(*b*) A control strategy that uses multiple sensors to measure the system's output
(*c*) A control strategy that adapts to changing environments
(*d*) A control strategy that is insensitive to uncertainties and variations

Q.28. What is the role of a control algorithm in control systems?

(*a*) To provide feedback to the system
(*b*) To adjust the controller's parameters
(*c*) To compute the control action based on input and feedback information
(*d*) To estimate the system's state

Q.29. What is the role of a relay in control systems?

(*a*) To provide feedback to the system
(*b*) To switch the control system between different modes of operation
(*c*) To measure the system's output
(*d*) To convert the control signal into physical action

Q.30. What is the minimum-phase property in control systems?

(*a*) The system has no poles with negative real parts, ensuring stability
(*b*) The system's output is minimized for a given input
(*c*) The system has no zeros with positive real parts, ensuring stability
(*d*) The system's output is maximized for a given input

Q.31. What is the role of dead band in control systems?

(*a*) To prevent overshooting in the control action
(*b*) To prevent the system from responding to small changes in the input
(*c*) To improve the system's stability
(*d*) To compensate for disturbances in the system

Q.32. What is robust control?

(*a*) A control strategy that adjusts its parameters based on the system's behavior
(*b*) A control strategy that is insensitive to uncertainties and variations
(*c*) A control strategy that uses multiple actuators to control the system
(*d*) A control strategy that adapts to changing environments

Q.33. What does the Nyquist stability criterion determine?

(*a*) The stability margin of the control system
(*b*) The stability of a control system by examining the frequency response
(*c*) The time response of the control system
(*d*) The accuracy of the control system's output

Q.34. What does the integral windup prevention in a PID controller aim to prevent?

(*a*) Overshoot (*b*) Instability
(*c*) Integral windup (*d*) Steady-state error

Q.35. What is the minimum-phase property in control systems?

(*a*) The system's output is minimized for a given input
(*b*) The system has no zeros with positive real parts, ensuring stability
(*c*) The system has no poles with negative real parts, ensuring stability
(*d*) The system's output is maximized for a given input

Q.36. What is the Nyquist stability criterion?

(*a*) A method for analyzing poles of a control system
(*b*) A method for determining the gain and phase margins of a control system
(*c*) A graphical method for analyzing the stability of a control system
(*d*) A method for analyzing the bandwidth of a control system

Q.37. What is adaptive control?

(*a*) A control strategy that adjusts its parameters based on the system's behavior
(*b*) A control strategy that uses multiple sensors to measure the system's output
(*c*) A control strategy that adapts to changing environments
(*d*) A control strategy that is insensitive to uncertainties and variations

Q.38. What is the role of a control algorithm in control systems?

(*a*) To provide feedback to the system (*b*) To adjust the controller's parameters
(*c*) To compute the control action based on input and feedback information
(*d*) To estimate the system's state

Q.39. What is the role of a relay in control systems?

(*a*) To provide feedback to the system
(*b*) To switch the control system between different modes of operation
(*c*) To measure the system's output
(*d*) To convert the control signal into physical action

Q.40. What is the role of a state observer in control systems?

(*a*) To measure the system's output (*b*) To adjust the controller's parameters

(*c*) To estimate the unmeasured state variables of a system

(*d*) To switch the control system between different modes of operation

Q.41. What is the role of a compensator in control systems?

(*a*) To convert the control signal into physical action

(*b*) To provide feedback to the system

(*c*) To modify the system's transfer function to achieve desired performance

(*d*) To measure the system's output

Q.42. What is the gain margin in control systems?

(*a*) A measure of the system's sensitivity to disturbances

(*b*) A measure of how much the system's gain can be increased before it becomes unstable

(*c*) A measure of the system's ability to achieve desired performance

(*d*) A measure of the system's output accuracy

Q.43. What is the role of a sensor in a control system?

(*a*) To convert the control signal into physical action

(*b*) To provide feedback to the controller (*c*) To measure the system's input

(*d*) To regulate the system's output

Q.44. What is the role of an actuator in a control system?

(*a*) To measure the system's output

(*b*) To convert the control signal into physical action

(*c*) To regulate the system's output

(*d*) To provide feedback to the controller

Q.45. What is system identification in control theory?

(*a*) The process of estimating the system's state from noisy measurements

(*b*) The process of determining the mathematical model of a system from input-output data

(*c*) The process of placing poles and zeros in the transfer function

(*d*) The process of selecting the controller's parameters

Q.46. What is a state observer in control systems?

(*a*) A device used to measure the system's output

(*b*) A controller that adjusts its parameters based on the system's behavior

(*c*) A mechanism used to estimate the unmeasured state variables of a system

(*d*) A compensator used to modify the system's transfer function

Q.47. What is the role of dead band in control systems?

(*a*) To prevent overshooting in the control action

(*b*) To prevent the system from responding to small changes in the input

(*c*) To improve the system's stability

(*d*) To compensate for disturbances in the system

Q.48. What is robust control?

(*a*) A control strategy that adjusts its parameters based on the system's behavior

(*b*) A control strategy that is insensitive to uncertainties and variations

(*c*) A control strategy that uses multiple actuators to control the system

(*d*) A control strategy that adapts to changing environments

Q.49. What does the Nyquist stability criterion determine?

(*a*) The stability margin of the control system

(*b*) The stability of a control system by examining the frequency response

(*c*) The time response of the control system

(*d*) The accuracy of the control system's output

Q.50. What does the integral windup prevention in a PID controller aim to prevent?

(*a*) Overshoot (*b*) Instability

(*c*) Integral windup (*d*) Steady-state error

Key to Multiple Choice Questions

1. (*c*)	**2.** (*c*)	**3.** (*a*)	**4.** (*c*)	**5.** (*b*)	**6.** (*b*)	**7.** (*a*)	**8.** (*b*)	**9.** (*c*)
10. (*c*)	**11.** (*c*)	**12.** (*b*)	**13.** (*b*)	**14.** (*b*)	**15.** (*b*)	**16.** (*b*)	**17.** (*b*)	**18.** (*c*)
19. (*b*)	**20.** (*c*)	**21.** (*b*)	**22.** (*b*)	**23.** (*b*)	**24.** (*c*)	**25.** (*b*)	**26.** (*c*)	**27.** (*a*)
28. (*c*)	**29.** (*b*)	**30.** (*c*)	**31.** (*b*)	**32.** (*b*)	**33.** (*b*)	**34.** (*c*)	**35.** (*b*)	**36.** (*c*)
37. (*a*)	**38.** (*c*)	**39.** (*b*)	**40.** (*c*)	**41.** (*c*)	**42.** (*b*)	**43.** (*b*)	**44.** (*b*)	**45.** (*b*)
46. (*c*)	**47.** (*b*)	**48.** (*b*)	**49.** (*b*)	**50.** (*c*)				

Projects Works

1. **PID Controller Tuning for Robotic Arm:** Design a PID controller for a robotic arm to achieve precise and smooth motion control. Implement various tuning methods like Ziegler-Nichols, Cohen-Coon, or Genetic Algorithms to optimize the controller's parameters for different robotic tasks.
2. **Trajectory Tracking for Mobile Robot:** Develop a control system that enables a mobile robot to follow a predefined trajectory accurately. Implement path planning algorithms and control techniques like Model Predictive Control (MPC) to handle constraints and environmental disturbances.
3. **Quadcopter Stabilization and Navigation:** Create a control system for stabilizing and navigating a quadcopter. Implement sensor fusion techniques using an Extended Kalman Filter (EKF) to estimate the drone's state and use a feedback controller to maintain stability during flight.
4. **Autonomous Drone Swarm:** Design a control system for a swarm of drones to work collaboratively on tasks like mapping an area or forming specific formations. Implement decentralized control algorithms that enable the drones to communicate and coordinate their actions.

5. **Robotic Manipulator Control for Object Manipulation:** Develop a control system for a robotic manipulator to grasp and manipulate objects of different shapes and sizes. Use computer vision and force feedback sensors to enhance the robot's grasp stability and adapt its manipulation strategy.
6. **Real-time Vision-Based Object Tracking:** Implement a control system that allows a robot to track and follow a moving object using real-time vision processing. Use techniques like optical flow or visual servoing to maintain the robot's position relative to the target.
7. **Adaptive Control for Flexible-Joint Robots:** Develop an adaptive control strategy for robots with flexible joints, enabling them to compensate for uncertain dynamics and maintain accuracy during motion.
8. **Control of Multi-Agent Systems:** Investigate the coordination and control of multi-robot systems to accomplish complex tasks like swarm exploration, formation control, or cooperative transportation.
9. **Intelligent Warehouse Automation:** Design a control system for an automated warehouse that manages inventory movement, optimizing the usage of automated guided vehicles (AGVs) and robots to handle storage and retrieval tasks efficiently.
10. **Human-Robot Interaction Control:** Develop a control system that allows robots to safely and intuitively interact with humans in dynamic environments, such as collaborative assembly lines or service robots in public spaces.
11. **Control of Unmanned Surface Vehicles (USVs):** Create a control system for USVs, enabling them to perform autonomous missions like environmental monitoring, surveillance, or marine research.
12. **Control of Underwater Remotely Operated Vehicles (ROVs):** Develop a control system for ROVs to execute tasks like underwater inspection, maintenance, or exploration of submerged structures.
13. **Robotic Swarm Exploration of Unknown Environments:** Implement control algorithms for a swarm of robots to explore unknown terrains efficiently and map their surroundings collaboratively.
14. **Adaptive Control for Rehabilitation Robotics:** Design a control system for rehabilitation robots that adapt to patients' needs and motor capabilities, providing personalized and effective therapeutic interventions.
15. **Distributed Control for Smart Grids:** Develop a distributed control system for smart grids that optimizes power generation, distribution, and consumption to achieve energy efficiency and grid stability.

07

Kinematics and Dynamics in the Study of Motion and the Behavior of Objects and Systems

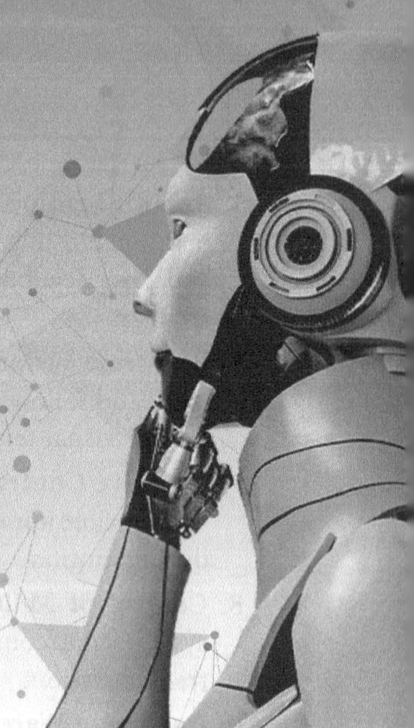

7.1 INTRODUCTION

Fundamental ideas in physics and engineering, particularly in the study of motion and the behavior of things and systems, include kinematics and dynamics. They serve as the foundation for comprehending how particles, rigid bodies, machines, and intricate mechanisms move and interact. In several fields, including robotics, mechanical engineering, aerospace engineering, and others, kinematics and dynamics are essential concepts. Without taking into account the forces or torques that create the motion, kinematics deals with the description and study of motion. It focuses on how objects move through space and time, including their position, speed, and acceleration. In kinematics, we utilize mathematical representations to explain the geometric features of motion, such as position vectors, velocity vectors, and acceleration vectors.

Dynamics, on the other hand, focuses on the investigation of the forces and torques that propel motion as well as how objects react to those forces. It entails using Newton's equations of motion and other concepts to understand how forces interact with an object's mass and geometry to modify how that thing moves. By studying dynamics, we may forecast system behavior in the future, investigate stability, and create control plans to achieve desired motion.

The design and analysis of complex mechanical systems, robotics, vehicles, and other technical systems are built on the foundation of kinematics and dynamics. Engineers and physicists may optimize designs, guarantee safety, and enhance total system performance by comprehending these ideas. Kinematics and dynamics are essential to understanding the behavior and underlying mechanics of a variety of physical phenomena, including the smooth motion of a robotic arm, aircraft flight, and the migration of planets in the solar system.

7.2 RIGID BODY MOTION

A fundamental idea in mechanics called rigid body motion defines how an entity moves as a whole without taking interior deformations or shape changes into account. In a rigid body motion, the object's points move in parallel while keeping their respective distances, orientations, and forms. Applications in engineering and physics including robotics, vehicle dynamics, and structural analysis can all benefit from this reduction.

We commonly employ vectors and matrices to express the position, orientation, velocity, and acceleration of the object in order to quantitatively explain rigid body motion.

Let's dissect the fundamental elements of the rigid body motion mathematical model:

1. **Translation:** Translation is the straight-line, non-rotational movement of an object from one point to another. The position vector r of a point P in a rigid body in a fixed reference frame can be used to characterize that point's location. The item moves during translation, changing the position vector r over time t.

   The translation can be represented mathematically as: $r(t) = r_0 + vt$, where:

   - r(t) is the position vector of point P at time t.
   - r_0 is the initial position vector of point P at t=0.
   - v is the translational velocity vector of the rigid body.

2. **Rotation:** Rotation describes the angular movement of the object about a fixed point or axis. It is often represented by a rotation matrix or quaternion. The orientation of the object can be defined using Euler angles or other coordinate systems.

   The rotation matrix R represents the orientation of the rigid body and is orthogonal, i.e., $R^T * R = I$, where I is the identity matrix. The rotation matrix can be used to transform points between the body frame and the fixed reference frame.

   The mathematical representation of rotation is usually more complex than translation and depends on the chosen coordinate system and rotation conventions.

3. **Rigid Body Transformation:** We combine translation and rotation into a single transformation matrix that connects the positions of rigid body points to the fixed reference frame in order to describe the full rigid body motion. The homogeneous transformation matrix or posture matrix is another name for this transformation matrix.

   The homogeneous transformation matrix T combines the rotation matrix R and the translation vector d : T = | R d | | 0 1 |

   The transformation can be applied to a point P in the rigid body to find its position in the fixed reference frame as follows: P_fixed = T * P_body

   where P_fixed is the position vector of point P in the fixed reference frame, P_body is the position vector of point P in the rigid body frame, and * denotes matrix multiplication.

4. **Velocity and Acceleration:** We can also examine the velocity and acceleration of individual rigid body locations to further our understanding of the motion. The derivative of a point P's position vector r with respect to time t determines the point's linear velocity, or v. Similar to this, the velocity vector's derivative with respect to time t is the linear acceleration, or a.

The rigid body's rate of change in orientation is indicated by the angular velocity, or. It can be shown as a skew-symmetric matrix or a vector.

Similar to the angular velocity, the angular acceleration is the derivative of time t. We can theoretically represent and examine the motion of rigid bodies in both two-dimensional and three-dimensional space by combining translation, rotation, and rigid body transformation. This mathematical foundation serves as the foundation for investigating more intricate mechanical systems, comprehending robot kinematics, and replicating the motion of engineering constructions like cars.

7.3 FORWARD AND INVERSE KINEMATICS

Robotics and biomechanics utilize the terms forward and inverse kinematics to explain the link between joint variables and the position or orientation of an end-effector (such as a human hand, a robot leg, or a robot gripper). For robotic systems, they are essential for motion planning, control, and trajectory development.

Let's explore each concept in detail:

***Forward Kinematics*:** By using the known joint variables of a robotic system, forward kinematics calculates the position and orientation of the end-effector (or a specific point on the end-effector) in the workspace. In other words, it entails determining the pose (position and orientation) of the end-effector based on the robot's joint angles or joint displacements.

Mathematically, forward kinematics is represented by a function that maps the joint variables (θ_1, θ_2, ..., θ_n) to the Cartesian coordinates (x, y, z) and orientation (roll, pitch, yaw) of the end-effector.

The forward kinematics function is denoted as: $T = FK(\theta_1, \theta_2, ..., \theta_n)$

where,

- T represents the homogeneous transformation matrix that describes the end-effector's pose in the workspace.
- FK is the forward kinematics function, which takes the joint variables θ_1, θ_2, ..., θ_n as inputs and computes the transformation matrix T.

The position and orientation of the end-effector with respect to a given reference frame are the outcomes of the forward kinematics. When particular joint angles are provided, it enables us to determine where the end-effector is situated in space.

The process of finding the joint variables (joint angles or joint displacements) that will lead to a desired end-effector position and orientation is known as inverse kinematics, on the other hand. Because it helps us to identify appropriate joint configurations to accomplish a particular position or orientation in the workspace, inverse kinematics is particularly crucial for motion planning.

Mathematically, inverse kinematics is represented by a function that maps the Cartesian coordinates (x, y, z) and orientation (roll, pitch, yaw) of the end-effector to the joint variables (θ_1, θ_2, ..., θ_n).

The inverse kinematics function is denoted as: $\theta = IK(x, y, z, roll, pitch, yaw)$

where,

- θ represents the joint variables (joint angles or joint displacements) required to achieve the desired end-effector pose.
- IK is the inverse kinematics function, which takes the Cartesian coordinates (x, y, z) and orientation (roll, pitch, yaw) of the end-effector as inputs and computes the joint variables θ.

Finding the inverse kinematics solution might be more difficult than finding the forward kinematics solution because, depending on the structure of the robot and the required end-effector pose, there may be several solutions or no solutions at all.

In conclusion, forward kinematics enables us to find the joint variables required to accomplish a desired end-effector posture, whereas inverse kinematics enables us to determine the pose of the end-effector from known joint variables. Both ideas are crucial for understanding and using robotic systems since they serve as the foundation for control, motion planning, and trajectory generation.

7.4 NEWTON-EULER EQUATIONS

The motion of a rigid body in three dimensions is described by a set of fundamental equations called the Newton-Euler equations. These equations are derived from Euler's equations for rotational motion and Newton's second law of motion. They offer a methodical approach for evaluating the forces and torques exerted on a rigid body in order to ascertain its translational and rotational motion.

The rigid body's linear and angular accelerations, as well as the external forces and torques acting on it, are commonly expressed in terms of the Newton-Euler equations. The rotational equations and the translational equations make up two sets of equations.

***Translational Equations (Linear Motion)*:** For a rigid body in translation, the translational equations relate the linear acceleration of the body's center of mass to the total force acting on it. These equations can be expressed as:

$$\Sigma F = m * a$$

where,

- ΣF is the vector sum of all external forces acting on the rigid body.
- m is the mass of the rigid body.
- a is the linear acceleration of the body's center of mass.

The translational equations are analogous to Newton's second law of motion ($F = ma$) for a point mass but are applied to the entire rigid body.

***Rotational Equations (Angular Motion)*:** For a rigid body in rotation, the rotational equations relate the angular acceleration of the body to the total torque acting on it. These equations can be expressed as:

$$\Sigma\tau = I * \alpha$$

where,

- $\Sigma\tau$ is the vector sum of all external torques acting on the rigid body.
- I is the body's inertia tensor, which represents its resistance to rotational motion.
- α is the angular acceleration of the rigid body.

The rotational equations are analogous to Euler's equations ($\tau = I * \alpha$) for a point mass but are applied to the entire rigid body.

***Combined Equations*:** The translational and rotational equations are merged into a single set of equations in order to study a rigid body's whole range of motion. These merged equations give a thorough account of the motion of the body, taking into account both its linear and angular accelerations as well as the associated outside forces and torques.

The Newton-Euler equations are frequently used to model and analyse the dynamics of complex systems involving rigid bodies in robotics, aeronautical engineering, mechanical engineering, and other domains. They are an effective instrument for researching mechanical systems' motion, stability, and control in three dimensions.

7.4.1 Newton-Euler Equations as Applied in Robotics

In robotics, modeling and analyzing the dynamics of robotic systems rely heavily on the Newton-Euler equations. They serve as the foundation for robot control and trajectory planning and are used to characterize the motion of robots, ascertain the forces and torques operating on their links and joints. Robots are frequently viewed as interconnected rigid bodies in rigid body dynamics, where the Newton-Euler equations are particularly helpful.

The motion and forces/torques at each joint and link are calculated for a robot with numerous links and joints using the Newton-Euler equations, which are then applied recursively along the robot's kinematic chain. Forward recursion and backward recursion are the two key steps that commonly make up the process.

Forward Recursion: The forward recursion stage uses the equations to calculate the linear and angular accelerations of each link's centre of mass from the robot's base to end-effector.

For each link i:

- The linear acceleration of the center of mass (a_c_i) is computed using the translational equations: ΣF_i = m_i * a_c_i, where m_i is the mass of link i and ΣF_i is the sum of external forces acting on link i.
- The angular acceleration (α_i) is determined using the rotational equations: Στ_i = I_i * α_i, where I_i is the inertia tensor of link i and Στ_i is the sum of external torques acting on link i.

The forward recursion process is performed sequentially from the base link to the end-effector, propagating the accelerations up the kinematic chain.

Backward Recursion: To calculate the joint forces and torques necessary to produce the accelerations obtained in the forward recursion, the equations are applied from the end-effector back to the base link in the backward recursion step.

For each joint i:

- The joint force (F_i) is computed using the translational equations: F_i = m_i * a_c_i – ΣF_i, where m_i is the mass of link i, a_c_i is the linear acceleration of the link's center of mass, and ΣF_i is the sum of external forces acting on link i.
- The joint torque (τ_i) is determined using the rotational equations: τ_i = I_i * α_i – Στ_i, where I_i is the inertia tensor of link i, α_i is the angular acceleration, and Στ_i is the sum of external torques acting on link i.

In order to calculate the forces and torques at each joint, the backward recursion procedure is carried out sequentially from the end-effector back to the base link. The Newton-Euler equations can be used to analyze the dynamics of robotic systems, forecast how they will respond to various control inputs, and improve the design of robotic systems for performance and stability. For robot control, motion planning, and simulation in a variety of applications, including industrial automation, autonomous driving, robotic arms, and humanoid robots, this knowledge is crucial.

7.5 LAGRANGE'S EQUATIONS

The strong set of equations known as Lagrange's equations is used in classical mechanics to explain the motion of a rigid body or system of particles. They offer a different way to express Newton's laws of motion and are particularly helpful in cases where constrained systems and generalized coordinates are involved. Joseph-Louis Lagrange first presented the equations in 1788 as a part of his work on analytical mechanics, which sought to offer a more orderly and elegant approach to classical mechanics.

The principle of least action, which asserts that a mechanical system moves along a route that minimises the action integral, is the source of Lagrange's equations. The difference between the system's kinetic and potential energies is known as the Lagrangian function, and it is this function's integral over time that is referred to as the action integral. For a system with n degrees of freedom and generalized coordinates q_1, q_2, ..., q_n, the Lagrangian function $L(q, \dot{y})$ is defined as:

$$L(q, \dot{y}) = T(q, \dot{y}) - V(q)$$

where,

T $(q, \dot{y})$ is the kinetic energy of the system as a function of the generalized coordinate q and their time derivatives $\dot{y}$.

V (q) is the potential energy of the system as a function of the generalized coordinate q.

The time derivative of the Lagrangian with respect to the generalized coordinate q and their time derivatives $\dot{y}$ is denoted as $\partial L/\partial q_i$ and $\partial L/\partial \dot{y}_i$, respectively.

The Lagrange's equations for a system with n degrees of freedom are given by:

$$d/dt\,(\partial L/\partial \dot{y}_i) - (\partial L/\partial q_i) = Q_i$$

where,

d/dt is the total derivative with respect to time.

Q_i represents the generalized force acting on the system along the coordinate q_i.

The motion of the system is described by a set of n second-order ordinary differential equations known as the Lagrange equations. Because we can express the constraints directly and derive the equations of motion more simply thanks to the use of generalized coordinates, these equations are particularly helpful for solving problems requiring constraints. Numerous physics and engineering disciplines, such as robotics, classical mechanics, celestial mechanics, and control theory, make extensive use of Lagrange's equations. They offer a potent mathematical foundation for deciphering and comprehending the dynamics of intricate mechanical systems.

7.5.1 Lagrange's Equations in Robotics and Automation

Lagrange's equations are crucial for modeling and analyzing the dynamics of robotic systems in robotics and automation. They offer a methodical and beautiful way to derive the equations of motion for mobile robots and robot manipulators, enabling the creation of control plans, motion simulations, and equations of motion.

In the context of robotics, the generalized coordinates q_i in Lagrange's equations represent the configuration variables of the robot, such as joint angles, joint displacements, or even the robot's pose in the workspace. The Lagrangian function L $(q, \dot{y})$ accounts for the kinetic and potential energy of the robot.

Robot Manipulators: Lagrange's equations are used to generate the equations of motion for robot manipulators (serial or parallel) that connect joint torques or forces to joint locations, velocities, and accelerations. This is essential for control design because it enables us to identify the torques or forces required to produce the desired movements and trajectories.

The sum of the kinetic energies of individual links, which are based on their masses, velocities, and inertia characteristics, makes up the kinetic energy term T (q,y) in the Lagrangian. The potential energy term V(q) denotes the energy that the robot might experience as a result of gravity or any other external forces operating on it. The joint space dynamic model, can be obtained by applying Lagrange's equations. This model aids in creating precise and reliable robot mobility, minimizing vibrations, and tracking intended trajectories.

Mobile Robots: Lagrange's equations can be utilized to develop the dynamic equations that explain the movement of mobile robots in their surroundings. The robot's position (x, y), orientation (θ), and generalized coordinates (q_i) may be included. The Lagrangian captures the kinetic energy related to the robot's linear and rotational velocities as well as the potential energy brought on by gravity or terrain fluctuations. The equations of motion that connect the forces or torques applied to the robot (such as wheel torques or thruster forces) to the robot's linear and angular accelerations can be obtained by applying Lagrange's equations to mobile robotics. This knowledge is essential for creating control algorithms that allow the robot to travel freely and complete desired tasks, such as feedback control or model predictive control.

Generally speaking, Lagrange's equations offer a strong and beautiful framework for analyzing the dynamics of robotic systems, allowing researchers and engineers in robotics and automation to create sophisticated control schemes, enhance system performance, and simulate robot behavior accurately in complex environments.

7.6 ROBOT ARM KINEMATICS

The study of the geometric relationships between the various parts of a robot arm and the motion of its end-effector in three-dimensional space is known as robot arm kinematics, which is a key area of robotics. It includes comprehending how the position and orientation of the robot's end-effector, which is frequently a gripper or a tool utilised for a certain task, are impacted by joint angles or joint displacements. In general, there are two forms of robot arm kinematics: forward kinematics and inverse kinematics.

Forward Kinematics: By using known joint angles or joint displacements, forward kinematics is the process of figuring out the location and orientation of the robot's end-effector. In other words, it entails determining the pose (position and orientation) of the end-effector using the robot's joint variables.

It is possible to compute the transformation matrix Ti, which represents the rigid body transformation from the base frame to the end-effector frame, for a serial-link robot arm with n joints. The Denavit-Hartenberg (DH) convention or other coordinate transformation techniques are used to obtain the transformation matrix Ti for each joint i. We may determine the pose of the end-effector in the robot's base frame by progressively multiplying the transformation matrices of all joints. The position and orientation of the end-effector in the robot's base frame are represented by the final transformation matrix T.

We can forecast the location of the robot's end-effector in space for a given set of joint angles using forward kinematics. It is necessary for activities like path generation, workspace analysis, and trajectory planning.

Inverse Kinematics: The reverse of forward kinematics is inverse kinematics. It entails figuring out the joint angles or joint displacements that will lead to the end-effector of the robot being in the required posture (position and orientation).

For robots with several degrees of freedom or intricate kinematic structures, inverse kinematics can be more difficult than forward kinematics. There may be several solutions or none at all in some circumstances. Numerous techniques, including closed-form solutions for particular robot configurations, geometric approaches, and numerical optimization, can be employed to obtain the inverse kinematics solution. For the control and motion planning of robots, inverse kinematics is essential. It enables us to specify the intended location and orientation of the end-effector while planning and carrying out tasks in the robot's workspace.

In conclusion, the study of robot arm kinematics is essential to robotics because it provides the mathematical framework for comprehending and directing the motion of robot arms. Researchers and engineers can create effective and precise robot arm systems capable of carrying out a variety of tasks in a variety of applications, including manufacturing, assembly, automation, and research, by mastering forward and inverse kinematics.

7.6.1 Robot arm Kinematics Modeling

In order to simulate the kinematics of a robot arm, joint angles (or joint displacements) must be related mathematically to the location and orientation of the robot's end-effector in three dimensions. Planning and simulating the movements of robot arms in various jobs and settings, as well as comprehending and regulating their motion, all depend on this modeling. Forward kinematics and inverse kinematics are the two primary categories of kinematic models used in robot arm kinematics.

Forward Kinematics Modeling: In forward kinematics modeling, the end-effector of the robot is positioned and oriented according to the joint variables (joint angles or joint displacements). Calculating the transformation matrix that depicts the end-effector's pose in the robot's base frame is the objective. A popular technique for simulating the forward kinematics of robot arms is the Denavit-Hartenberg (DH) convention. Each joint's coordinate frames must be assigned, the joint parameters (link lengths, joint angles, and link offsets) must be specified, and the transformation matrices between succeeding joints must be derived. The overall transformation matrix, which depicts the end-effector's attitude in relation to the robot's base frame, can be obtained by successively multiplying the transformation matrices.

For each given joint configuration, we may forecast the position and orientation of the end-effector using the forward kinematics model, which provides the mathematical link between the joint variables and the end-effector posture. It is essential for workspace analysis, robot motion planning, and trajectory development.

Inverse Kinematics Modeling: Forward kinematics modeling is the opposite of inverse kinematics modelling. It entails figuring out the joint variables (joint angles or joint displacements) necessary to get the end-effector of the robot into the desired posture (position and orientation). Given that there may be several solutions or none at all, depending on the kinematic structure of the robot and the job requirements, inverse kinematics is frequently more difficult than forward kinematics. The inverse kinematics problem can be resolved using a variety of techniques, such as closed-form solutions for certain robot configurations, geometric approaches, and numerical optimization.

For the control and motion planning of robots, inverse kinematics modeling is essential. It enables us to specify the desired end-effector position and orientation and determine the joint angles required to realize that configuration.

In practice, robot arm kinematics modeling involves the following steps:

1. Identify the kinematic framework of the robot and give each joint a coordinate frame.
2. Based on the physical properties of the robot, derive the Denavit-Hartenberg parameters or other joint parameters.
3. To calculate the transformation matrix from the base frame to the end-effector frame, develop the forward kinematics equations.
4. Use algorithms to help the robot arm reach the appropriate locations and orientations by resolving the inverse kinematics problem.

Researchers and engineers can develop, control, and optimize robotic systems for a variety of applications, including industrial automation, manufacturing, assembly, surgical robotics, and more by precisely modeling the kinematics of robot arms.

7.6.2 Forward Kinematics for Joint Angles to End-Effector Position and Orientation

Based on the known joint angles (or joint displacements), forward kinematics is the process of establishing the location and orientation of the robot's end-effector in the workspace. It entails establishing a mathematical connection between the joint variables and the end-effector's pose (position and orientation). We employ transformation matrices to represent the rigid body transformations between neighboring links in the robot's kinematic chain in order to link the joint angles to the end-effector position and orientation. Robotics commonly employs the Denavit-Hartenberg (DH) standard to model transformations.

Let's consider a serial-link robot manipulator with n joints, where each joint is indexed from 1 to n. The robot's base frame is denoted as frame 0, and the end-effector frame is denoted as frame n + 1. The transformation matrices T_i represent the transformations from frame i to frame i + 1, and they depend on the joint variables (θ_1, θ_2, ..., θ_n) and the robot's physical parameters (link lengths, link offsets, and joint angles).

The transformation matrix T_i can be expressed as:

$$T_i = [\cos\theta_i - \sin\theta_i\cos\alpha_i \; \sin\theta_i\sin\alpha_i \; a_i\cos\theta_i]$$

$$[\sin\theta_i \; \cos\theta_i\cos\alpha_i \; -\cos\theta_i\sin\alpha_i \; a_i\sin\theta_i] \; [0 \; \sin\alpha_i \; d_i] \; [\,0\;0\;1\,]$$

where,

θ_i is the joint angle of joint i.

α_i is the link twist angle between joint i and joint i+1.

a_i is the link length between joint i and joint i+1.

d_i is the link offset (or link displacement) between joint i and joint i+1.

The transformation matrix T_0 represents the transformation from the robot's base frame to the first joint, and it is usually a constant value. The transformation matrix T_{n+1} represents the transformation from the last joint to the end-effector frame.

To compute the overall transformation from the robot's base frame to the end-effector frame (T_base_to_ee), we multiply the individual transformation matrices for all joints:

$$T\_base\_to\_ee = T_1 * T_2 * . * T_{n+1}$$

The resulting T_base_to_ee matrix represents the end-effector's pose (position and orientation) relative to the robot's base frame. The position (x, y, z) is represented by the elements in the fourth column of T_base_to_ee, and the orientation is encoded in the rotational elements of the upper 3 × 3 sub matrix. By knowing the joint angles (θ_1, θ_2, ..., θ_n) and the robot's kinematic parameters (a_i, α_i, d_i), we can compute the end-effector's pose using the forward kinematics equations. This enables us to predict the position and orientation of the end-effector for any given joint configuration and facilitates robot motion planning and control.

7.6.3 Process of determining the Joint Angles of a Robot Manipulator

The method of calculating a robot manipulator's joint angles (or joint displacements) from a desired end-effector position and orientation in the workspace is known as inverse kinematics. It entails identifying the joint parameters that will enable the robot's end-effector to assume a particular posture (position and orientation). The inverse kinematics problem might be more difficult to solve than the forward kinematics problem because, depending on the kinematic structure of the robot and the job requirements, it may have several solutions or none at all. The inverse kinematics challenge will be approached differently by various robots with varied degrees of freedom and kinematic configurations.

Here are some common techniques used for solving the inverse kinematics problem:

1. **Geometric approaches:** For some straightforward robot manipulators, the inverse kinematics can be solved analytically using geometric approaches. These techniques entail deriving closed-form answers for the joint angles through the use of geometrical connections and trigonometry.

   The Law of cosines and Law of sines can be used to solve the inverse kinematics, for instance, in a planar 2-link robot arm. Geometric solutions can also be found for some particular robot configurations, such as delta robots or SCARA robots.
2. **Numerical approaches:** Inverse kinematics for increasingly sophisticated robot manipulators are frequently solved using numerical approaches. Numerical solutions can be found by using iterative techniques like the Jacobian transposition method or the Newton-Raphson method.

   In numerical approaches, a first-guess estimate of the joint angles is given, and the algorithm iteratively improves the answer until it converges to the desired end-effector position within an allowable tolerance. These approaches are flexible and can deal with a variety of robot configurations, but they call for repetitive calculations and could end up in local minima.
3. **Task-Specific Optimization:** In some circumstances, the inverse kinematics solution can be discovered using task-specific optimization techniques. These techniques take into account extra restrictions and goals relevant to the robot's task, such as joint limits, workspace restrictions, or reducing energy usage.

   With infinitely many ways to arrive at a given position and orientation, the inverse kinematics problem might be more difficult for robots with redundant degrees of freedom (more joints than are required to reach a point).

In general, the inverse kinematics issue solution is a crucial part of planning and controlling robot movements. Robots may now carry out precise and accurate activities in a variety of applications thanks to our ability to describe desired end-effector poses and calculate the joint angles required to accomplish those configurations.

7.7 ROBOTIC DYNAMIC MODELING

In robotic dynamic modeling, the motion of robotic systems under the influence of forces and torques is mathematically described. For a variety of robotic applications, such as control design, motion planning, trajectory generation, and simulation, it is a critical step in understanding how robots behave.

Forward dynamics and inverse dynamics are the two primary categories of dynamic modeling for robotic systems.

***Forward Dynamics Modeling*:** A robot's motion can be predicted using forward dynamics modeling based on the applied forces and torques at its joints. Calculating the end-effector motion as well as the link accelerations and velocities of the robot is the objective.

The second rule of motion, which states that the total force acting on an object is equal to its mass times its acceleration (F = m * a), is the foundation for the forward dynamics model. A robotic system's forces and acceleration are represented by the joint forces or torques and the angular and linear acceleration of each connection, respectively.

The kinematic structure, mass distribution, and link inertia characteristics of the robot are all taken into account by the forward dynamics model. In order to explain the motion of the robot in terms of the joint forces or torques applied, a set of differential equations must be created. We can find the time-evolution of the robot's joint angles, joint velocities, and end-effector posture by solving these equations.

***Inverse Dynamics Modeling*:** Forward dynamics is processed backwards using inverse dynamics modeling. It entails figuring out the joint forces or torques needed to make the robot move in the appropriate way. By rearranging the forward dynamics equations to solve for the joint forces or torques, the inverse dynamics model is created. It enables us to figure out the control inputs required to get the end-effector of the robot moving in the desired directions.

Robot control requires the use of inverse dynamics modeling, which helps us create controllers that produce the proper joint torques to follow particular trajectories or complete particular tasks. When precise force interactions with the environment are necessary for force management and impedance control, inverse dynamics is also essential.

Dynamic modeling of robotic systems in practice can be challenging, particularly for robots with numerous degrees of freedom and intricate kinematic designs. Further complicating the modeling process are frictional effects, actuator dynamics, and external disturbances that may be included in dynamic models. Roboticists frequently employ computational methods like Newton-Euler formulations, Lagrange-Euler formulations, or Kane's method to carry out dynamic modeling. The systematic derivation of the dynamic equations of motion for various types of motion is made possible by these techniques.

Researchers and engineers may create sophisticated control techniques, enhance robot performance, and guarantee the stability and safety of robotic operations in practical applications by precisely modeling the dynamics of robotic systems.

7.8 ROBOTIC TRAJECTORY PLANNING

Robotic trajectory planning is the process of creating a series of desired poses for a robot's end-effector or joints with the goal of moving the robot smoothly and effectively from its current position to a target location while avoiding obstacles and respecting any task-specific constraints.

In robotics, trajectory planning is crucial, especially in jobs requiring accurate and fluid motion, like pick-and-place operations, welding, painting, and assembly work. It entails determining a workable and ideal path that the robot can take to accomplish its goal.

There are several key components in robotic trajectory planning:

1. **Path Planning:** Path planning entails determining a path that avoids collisions from the robot's starting point to its destination position. This procedure frequently takes into account the workspace of the robot, the presence of impediments, and any environmental restrictions. A* (A-star), Dijkstra's algorithm, and rapidly exploring random trees (RRT) are common path planning techniques.
2. **Interpolation:** Following the discovery of a collision-free path, interpolation methods are applied to create a smooth trajectory from the discrete locations on the path. Linear interpolation, cubic splines, and Bezier curves are common interpolation techniques.
3. **Joint Space vs. Cartesian Space Planning:** Trajectory planning can be done in either Cartesian Space (space of end-effector poses) or Joint Space (space of joint angles or joint displacements). In Cartesian space planning, the trajectory is planned for the position and orientation of the end-effector, but in joint space planning, the trajectory is directly planned for the robot's joints.
4. **Time Parameterization:** To regulate the robot's mobility at a specified speed while adhering to velocity and acceleration limits, the created trajectory must be time-parameterized. For time parameterization, trapezoidal profiles or S-curve profiles are frequently utilized.

5. **Restrictions:** Task-specific restrictions for trajectory planning may include avoiding singularities, joint limits, or adhering to the maximum joint velocities and accelerations. To ensure the robot's performance and safety, certain limits must be taken into account.
6. **Online vs. Offline Trajectory Planning:** Trajectory planning can be done online (in real-time) or offline (pre-computed). When planning offline, the full trajectory is generated prior to execution; when planning online, the trajectory is planned as-needed while the robot is in operation.
7. **Dynamic vs. Kinematic Planning:** In dynamic trajectory planning, the effects of pressures and torques during motion are taken into account by using the robot's dynamic model. The robot's kinematic model is utilized in kinematic trajectory planning, with the assumption that the robot moves without taking the forces at play into account.

The performance, energy efficiency, and safety of a robot while performing a task are all greatly impacted by effective trajectory planning. Particularly for complicated robotic systems and dynamic surroundings, it necessitates a delicate balancing act between computational complexity and real-time restrictions. To enhance robotic trajectory planning skills, cutting-edge methodologies like optimization, learning-based approaches, and reinforcement learning are also being investigated.

NUMERICAL PROBLEMS WITH SOLUTIONS

Problem 1. A car accelerates from rest at 2 m/s^2 for 5 seconds. Calculate its final velocity.

Solution: Given: Initial velocity (u) = 0 m/s, Acceleration (a) = 2 m/s^2, Time (t) = 5 seconds. Final velocity (v) = u + at; v = 0 + (2 m/s^2 × 5 seconds), i.e., v = 10 m/s

Problem 2. An object is thrown vertically upward with an initial velocity of 15 m/s. How high does it rise?

Solution: Given: Initial velocity (u) = 15 m/s, Final velocity (v) = 0 m/s (at maximum height), Acceleration (a) = −9.81 m/s^2 (acceleration due to gravity). Using the kinematic equation: $v^2 = u^2 + 2as$ $0 = (15 \text{ m/s})^2 + 2 \times (-9.81 \text{ m/s}^2) \times s$. Solving for s (height): $s = (15 \text{ m/s})^2/(2 \times 9.81 \text{ m/s}^2) \approx 11.48$ m

Problem 3. A train accelerates from rest to a speed of 72 km/h in 30 seconds. Calculate its acceleration.

Solution: Given: Initial velocity (u) = 0 m/s, Final velocity (v) = 72 km/h = 20 m/s, Time (t) = 30 seconds. Acceleration (a) = (v – u)/t, a = (20 m/s – 0 m/s)/30 seconds, a ≈ 0.67 m/s^2

Problem 4. A ball is thrown horizontally from the top of a cliff with a speed of 25 m/s. Calculate the time it takes for the ball to reach the ground.

Solution: Given: Initial vertical velocity (u) = 0 m/s (thrown horizontally), Acceleration due to gravity (a) = 9.81 m/s^2, Vertical displacement (s) = height of the cliff Using the kinematic equation: $s = ut + (1/2)at^2$, $h = 0 + (1/2) \times 9.81 \text{ m/s}^2 \times t^2$. Solving for t (time of flight): $t^2 = (2h)/9.81 \text{ m/s}^2$ $t = \sqrt{[(2 \times \text{height})/9.81 \text{ m/s}^2]}$

Problem 5. A car travels at a constant speed of 80 km/h for 2 hours. Calculate the total distance traveled by the car.

Solution: Given: Speed (v) = 80 km/h = 22.22 m/s (converted to m/s), Time (t) = 2 hours = 7200 seconds (converted to seconds). Distance (d) = v × t d = 22.22 m/s × 7200 seconds, d ≈ 160,000 meters = 160 km

Problem 6. An object is moving in a circle of radius 3 meters with a constant speed of 5 m/s. Calculate its centripetal acceleration.

Solution: Given: Radius (r) = 3 meters, Speed (v) = 5 m/s, Centripetal acceleration (*a*) = v^2/r a = $(5 \text{ m/s})^2/3$ meters a ≈ 8.33 m/s^2

Problem 7. A stone is dropped from the top of a tower 80 meters high. Calculate the time it takes for the stone to reach the ground.

Solution: Given: Initial vertical velocity (u) = 0 m/s (stone is dropped), Acceleration due to gravity (*a*) = 9.81 m/s^2, Vertical displacement (s) = 80 meters (height of the tower). Using the kinematic equation: s = ut + $(1/2)at^2$ 80 meters = 0 + (1/2) × 9.81 m/s^2 × t^2. Solving for t (time of flight): t^2 = (2 × 80 meters)/9.81 m/s^2 t = $\sqrt{[(2 \times 80 \text{ meters})/9.81 \text{ m/s}^2]}$

Problem 8. A cyclist accelerates from rest to a speed of 10 m/s in 5 seconds. Calculate the cyclist's acceleration.

Solution: Given: Initial velocity (u) = 0 m/s, Final velocity (v) = 10 m/s, Time (t) = 5 seconds. Acceleration (*a*) = (v – u)/t a = (10 m/s – 0 m/s)/5 seconds a = 2 m/s^2

Problem 9. A car travels at a constant speed of 60 km/h. How far will the car travel in 3 hours?

Solution: Given: Speed (v) = 60 km/h = 16.67 m/s (converted to m/s), Time (t) = 3 hours = 10,800 seconds (converted to seconds). Distance (*d*) = v × t d = 16.67 m/s × 10,800 seconds. d ≈ 180,000 meters = 180 km

Problem 10. A ball is thrown vertically upward with an initial velocity of 20 m/s. Calculate the maximum height it reaches.

Solution: Given: Initial velocity (u) = 20 m/s, Final velocity (v) = 0 m/s (at maximum height), Acceleration (*a*) = –9.81 m/s^2 (acceleration due to gravity). Using the kinematic equation: $v^2 = u^2 + 2as$ 0 = $(20 \text{ m/s})^2$ + 2 × (–9.81 m/s^2) × s. Solving for s (height): s = $(20 \text{ m/s})^2/(2 \times 9.81 \text{ m/s}^2)$

Problem 11. An object is released from rest and falls freely for 3 seconds. Calculate its final velocity.

Solution: Given: Initial velocity (u) = 0 m/s, Time (t) = 3 seconds, Acceleration (*a*) = 9.81 m/s^2 (acceleration due to gravity). Final velocity (v) = u + at v = 0 + (9.81 m/s^2 × 3 seconds)

Problem 12. A car accelerates uniformly from 10 m/s to 30 m/s in 5 seconds. Calculate its acceleration.

Solution: Given: Initial velocity (u) = 10 m/s, Final velocity (v) = 30 m/s, Time (t) = 5 seconds. Acceleration (*a*) = (v – u)/t

Problem 13. A car accelerates from rest at 2 m/s² for 5 seconds. Calculate its final velocity.

Solution: Given: Initial velocity (u) = 0 m/s, Acceleration (*a*) = 2 m/s^2, Time (t) = 5 seconds. Final velocity (v) = u + at v = 0 + (2 m/s^2 × 5 seconds) v = 10 m/s

Problem 14. An object is thrown vertically upward with an initial velocity of 15 m/s. How high does it rise?

Solution: Given: Initial velocity (u) = 15 m/s, Final velocity (v) = 0 m/s (at maximum height), Acceleration (*a*) = –9.81 m/s^2 (acceleration due to gravity). Using the kinematic equation: $v^2 = u^2 + 2as$ 0 = $(15 \text{ m/s})^2$ + 2 × (–9.81 m/s^2) × s. Solving for s (height): s = $(15 \text{ m/s})^2/(2 \times 9.81 \text{ m/s}^2)$ ≈ 11.48 m

Problem 15. A train accelerates from rest to a speed of 72 km/h in 30 seconds. Calculate its acceleration.

Solution: Given: Initial velocity (u) = 0 m/s, Final velocity (v) = 72 km/h = 20 m/s, Time (t) = 30 seconds. Acceleration (*a*) = (v – u)/t a = (20 m/s – 0 m/s)/30 seconds, a ≈ 0.67 m/s^2

Problem 16. A ball is thrown horizontally from the top of a cliff with a speed of 25 m/s. Calculate the time it takes for the ball to reach the ground.

Solution: Given: Initial vertical velocity (u) = 0 m/s (thrown horizontally). Acceleration due to gravity (*a*) = 9.81 m/s^2, Vertical displacement (s) = height of the cliff. Using the kinematic equation: s = ut + (1/2) at^2 h = 0 + (1/2) × 9.81 m/s^2 × t^2. Solving for t (time of flight): t^2 = (2h)/9.81 m/s^2 t = √[(2 × height)/9.81 m/s^2]

Problem 17. A car travels at a constant speed of 80 km/h for 2 hours. Calculate the total distance traveled by the car.

Solution: Given: Speed (v) = 80 km/h = 22.22 m/s (converted to m/s), Time (t) = 2 hours = 7200 seconds (converted to seconds). Distance (*d*) = v × t d = 22.22 m/s × 7200 seconds, d ≈ 160,000 meters = 160 km

Problem 18. An object is moving in a circle of radius 3 meters with a constant speed of 5 m/s. Calculate its centripetal acceleration.

Solution: Given: Radius (r) = 3 meters, Speed (v) = 5 m/s. Centripetal acceleration (*a*) = v^2/r a = (5 m/s)^2/3 meters, a ≈ 8.33 m/s^2

Problem 19. A stone is dropped from the top of a tower 80 meters high. Calculate the time it takes for the stone to reach the ground.

Solution: Given: Initial vertical velocity (u) = 0 m/s (stone is dropped), Acceleration due to gravity (*a*) = 9.81 m/s^2, Vertical displacement (s) = 80 meters (height of the tower). Using the kinematic equation: s = ut + (1/2)at^2 80 meters = 0 + (1/2) × 9.81 m/s^2 × t^2. Solving for t (time of flight): t^2 = (2 × 80 meters)/9.81 m/s^2 t = √[(2 × 80 meters)/9.81 m/s^2]

Problem 20. A cyclist accelerates from rest to a speed of 10 m/s in 5 seconds. Calculate the cyclist's acceleration.

Solution: Given: Initial velocity (u) = 0 m/s, Final velocity (v) = 10 m/s, Time (t) = 5 seconds. Acceleration (*a*) = (v – u)/t a = (10 m/s – 0 m/s)/5 seconds, a = 2 m/s^2

Problem 21. A car travels at a constant speed of 60 km/h. How far will the car travel in 3 hours?

Solution: Given: Speed (v) = 60 km/h = 16.67 m/s (converted to m/s), Time (t) = 3 hours = 10,800 seconds (converted to seconds). Distance (*d*) = v × t d = 16.67 m/s × 10,800 seconds, d ≈ 180,000 meters = 180 km

Problem 22. A ball is thrown vertically upward with an initial velocity of 20 m/s. Calculate the maximum height it reaches.

Solution: Given: Initial velocity (u) = 20 m/s, Final velocity (v) = 0 m/s (at maximum height), Acceleration (*a*) = –9.81 m/s^2 (acceleration due to gravity). Using the kinematic equation: v^2 = u^2 + 2as 0 = (20 m/s)^2 + 2 × (–9.81 m/s^2) × s. Solving for s (height): s = (20 m/s)^2/(2 × 9.81 m/s^2)

Problem 23. An object is released from rest and falls freely for 3 seconds. Calculate its final velocity.

Solution: Given: Initial velocity (u) = 0 m/s, Time (t) = 3 seconds, Acceleration (*a*) = 9.81 m/s^2 (acceleration due to gravity), Final velocity (v) = u + at v = 0 + (9.81 m/s^2 × 3 seconds)

Problem 24. A car accelerates uniformly from 10 m/s to 30 m/s in 5 seconds. Calculate its acceleration.

Solution: Given: Initial velocity (u) = 10 m/s, Final velocity (v) = 30 m/s, Time (t) = 5 seconds. Acceleration (*a*) = (v – u)/t = 4 m/s^2.

Short Questions and Answers

Q.1. What is kinematics?

Ans. Kinematics is the study of motion without considering the forces causing the motion.

Q.2. What is dynamics?

Ans. Dynamics is the study of motion, taking into account the forces causing the motion.

Q.3. What is displacement?

Ans. Displacement is the change in position of an object, usually measured in meters (m).

Q.4. What is velocity?

Ans. Velocity is the rate of change of displacement with time and has both magnitude and direction.

Q.5. What is speed?

Ans. Speed is the rate of change of distance with time and only has magnitude.

Q.6. What is the SI unit of velocity?

Ans. The SI unit of velocity is meters per second (m/s).

Q.7. What is acceleration?

Ans. Acceleration is the rate of change of velocity with time.

Q.8. What is the SI unit of acceleration?

Ans. The SI unit of acceleration is meters per second squared (m/s^2).

Q.9. What is uniform motion?

Ans. Uniform motion is when an object moves with constant velocity (constant speed and direction).

Q.10. What is non-uniform motion?

Ans. Non-uniform motion is when an object's velocity changes over time.

Q.11. What is projectile motion?

Ans. Projectile motion is the motion of an object launched into the air and subject only to the force of gravity.

Q.12. What is a scalar quantity?

Ans. A scalar quantity is a physical quantity that has only magnitude and no direction, e.g., distance, speed, time.

Q.13. What is a vector quantity?

Ans. A vector quantity is a physical quantity that has both magnitude and direction, e.g., displacement, velocity, acceleration.

Q.14. How do you calculate average velocity?

Ans. Average velocity is calculated by dividing the total displacement by the total time taken.

Q.15. What is instantaneous velocity?

Ans. Instantaneous velocity is the velocity of an object at a specific instant in time.

Q.16. How do you calculate acceleration from velocity-time graph?

Ans. Acceleration is the slope of the velocity-time graph.

Q.17. What is the first law of motion?

Ans. The first law of motion, also known as Newton's first law, states that an object will remain at rest or in uniform motion unless acted upon by an external force.

Q.18. What is the second law of motion?

Ans. The second law of motion, also known as Newton's second law, states that the acceleration of an object is directly proportional to the net force acting on it and inversely proportional to its mass. The formula is F = ma.

Q.19. What is the third law of motion?

Ans. The third law of motion, also known as Newton's third law, states that for every action, there is an equal and opposite reaction.

Q.20. What is friction?

Ans. Friction is the force that opposes the relative motion or attempted motion of two surfaces in contact.

Q.21. What is static friction?

Ans. Static friction is the frictional force that prevents an object from moving when a force is applied to it.

Q.22. What is kinetic friction?

Ans. Kinetic friction is the frictional force between two surfaces in relative motion.

Q.23. What is the coefficient of friction?

Ans. The coefficient of friction is a dimensionless constant that relates the force of friction between two surfaces to the normal force pressing them together.

Q.24. What is centripetal acceleration?

Ans. Centripetal acceleration is the acceleration directed towards the center of the circular path of an object moving in a circle.

Q.25. What is centripetal force?

Ans. Centripetal force is the force that keeps an object moving in a circular path and directed towards the center of the circle.

Q.26. What is gravitational force?

Ans. Gravitational force is the force of attraction between any two objects with mass.

Q.27. What is weight?

Ans. Weight is the force experienced by an object due to gravity and is equal to the mass of the object multiplied by the acceleration due to gravity.

Q.28. What is the SI unit of force?

Ans. The SI unit of force is the newton (N).

Q.29. What is Newton's universal law of gravitation?

Ans. Newton's universal law of gravitation states that every particle in the universe attracts every other particle with a force that is directly proportional to the product of their masses and inversely proportional to the square of the distance between them.

Q.30. What is Hooke's law?

Ans. Hooke's law states that the force needed to extend or compress a spring is directly proportional to the displacement or change in length of the spring.

Q.31. What is the principle of conservation of momentum?

Ans. The principle of conservation of momentum states that the total momentum of a system of objects remains constant if no external forces act on the system.

Q.32. What is elastic collision?

Ans. An elastic collision is a collision where both kinetic energy and momentum are conserved.

Q.33. What is inelastic collision?

Ans. An inelastic collision is a collision where kinetic energy is not conserved.

Q.34. What is the law of conservation of energy?

Ans. The law of conservation of energy states that energy cannot be created or destroyed, only transformed from one form to another.

Q.35. What is potential energy?

Ans. Potential energy is the energy an object possesses due to its position or configuration.

36. What is kinetic energy?

Ans. Kinetic energy is the energy an object possesses due to its motion.

Q.37. What is work?

Ans. Work is the product of force and displacement in the direction of the force.

Q.38. What is power?

Ans. Power is the rate at which work is done or the rate at which energy is transferred or converted.

Q.39. What is mechanical advantage?

Ans. Mechanical advantage is the ratio of the force output of a machine to the force input.

Q.40. What is a simple machine?

Ans. A simple machine is a mechanical device that changes the magnitude or direction of a force and includes tools like levers, pulleys, and inclined planes.

Q.41. What is torque?

Ans. Torque is the rotational analog of force and is the product of force and the lever arm (distance from the pivot).

Q.42. What is angular momentum?

Ans. Angular momentum is the rotational analog of linear momentum and is the product of rotational inertia and angular velocity.

Q.43. What is angular velocity?

Ans. Angular velocity is the rate of change of angular displacement with time.

Q.44. What is the law of conservation of angular momentum?

Ans. The law of conservation of angular momentum states that the total angular momentum of a system remains constant in the absence of external torques.

Q.45. What is the moment of inertia?

Ans. The moment of inertia is a measure of an object's resistance to changes in its rotational motion.

Q.46. What is rotational kinetic energy?

Ans. Rotational kinetic energy is the kinetic energy associated with the rotation of an object about an axis.

Q.47. What is the center of mass?

Ans. The center of mass is the point where the entire mass of an object can be considered to be concentrated.

Q.48. What is a system in dynamics?

Ans. A system in dynamics refers to a group of objects or particles under consideration.

Q.49. What is linear momentum?

Ans. Linear momentum is the product of an object's mass and its linear velocity.

Q.50. What is the principle of conservation of linear momentum?

Ans. The principle of conservation of linear momentum states that the total linear momentum of an isolated system remains constant in the absence of external forces.

Multiple Choice Questions and Answers

Q.1. What is kinematics?

(*a*) The study of motion with consideration of forces
(*b*) The study of motion without consideration of forces
(*c*) The study of motion at rest
(*d*) The study of motion in a vacuum

Q.2. What is dynamics?

(*a*) The study of motion with consideration of forces
(*b*) The study of motion without consideration of forces
(*c*) The study of motion at rest
(*d*) The study of motion in a vacuum

Q.3. Which of the following is a scalar quantity?

(*a*) Velocity (*b*) Acceleration (*c*) Displacement (*d*) Force

Q.4. Which of the following is a vector quantity?

(*a*) Speed (*b*) Time (*c*) Distance (*d*) Displacement

Q.5. What is the SI unit of velocity?

(*a*) Meters per second squared (m/s^2)
(*b*) Meters per second (m/s)
(*c*) Kilometers per hour (km/h)
(*d*) Meters (m)

Q.6. What is the SI unit of acceleration?

(*a*) Meters per second squared (m/s^2)
(*b*) Meters per second (m/s)
(*c*) Kilometers per hour (km/h)
(*d*) Meters (m)

Q.7. What is the first law of motion?

(*a*) Every action has an equal and opposite reaction.
(*b*) Force is equal to mass times acceleration (F = ma).
(*c*) An object in motion will stay in motion, and an object at rest will stay at rest unless acted upon by an external force.

(*d*) The rate of change of momentum of an object is directly proportional to the net force acting on it.

Q.8. What is the second law of motion?

(*a*) Every action has an equal and opposite reaction.

(*b*) Force is equal to mass times acceleration (F = ma).

(*c*) An object in motion will stay in motion, and an object at rest will stay at rest unless acted upon by an external force.

(*d*) The rate of change of momentum of an object is directly proportional to the net force acting on it.

Q.9. What is the third law of motion?

(*a*) Every action has an equal and opposite reaction.

(*b*) Force is equal to mass times acceleration (F = ma).

(*c*) An object in motion will stay in motion, and an object at rest will stay at rest unless acted upon by an external force.

(*d*) The rate of change of momentum of an object is directly proportional to the net force acting on it.

Q.10. What is the force that keeps an object moving in a circular path called?

(*a*) Centrifugal force (*b*) Centripetal force (*c*) Gravitational force (*d*) Frictional force

Q.11. What is the formula for calculating average velocity?

(*a*) Average Velocity = (Final Velocity – Initial Velocity)/Time

(*b*) Average Velocity = Distance/Time

(*c*) Average Velocity = (Final Velocity + Initial Velocity)/Time

(*d*) Average Velocity = Distance × Time

Q.12. What is the acceleration of an object in free fall near the surface of the Earth?

(*a*) 9.81 m/s^2 upward (*b*) 9.81 m/s^2 downward

(*c*) 0 m/s^2 (*d*) 9.81 m/s

Q.13. Which of the following is true about projectile motion?

(*a*) The horizontal component of velocity remains constant.

(*b*) The vertical component of velocity remains constant.

(*c*) The acceleration is only in the vertical direction.

(*d*) The acceleration is only in the horizontal direction.

Q.14. What is the law of conservation of momentum?

(*a*) Momentum is always conserved in any collision.

(*b*) The total momentum of a system is zero.

(*c*) The total momentum of a system remains constant if no external forces act on the system.

(*d*) Momentum is never conserved in any collision.

Q.15. Which type of collision conserves both momentum and kinetic energy?

(*a*) Elastic collision (*b*) Inelastic collision

(*c*) Perfectly inelastic collision (*d*) None of the above

Q.16. What is potential energy?

(*a*) The energy an object possesses due to its motion

(*b*) The energy an object possesses due to its position or configuration

(*c*) The energy an object possesses due to its mass

(*d*) The energy an object possesses due to its temperature

Q.17. What is the work done on an object if the force and displacement are in the same direction?

(*a*) Positive (*b*) Negative

(*c*) Zero (*d*) None of the above

Q.18. What is power?

(*a*) The rate of change of momentum (*b*) The rate of change of velocity

(*c*) The rate at which work is done or energy is transferred

(*d*) The rate at which distance is covered

Q.19. What is a simple machine?

(*a*) A machine with a simple design

(*b*) A machine that operates without any external forces

(*c*) A machine with few or no moving parts that changes the magnitude or direction of a force

(*d*) A machine that operates only in a vacuum

Q.20. What is torque?

(*a*) The rotational analog of acceleration (*b*) The rotational analog of velocity

(*c*) The rotational analog of force (*d*) The rotational analog of distance

Q.21. What is thc law of conservation of angular momentum?

(*a*) Angular momentum is always conserved in any collision.

(*b*) The total angular momentum of a system remains constant in the absence of external torques.

(*c*) The total angular momentum of a system is zero.

(*d*) Angular momentum is never conserved in any collision.

Q.22. What is the moment of inertia?

(*a*) A measure of an object's resistance to linear motion

(*b*) A measure of an object's resistance to rotational motion

(*c*) A measure of an object's weight

(*d*) A measure of an object's velocity

Q.23. What is the center of mass?

(*a*) The point where an object's weight is concentrated

(*b*) The point where an object's mass is concentrated

(*c*) The point where the entire mass of an object can be considered to be concentrated
(*d*) The point where an object is in equilibrium

Q.24. What is linear momentum?
(*a*) The product of an object's mass and its linear velocity
(*b*) The product of an object's mass and its angular velocity
(*c*) The product of an object's weight and its acceleration
(*d*) The product of an object's displacement and its linear velocity

Q.25. Which of the following statements about linear momentum is true?
(*a*) Linear momentum is a scalar quantity.
(*b*) Linear momentum is always conserved in any collision.
(*c*) Linear momentum is equal to force times acceleration.
(*d*) Linear momentum depends on the direction of motion.

Q.26. Which of the following statements about angular momentum is true?
(*a*) Angular momentum is a vector quantity.
(*b*) Angular momentum is always conserved in any collision.
(*c*) Angular momentum is equal to force times acceleration.
(*d*) Angular momentum depends on the mass of the object.

Q.27. The acceleration due to gravity on the surface of the Earth is approximately:
(*a*) 9.81 m/s^2 (*b*) 6.67×10^{-11} $m^3/kg/s^2$
(*c*) 3.00×10^8 m/s (*d*) 1.00 m/s

Q.28. An object is dropped from a certain height above the ground. Neglecting air resistance, what is the acceleration of the object during free fall?
(*a*) Zero (*b*) 9.81 m/s^2 downward
(*c*) 9.81 m/s^2 upward (*d*) 4.90 m/s^2 upward

Q.29. Which of the following equations represents the relationship between force (F), mass (m), and acceleration (*a*)?
(*a*) F = m × a (*b*) F = m/a (*c*) F = a/m (*d*) F = a + m

Q.30. An object is thrown horizontally with an initial velocity v from the top of a cliff. What will be the shape of its trajectory?
(*a*) Straight line (*b*) Parabola (*c*) Circle (*d*) Ellipse

Q.31. A ball is thrown vertically upward. What is the direction of its acceleration at the highest point of its trajectory?
(*a*) Upward (*b*) Downward (*c*) Zero
(*d*) It depends on the initial velocity of the ball.

Q.32. Which of the following is true for an object moving in a circle at a constant speed?
(*a*) The object is at rest. (*b*) The object's acceleration is zero.
(*c*) The object's velocity is zero. (*d*) The object's net force is zero.

Q.33. A car accelerates from rest to 30 m/s in 10 seconds. What is its acceleration?

(*a*) 0.3 m/s^2 (*b*) 3 m/s^2 (*c*) 30 m/s^2 (*d*) 300 m/s^2

Q.34. An object moves with a constant velocity of 5 m/s for 10 seconds. What is its displacement during this time?

(*a*) 10 m (*b*) 50 m (*c*) 5 m (*d*) 100 m

Q.35. Which of the following is true for a body in uniform circular motion?

(*a*) The body is at rest.
(*b*) The body's velocity is constant.
(*c*) The body's acceleration is constant.
(*d*) The body's speed is constant.

Q.36. Which of the following quantities can never be negative?

(*a*) Velocity (*b*) Displacement (*c*) Speed (*d*) Acceleration

Q.37. An object is moving at a constant velocity. What can you say about its acceleration?

(*a*) The acceleration is zero.
(*b*) The acceleration is constant.
(*c*) The acceleration is positive.
(*d*) The acceleration is negative.

Q.38. The time it takes for an object to return to its starting position in circular motion is called:

(*a*) Period (*b*) Frequency (*c*) Amplitude (*d*) Wavelength

Q.39. Which of the following is the correct expression for gravitational force between two objects with masses m_1 and m_2, separated by a distance r?

(*a*) $F = G \times m_1 \times m_2/r^2$
(*b*) $F = G/m_1 \times m_2/r^2$
(*c*) $F = G \times m_1/m_2 \times r^2$
(*d*) $F = G \times m_2/m_1 \times r^2$

Q.40. If you double the distance between two objects, how does the gravitational force between them change?

(*a*) The force is halved.
(*b*) The force is doubled.
(*c*) The force remains the same.
(*d*) The force becomes zero.

Q.41. Which of the following represents the relationship between kinetic energy (KE), mass (m), and velocity (v)?

(*a*) $KE = m \times v$
(*b*) $KE = m \times v^2$
(*c*) $KE = 1/2 \times m \times v$
(*d*) $KE = 1/2 \times m \times v^2$

Q.42. Which of the following is an example of an inelastic collision?

(*a*) Two billiard balls collide and bounce off each other.
(*b*) A ball is dropped and bounces back up after hitting the ground.
(*c*) Two clay balls collide and stick together.
(*d*) A ball rolls on a smooth surface without any impact.

Q.43. What happens to the total momentum of a system of objects if no external forces act on the system?

(*a*) It increases.
(*b*) It decreases.
(*c*) It remains constant.
(*d*) It becomes zero.

Q.44. Which of the following types of energy is associated with an object in motion?

(*a*) Kinetic energy (*b*) Potential energy
(*c*) Gravitational potential energy (*d*) Elastic potential energy

Q.45. A car has a mass of 1000 kg and is moving with a velocity of 20 m/s. What is its kinetic energy?

(*a*) 10,000 J (*b*) 40,000 J (*c*) 100,000 J (*d*) 200,000 J

Q.46. What is the unit of power?

(*a*) Joules (J) (*b*) Watts (W) (*c*) Newtons (N) (*d*) Kilograms (kg)

Q.47. Which of the following is NOT an example of a simple machine?

(*a*) Lever (*b*) Pulley (*c*) Screw (*d*) Gear

Q.48. The moment of inertia of an object depends on its:

(*a*) Mass and velocity (*b*) Shape and mass distribution
(*c*) Weight and size (*d*) Density and temperature

Q.49. Which of the following statements about torque is true?

(*a*) Torque depends only on force. (*b*) Torque depends only on distance.
(*c*) Torque depends on both force and distance.
(*d*) Torque depends on neither force nor distance.

Q.50. If the mass of an object is doubled, how does its momentum change, assuming the velocity remains constant?

(*a*) Momentum is halved. (*b*) Momentum is doubled.
(*c*) Momentum remains the same. (*d*) Momentum becomes zero.

Key to Multiple Choice Questions

1. (*b*)	**2.** (*a*)	**3.** (*c*)	**4.** (*d*)	**5.** (*b*)	**6.** (*a*)	**7.** (*c*)	**8.** (*b*)	**9.** (*a*)
10. (*b*)	**11.** (*a*)	**12.** (*b*)	**13.** (*a*)	**14.** (*c*)	**15.** (*a*)	**16.** (*b*)	**17.** (*a*)	**18.** (*c*)
19. (*c*)	**20.** (*c*)	**21.** (*b*)	**22.** (*b*)	**23.** (*c*)	**24.** (*a*)	**25.** (*d*)	**26.** (*a*)	**27.** (*a*)
28. (*b*)	**29.** (*a*)	**30.** (*b*)	**31.** (*b*)	**32.** (*b*)	**33.** (*b*)	**34.** (*b*)	**35.** (*c*)	**36.** (*c*)
37. (*a*)	**38.** (*a*)	**39.** (*a*)	**40.** (*a*)	**41.** (*d*)	**42.** (*c*)	**43.** (*c*)	**44.** (*a*)	**45.** (*c*)
46. (*b*)	**47.** (*d*)	**48.** (*b*)	**49.** (*c*)	**50.** (*c*)				

Projects Works

1. **Robotic Arm Control:** Design and implement a robotic arm that can move in a controlled manner and perform precise tasks using kinematic principles.
2. **Dynamic Trajectory Planning:** Develop algorithms for dynamic trajectory planning of robots to efficiently navigate through complex environments and avoid collisions.
3. **Inverse Kinematics Solver:** Create an inverse kinematics solver for a robotic arm to calculate the joint angles required to reach a specific end-effector position.

4. **Mobile Robot Path Planning:** Implement a path planning algorithm for a mobile robot to move from one point to another while avoiding obstacles and optimizing the path.
5. **Dynamic Balancing Robot:** Build a robot that can balance itself dynamically using feedback control systems and principles of dynamics.
6. **Robotic Gripper Design:** Design and construct a robotic gripper that can adaptively grasp objects of various shapes and sizes based on their dynamics.
7. **Robot Arm Dynamics Simulation:** Develop a dynamic simulation of a robotic arm to study its motion and behavior under different loading conditions.
8. **Human-Robot Interaction Dynamics:** Investigate the dynamics of human-robot interaction, including force sensing and compliance control to enable safe collaboration between humans and robots.
9. **Mobile Robot Localization and Mapping:** Implement algorithms for simultaneous localization and mapping (SLAM) to enable a robot to create a map of its environment and localize itself within it.
10. **Robot Gait Optimization:** Optimize the gait of a legged robot using dynamic modeling and control techniques to achieve efficient and stable locomotion.
11. **Robot Control with Redundant Degrees of Freedom:** Explore methods to control robots with redundant degrees of freedom to enhance their flexibility and improve task performance.
12. **Robot Manipulator Dynamics Identification:** Develop techniques to identify the dynamic parameters of a robot manipulator to improve control accuracy and stability.
13. **Adaptive Control for Robots:** Design adaptive control algorithms for robots that can adjust their parameters based on changes in the environment or the robot's characteristics.
14. **Dynamic Obstacle Avoidance:** Create a real-time obstacle avoidance system for a robot that can handle dynamic obstacles and adapt its trajectory accordingly.
15. **Dynamic Interaction with Soft Robotics:** Investigate the dynamics of soft robots and their interaction with the environment, exploring applications in delicate manipulation and human-robot interaction.
16. **Robot Arm Compliance Control:** Implement compliance control algorithms for a robotic arm to make it safe and adaptable when interacting with humans or delicate objects.
17. **Dynamic Multi-Robot Coordination:** Develop strategies for dynamic coordination of multiple robots to perform cooperative tasks efficiently.
18. **Energy Efficiency in Robot Motion:** Optimize robot motion planning and control to minimize energy consumption, especially for battery-powered robots.
19. **Human-Robot Collaborative Control:** Design control schemes that allow humans to work collaboratively with robots, focusing on seamless motion coordination and shared workspace.
20. **Adaptive Robot Grasping:** Develop a grasping system for robots that can adapt its grip force based on the object's weight and fragility, using force feedback control.
21. **Dynamic Legged Robot:** Build a dynamic legged robot capable of running and jumping, leveraging principles of dynamics to achieve agile locomotion.
22. **Cooperative Robot Manipulation:** Investigate strategies for multiple robots to cooperate in manipulating large objects collaboratively, utilizing kinematic coordination and force distribution.

23. **Robot Soccer Control:** Develop control algorithms for robots playing soccer, emphasizing dynamic maneuvering, and strategic decision-making.
24. **Reactive Control for Mobile Robots:** Implement reactive control approaches that allow mobile robots to respond swiftly to unexpected changes in the environment.
25. **Dynamic Object Tracking:** Create a system that enables a robot to dynamically track moving objects in its environment using sensor feedback and predictive modeling.
26. **Robot Swarm Dynamics:** Study the dynamics of robot swarms, analyzing emergent behaviors and self-organization principles.
27. **Dynamics-based Robot Dance:** Choreograph a dance routine for a humanoid robot, utilizing dynamic motions and fluid movements.
28. **Optimal Robot Configuration Planning:** Develop algorithms for finding the optimal robot configuration for specific tasks by considering dynamic and environmental constraints.
29. **Robot Racing Control:** Design control strategies for autonomous robot racing, optimizing speed and stability around curves and obstacles.
30. **Reactive Grasping in Unstructured Environments:** Implement reactive grasping techniques for a robot to efficiently pick up objects in cluttered and unstructured environments.

08

Machine Learning and Artificial Intelligence as applied in Robotics and Automation

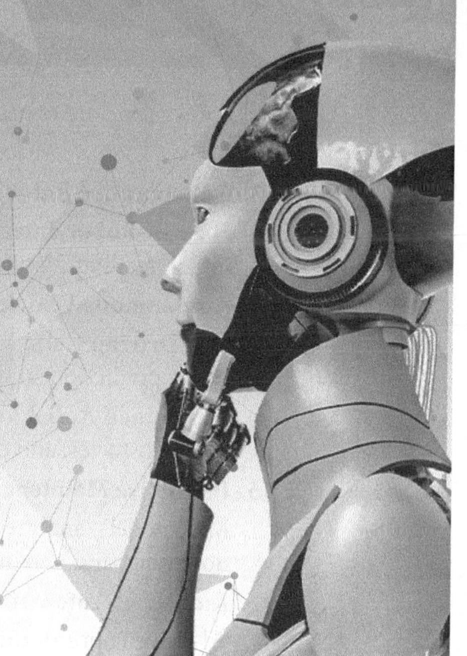

8.1 INTRODUCTION

Robotics and automation have undergone a revolution thanks to machine learning (ML) and artificial intelligence (AI), which have made it possible for machines to carry out jobs that were previously regarded to be solely the province of humans. Even though ML and AI have significantly advanced robotics and automation, there are still difficulties. There is ongoing study and development in the areas of ensuring safety, dealing with ethical issues, managing uncertainty in real-world settings, and enhancing generalization and transfer learning.

8.2 ML AND AI APPLIED IN ROBOTICS AND AUTOMATION

Here's a breakdown of how ML and AI are applied in robotics and automation:

1. **Perception and Sensing:**

   Computer Vision: ML algorithms are used to interpret visual data from cameras and sensors, enabling robots to properly pick up and place objects, recognize objects in their environment, and navigate.

   Processing of LiDAR and Radar Data: ML aids in the interpretation of data from LiDAR and Radar sensors, assisting in mapping and localization for autonomous navigation.

   Path Planning: AI methods like A* search, RRT (Rapidly-exploring Random Trees), and neural networks can optimize robot trajectories in complex settings while avoiding obstacles.

   Control Systems: Robotic systems are controlled using reinforcement learning (RL) and deep learning, allowing them to discover the best control methods through trial and error.

2. **Manipulation and Grasping:**

   Grasping: ML algorithms help robots determine the best ways to grasp objects of various shapes and sizes, improving efficiency in picking and placing tasks.

   Force Feedback: ML-driven force sensing allows robots to manipulate delicate objects with a human-like sense of touch.

3. **Collaborative Robotics (Cobots):**

   Human-Robot Interaction: AI enables robots to understand human gestures, speech, and intent, facilitating safer and more intuitive collaboration between humans and robots.

4. **Autonomous Vehicles:**

   Self-Driving Cars: ML algorithms are crucial for tasks such as object detection, lane keeping, traffic prediction, and decision-making in autonomous vehicles.

   Unmanned Aerial Vehicles (UAVs): ML-powered UAVs can navigate complex environments, avoid obstacles, and optimize flight paths.

5. **Predictive Maintenance:**

   ML models analyze sensor data from machines to predict maintenance needs, reducing downtime and maximizing operational efficiency.

6. **Industrial Automation:**

   AI-driven optimization algorithms help in scheduling and resource allocation for manufacturing processes, improving efficiency and reducing costs.

7. **Natural Language Processing (NLP) for Human-Robot Communication:**

   NLP techniques allow humans to communicate with robots using natural language, making it easier to control and coordinate tasks.

8. **Simulation and Training:**

   ML is used to create realistic simulations for training robots, allowing them to learn and improve their skills in virtual environments before applying them in the real world.

9. **Adaptive Systems:**

   ML allows robots to adapt to changing environments and tasks, enhancing their versatility and usefulness.

8.2.1 Perception and Sensing of ML and AI

Robotics and automation depend heavily on perception and sensing to help robots comprehend and interact with their surroundings. Robotics and automated systems can be improved in this area by using machine learning (ML) and artificial intelligence (AI). ML and AI are used in robotics and automation to improve perception and sensing in the following ways:

1. **Object detection and recognition:** ML algorithms, in particular convolution neural networks (CNNs), are used to find and identify objects in photos and videos taken by cameras and sensors.

   Semantic Segmentation: By using ML to classify each pixel in an image, semantic segmentation approaches help robots recognize the layout of an environment and distinguish between various items and regions.

2. **LiDAR and Radar Data Processing:** ML models are employed to process data from LiDAR and radar sensors, enabling accurate mapping, localization, and object detection even in challenging conditions.

3. **Depth Estimation:** Robots can perceive depth with the use of ML-based techniques, which use stereo vision or monocular depth estimates to assist with navigation and manipulation tasks.

4. **Gesture and Pose Recognition:** Robots can comprehend and react to human commands and interactions thanks to machine learning (ML) algorithms, which analyze human movements and poses from photographs or sensor data.

5. **Speech and Audio Processing:** Speech recognition and natural language processing (NLP) technologies allow robots to converse verbally with people and comprehend spoken orders.
6. **Haptic Sensing and Touch Recognition:** Robots can communicate orally with people and understand spoken orders thanks to speech recognition and natural language processing (NLP) technologies.
7. **Sensor Fusion:** ML algorithms fuse data from multiple sensors, such as cameras, LiDAR, and IMUs (Inertial Measurement Units), to create a comprehensive understanding of the environment, improving accuracy and robustness.
8. **Anomaly Detection:** In order to support predictive maintenance and the early identification of failures, machine learning (ML) models are used to discover abnormalities or departures from predicted patterns in sensor data.
9. **Simultaneous Localization and Mapping (SLAM):** Robots can construct maps of uncharted regions using SLAM techniques while also tracking their own positions, enabling autonomous navigation.
10. **Attention Mechanisms:** Robots' processing and decision-making are optimised when they can focus on pertinent areas of their surroundings thanks to AI-based attention processes.
11. **Transfer Learning and Generalization:** ML techniques like transfer learning enable robots to generalize knowledge from one environment to another, enhancing their adaptability to new scenarios.

Robots are now able to perceive and interact with their environment more intelligently and successfully because to the integration of ML and AI in perception and sensing systems. To ensure the secure and dependable operation of robotic systems, however, issues including data quality, resilience, real-time processing, and ethical considerations relating to privacy and bias must be properly addressed.

8.2.2 Motion Planning and Control

Choosing how robots and autonomous systems should move and interact with their surroundings is the basic task of motion planning and control in robotics and automation. Robotic motion planning and control abilities have been greatly improved by machine learning (ML) and artificial intelligence (AI) techniques. ML and AI are used in this situation as follows:

1. **Path Planning:**

   Dijkstra's algorithm and a search:* When attempting to discover the shortest route between two points while avoiding obstacles, these traditional procedures are frequently used.

   Exploring Random Trees Quickly (RRT): RRT and its variations are ML-inspired methods for quickly locating and creating viable pathways through complicated situations.

2. **Trajectory Generation:**

   Techniques for optimization: AI-driven optimization algorithms can produce ideal paths that minimize time, energy use, or other metrics while taking restrictions into account.

   Reinforcement Learning (RL): By adjusting trajectories in response to feedback from the environment, RL can be used to learn optimal control policies by trial and error.

3. **Obstacle Avoidance:**

   Machine learning (ML) algorithms assess sensor data from sources like cameras and LiDAR to foresee potential collisions and adjust the robot's path as needed.

Potential Fields: AI techniques like potential fields create imaginary forces that guide the robot past obstacles and in the right direction.

4. **Dynamics Modeling and Control:**

   Inverse Kinematics and Dynamics: Machine learning algorithms have the capacity to learn the inverse kinematics and dynamics of robots, allowing for highly accurate control of manipulators and end-effectors.

   Model Predictive Control (MPC): To optimize control inputs across a finite time horizon and ensure stability and performance, MPC uses predictions from learnt models.

5. **Adaptive Control:**

   Systems with Adaptive Behavior: AI-based adaptive control modifies control parameters in response to changes in the external environment or the dynamics of the system, improving robustness and adaptability.

6. **Coordinated Multi-Robot Motion:**

   Swarm Robotics: ML-based techniques allow swarms of robots to efficiently coordinate their movements and behavior in order to accomplish shared goals.

7. **Human-Robot Interaction:**

   Safe and Intuitive Collaboration: ML and AI enable robots to anticipate and react to human motion and intentions, allowing for safe and natural interaction.

8. **Learning from Demonstrations:**

   Robots can learn motion patterns from human demonstrations through imitation learning, which helps them become more adept at carrying out tasks with movements that resemble those of a human.

9. **Real-time Control:**

   Neural network controllers, for example, are ML-driven control algorithms that can quickly adjust to unforeseen environmental changes.

10. **Control in Uncertain Environments:**

    Stochastic Control: AI approaches take into account environmental unpredictability to ensure that robots make decisions that strike a balance between performance and risk.

11. **Transfer Learning and Generalization:**

    Transferable Skills: Robots can pick up reusable control techniques in one setting and use them in other contexts, which speeds up learning in other settings.

Robots are now capable of navigating complicated surroundings, interacting safely with people, and maximizing their movements to complete a variety of tasks thanks to the integration of ML and AI into motion planning and control. Real-time computation problems, safety concerns, and adaptability to changing settings are still being researched and developed, nevertheless.

8.2.3 Manipulation and Grasping

Robots' ability to interact with and manipulate items in their environment is one of the most important features of robotics and automation. Robotic manipulation and gripping abilities have significantly improved because to machine learning (ML) and artificial intelligence (AI) technology. ML and AI are used in this situation as follows:

1. **Grasping Strategy:**

   Grasp Prediction: Machine learning (ML) models examine the forms and properties of the objects to forecast the most efficient gripping techniques, enhancing the success rate of choosing and placement activities.

   Gaining Knowledge from Experience: Robots can pick up effective grasping skills through trial and error, modifying their tactics in response to the results.

2. **Force and Tactile Sensing:**

   Tactile Feedback: By interpreting data from tactile sensors on robotic fingers, ML algorithms allow robots to sense the force and pressure used during grasping and manipulation.

   Learning Object Qualities: AI approaches use force feedback and tactile feedback to infer object qualities such as weight, texture, and stiffness, assisting in the development of more effective manipulation techniques.

3. **Dexterous Manipulation:**

   Finger Coordination: Robots can coordinate the movement of individual fingers to carry out sophisticated manipulation tasks thanks to ML-based control algorithms.

   Robots may learn sophisticated manipulation sequences, such as twisting, flipping, and rotating things, using AI approaches.

4. **In-hand Manipulation:**

   Robots with adaptive grasping may reposition or realign an object as necessary by adjusting their grasp on it while it is being handled.

   Robots can manipulate items that are subject to dynamic changes in shape or attributes thanks to AI-driven algorithms.

5. **Soft and Deformable Objects:**

   Grasping Soft Objects: Machine learning and artificial intelligence (AI) approaches help robots grasp and manipulate objects with variable and flexible shapes, like cloth or food.

6. **Tool Use and Tool Manipulation:**

   Robots may learn to use tools effectively for a variety of tasks thanks to ML algorithms, which increases their capabilities.

   Tool Interaction: By taking into account the dynamics and physics involved, AI enables robots to comprehend how to operate objects using tools.

7. **Human-Robot Collaboration:**

   Safe Interaction: By taking into account human intentions and safety, ML and AI enable robots to work with humans on activities involving shared manipulation.

8. **Learning from Human Demonstrations:**

   Imitation Learning: Using ML for skill transfer, robots can learn manipulation techniques by studying and mimicking human movements.

9. **Learning Object Affordances:**

   Affordance Detection: Using AI approaches, robots can identify the innate properties and affordances of objects, which inform manipulation tactics.

10. **Generalization Across Objects:**

    Transfer Learning: To increase flexibility, robots can apply the grasping and manipulating abilities they have mastered with one set of items to new, comparable objects.

11. **Real-time Feedback and Adjustment:**

    Sensor Integration: ML-driven systems analyze real-time sensor data to modify manipulation techniques and grab forces in response to object behavior.

Robots can handle a variety of objects, adapt to changing situations, and work securely with humans when ML and AI are integrated into manipulation and grasping activities. However, there is still a need for more study and development in the areas of dealing with complicated object geometries, managing uncertainties, and assuring dependable and precise grabbing.

8.2.4 Collaborative Robotics

A specialized area of robotics called "Collaborative Robotics," or "Cobots," focuses on creating robots that can productively and securely work alongside people in shared workspaces. Instead of replacing human workers, cobots are made to collaborate, help, and improve human tasks. Collaborative robotics is made feasible and efficient by machine learning (ML) and artificial intelligence (AI).

Here's how Cobots benefit from ML and AI applications:

1. **Safety and Human-Robot Interaction:**

   Cobots may dynamically adapt their behavior in order to recognise and anticipate human movements, intentions, and gestures. This helps to ensure safe collaboration.

   Cobots can keep a safe distance from people, avoid collisions, and react to unforeseen changes in the environment thanks to AI algorithms.

2. **Force and Tactile Feedback:**

   Cobots with tactile and force sensors can employ machine learning (ML) to understand and react to human touch and pressure, enhancing interactions.

   AI-driven systems analyze force data to decide how much force is necessary during physical encounters in order to avoid harm.

3. **Task Allocation and Cooperation:**

   ML algorithms support efficient job distribution between humans and Cobots by taking into account each entity's advantages and disadvantages.

   AI-based coordination techniques provide effective collaboration and allow Cobots to modify their behavior in response to the advancement of the human.

4. **Learning from Demonstration:**

   Cobots can mimic and learn from human behaviors to execute tasks, then use ML to adapt the learnt skills to other settings.

5. **Natural Language Processing (NLP):**

   Cobots can engage with people using natural language thanks to NLP techniques, which improve task performance and communication.

6. **Adaptive Behavior:**

   ML-based Cobots are capable of learning from their experiences and changing their behavior over time to more closely match human needs and expectations.

7. **Shared Control:**

   AI makes it possible for human operators and Cobots to share control, allowing people to direct the robot's operations while keeping control over important variables.

8. **Skill Transfer and Reprogramming:**
   By utilizing ML to transfer skills from one task to another, cobots may learn new tasks quickly and with less reprogramming work.
9. **Context Awareness:**
   Cobots can adapt properly to changes in the environment or work requirements thanks to ML and AI that enable them comprehend the context of the task.
10. **Gesture and Pose Recognition:**
    Cobots can comprehend and react to human directions and cues because AI systems scan visual data to recognize human gestures and poses.
11. **User Experience and Adaptation:**
    Cobots with AI capabilities may personalize support and change their behavior in response to user preferences, improving the user experience.

By bridging the gap between automation and human labor, cobots enable greater productivity, safer operations, and the ability to tackle jobs that call for human skill and judgment. While ML and AI have many advantages for collaborative robots, there are still a number of issues that require further research and development, including building intuitive human-robot interfaces, handling unpredictable human behavior, and guaranteeing effective safety systems.

8.2.5 Autonomous Vehicles Applied In Robotics and Automation

Robotics and automation are widely used in autonomous vehicles, often known as self-driving cars or driverless cars, which have the potential to revolutionize transportation systems. Artificial intelligence (AI) and machine learning (ML) are crucial for the functioning of autonomous cars. How they are used in this situation is as follows:

1. **Perception and Sensor Fusion:**
   To accurately understand the environment around the car, ML algorithms process data from a variety of sensors, including LiDAR, cameras, radar, and ultrasonic sensors.
   Sensor fusion creates a thorough and accurate depiction of the environment by combining input from several sources, increasing awareness.
2. **Object Detection and Recognition:**
   Convolutional neural networks (CNNs), a type of deep learning model, are used in artificial intelligence (AI) to detect and identify things including pedestrians, automobiles, signs, and road markings.
3. **Semantic Segmentation:**
   Each pixel in the camera photos is classified using ML-driven semantic segmentation, which enables the car to distinguish between various objects and road aspects.
4. **Localization and Mapping:**
   Autonomous cars use SLAM (simultaneous localization and mapping) algorithms to concurrently map their surroundings accurately and determine their own location inside those maps.
5. **Path Planning and Trajectory Generation:**
   AI-powered algorithms design the best routes and trajectories for the car while taking into account variables like traffic, weather, and road conditions.

ML approaches can adapt human driving behaviors to autonomous control, resulting in smoother and safer manoeuvres.

6. **Decision-Making:**

   In order to make judgments in real-time, such as changing lanes, merging, and yielding at crossings, AI algorithms analyze data from sensors, maps, and traffic conditions.

7. **Behavior Prediction:**

   ML models forecast other drivers' behavior, including that of pedestrians and other cars, allowing the autonomous vehicle to anticipate their movements and react appropriately.

8. **Traffic Prediction and Navigation:**

   AI-driven systems use historical and real-time traffic data analysis to optimize routes, avoiding traffic, and choosing the quickest paths.

9. **Human Interaction and Communication:**

   AI is used by autonomous vehicles to convey their intentions to other road users including pedestrians via signs, lighting, displays, and other devices.

10. **Sensor Calibration and Maintenance:**

    ML supports long-term calibration and accuracy maintenance of the vehicle's sensors, assuring reliable performance.

11. **Edge Computing and Real-time Processing:**

    ML models can be tailored for edge computing, making it possible to process sensor data in real-time without having to rely substantially on cloud resources.

12. **Adaptive Learning:**

    Through methods like reinforcement learning, autonomous cars can continuously learn from their experiences and enhance their performance over time.

Autonomous vehicle applications of ML and AI promise more practical, effective, and secure transportation options. Nevertheless, research and development are still being done in areas such assuring robustness against unforeseen circumstances, legal considerations, and resolving ethical issues associated to decision-making.

8.2.6 Predictive Maintenance

In order to predict and stop equipment breakdowns before they happen, predictive maintenance is a crucial application of robotics and automation that makes use of data analytics, machine learning (ML), and artificial intelligence (AI). This strategy aids in maintenance schedule optimization, downtime reduction, and improved industrial process efficiency.

Here's how predictive maintenance is applied in robotics and automation:

1. **Data Collection and Monitoring:**

   Sensors and monitoring systems gather data from robotic systems, machines, and equipment, including temperature, vibration, pressure, and other relevant parameters.

   Data is collected in real-time and transmitted to a central system for analysis.

2. **Data Preprocessing:**

   Preprocessing is done on the collected data to get rid of noise, outliers, and consistency issues, which ensures the correctness of the analysis that follows.

3. **Feature Extraction:**
   From the data, pertinent aspects are retrieved that offer perceptions into the condition and functionality of the machinery.
   Relevant features can be automatically selected and extracted using ML approaches.
4. **Anomaly Detection:**
   These models detect departures from the expected behavior, identifying probable abnormalities or symptoms of deterioration.
   ML models are trained on past data to learn the typical patterns and behaviors of equipment.
5. **Fault Prediction:**
   ML algorithms forecast when a piece of equipment is likely to fail in the future based on the abnormalities found.
   Different time horizons can be used to make predictions, which enables maintenance teams to make appropriate plans.
6. **Health Assessment:**
   A quantitative evaluation of the equipment's health is provided by ML-based systems, and this evaluation aids in resource allocation and maintenance task prioritization.
7. **Remaining Useful Life (RUL) Estimation:**
   ML models determine how long a piece of equipment can work without needing maintenance by estimating its remaining useful life.
8. **Maintenance Scheduling:**
   Predictive maintenance systems provide optimized maintenance schedules, making sure that maintenance tasks are carried out at the most advantageous and economical times.
9. **Reduced Downtime:**
   Predictive maintenance reduces unscheduled downtime by resolving possible problems before they become failures, hence improving overall operational efficiency.
10. **Cost Savings:**
    Predictive maintenance lowers the expenses of unplanned maintenance, production halts, and emergency repairs.
11. **Continuous Learning:**
    Machine learning algorithms may modify their predictive models over time and continuously learn from fresh data, increasing accuracy.
12. **Integration with Automation:**
    Automated reactions, such as scheduling repair chores or changing production procedures, can be triggered by predictive maintenance systems.

In sectors like manufacturing, energy, and transportation where downtime can result in considerable losses, predictive maintenance is especially beneficial. Its use in robotics and automation ensures the efficient operation of machinery and robotic systems while maximizing resource management and budgeting.

8.2.7 Industrial Automation to automate various Tasks and Processes

Automation of numerous jobs and processes in industrial settings is referred to as industrial automation. Advanced technologies like robots, control systems, and information technologies are used in this context. It tries to increase manufacturing and other industrial processes' effectiveness, productivity quality, and safety. Implementing industrial automation requires both machine learning (ML) and artificial intelligence (AI).

Here's how they are applied in this context:

1. **Manufacturing Process Optimization**: ML algorithms examine production data to find inefficiencies, bottlenecks, and opportunities for manufacturing process optimization.

   AI-driven algorithms continuously analyze workflows and modify settings to guarantee dependable, high-quality results.

2. **Quality Control and Inspection:** AI models may learn from past inspection data to improve accuracy and adapt to changing product standards. ML-powered visual inspection systems detect flaws, anomalies, and quality issues in real-time using image analysis and pattern recognition.
3. **Predictive Maintenance:** ML and AI analyze sensor data to predict equipment failures and maintenance requirements, reducing expensive downtime and prolonging equipment lifespan.
4. **Supply Chain Management:** AI algorithms optimize logistics, demand forecasting, and inventory management, cutting waste and boosting supply chain effectiveness.
5. **Automated Robotics:** Assembling, picking, and placing items, as well as material handling, are tasks that robots with AI capabilities can complete with great accuracy and effectiveness.
6. **Process Monitoring and Control:** To maintain strict control over numerous industrial processes and guarantee constant product quality, ML models examine real-time sensor data.
7. **Energy Efficiency:** By modifying procedures, tools, and resources in accordance with demand and cost, ML algorithms optimize energy usage in industrial plants.
8. **Personalization and Customization:** AI-driven systems enable mass customization by modifying production procedures to produce goods that are specifically catered to the tastes of each customer.
5. **Human-Machine Collaboration:** AI-equipped Cobots (collaborative robots) assist human workers while boosting efficiency and safety.
6. **Decision-Making and Data Analytics:** To deliver insights for enhanced decision-making in real-time, ML and AI systems evaluate massive datasets from numerous sources.
7. **Digital Twin Technology:** AI-powered digital twins replicate actual industrial systems and processes to allow for experimentation and optimization without interfering with ongoing operations.
8. **Cyber Security:** AI algorithms keep an eye on network activity and spot any security risks, protecting industrial systems from online attacks.
9. **Process Adaptability:** ML and AI make it possible for systems to change in response to a variety of variables, including alterations in the needs for production or raw material supply.
10. **Remote Monitoring and Control:** Operators may monitor and manage industrial processes remotely with the help of AI-driven remote monitoring systems, which boosts operational effectiveness.

Industrial automation enabled by ML and AI improves output, quality, and security while lowering costs and having a less negative impact on the environment. However, obstacles like system integration, workforce development, and data security still need to be taken into account for a successful deployment.

3.2.8 Natural Language Processing (NLP) for Human-Robot Communication

The goal of Natural Language Processing (NLP), a subfield of Artificial Intelligence (AI), is to make it possible for computers to comprehend, analyze, and produce human language. In robotics and automation, NLP is essential for enabling successful communication between people and robots. Using NLP for human-robot communication looks like this:

1. **Voice Commands and Speech Recognition:**

   NLP makes it possible for robots to comprehend verbal instructions from humans and act accordingly, enabling intuitive control and interaction.

   Speech recognition algorithms translate spoken words into text that the robot can analyse and interpret.

2. **Text-based Interaction:**

   Robots with NLP capabilities can comprehend and produce text messages, enabling communication through text-based displays or interfaces.

3. **Natural Language Understanding (NLU):**

   Because NLP models can interpret the context and intent of human language, they enable robots to comprehend the meaning of requests, questions, and orders.

4. **Contextual Understanding:**

   Robots are able to maintain coherent and meaningful relationships across several exchanges thanks to NLP algorithms that take the conversation's context into account.

5. **Sentiment Analysis:**

   NLP-based sentiment analysis enables robots to understand the emotional undertone of a discussion and react to it accordingly.

6. **Language Translation:**

   Language barriers can be overcome by robots communicating with people who speak different languages thanks to NLP-driven translation systems.

7. **Information Retrieval:**

   In response to user inquiries, NLP algorithms can obtain data from databases or the internet, giving consumers access to pertinent information.

8. **Human-Robot Collaboration:**

   NLP enables clear instructions and clarifications for robots working alongside humans on complex jobs.

9. **Personalization and Adaptation:**

   Robots with NLP capabilities can modify their responses and actions based on the preferences and past actions of specific users.

10. **Learning and Knowledge Acquisition:**

    Computer systems that use natural language processing (NLP) can pick up new language skills from interacting with people.

11. **User Assistance:**

    Natural language explanations, support, and direction from robots can improve user experience.

12. **Safety and Clarification:**

    Robots may ask people for clarification when given unclear instructions thanks to NLP, which lowers the likelihood of mistakes.

13. **Remote Operation and Control:**

    Natural language processing (NLP) makes it possible for remote operators to interact with and command robots.

14. **Multimodal Interaction:**

    NLP can be used in conjunction with other modalities, such as gestures and visual signals, to provide interactions that are more thorough and human-like.

Robots' capacity to comprehend and respond to human orders is improved by the incorporation of NLP into robotics and automation, allowing for more natural and user-friendly interactions. However, there is still room for more research and development in areas like resolving ambiguity, identifying intricate linguistic patterns, and accurately comprehending user intent.

8.2.9 Simulation and Training

Using virtual environments and educational resources, simulation and training are crucial elements of robotics and automation that improve the design, testing, and use of robotic systems. The accuracy of simulations and the efficiency of training are significantly improved by machine learning (ML) and artificial intelligence (AI). The use of simulation and training in robotics and automation is as follows:

1. **Virtual Prototyping:**

   Engineers and developers can build digital prototypes of robotic systems in simulation settings to test their performance and functionality before putting them into actual use.

   AI can improve simulation realism by simulating intricate physical dynamics and interactions.

2. **Algorithm Testing and Validation:**

   Prior to being applied to actual robots, ML and AI algorithms can be tested and validated in simulated situations, verifying their efficacy and security.

3. **Training AI Models:**

   Simulated settings offer plenty of data for training reinforcement learning agents, a type of AI model, to learn new skills ahead of time.

4. **Robot Behavior Testing:**

   Robots can be taught and tested in simulations to learn and perfect their behaviors, enhancing their ability to adapt and perform in real-world situations.

5. **Complex Scenario Testing:**

   In order to assess robot responses and develop tactics, simulations enable the production of complicated scenarios, such as uncommon and hazardous situations.

6. **Sensor and Perception Testing:**

   Robots can more precisely understand their surroundings thanks to the validation and optimization provided by simulated sensor data for perception algorithms.

7. **Adversarial Testing:**

   Robotic systems' resistance to threats and attacks is put to the test in simulated hostile situations.

8. **Optimizing Control Strategies:**
   Robots can learn and adapt to various control strategies thanks to AI algorithms that optimize control strategies in simulation settings.
9. **Human-Robot Interaction Training:**
   Robots can learn to interact with humans safely and effectively by practicing various gestures, speech patterns, and behaviors through simulation-based training.
10. **Transfer Learning:**
    ML approaches make it possible to fine-tune models developed in simulations in real-world situations, which reduces the quantity of data required for real-world training.
11. **Cost Reduction:**
    Why Simulation makes actual hardware and experimentation less expensive, allowing for iterative development and improvement without incurring exorbitant costs.
12. **Fast Iteration and Prototyping:**
    Simulated testing expedites the development process by enabling developers to swiftly iterate via design modifications and enhancements.
13. **Education and Skill Development:**
    Students and employees are trained in operating and programming robotic systems using simulators, improving their skill sets.
14. **Scenario Exploration:**
    By enabling the examination of various scenarios that would not be simple or safe to recreate in the actual world, simulations help robots gain new experiences.

In robotics and automation, simulations and training are essential tools that help researchers, engineers, and operators develop, test, and implement cutting-edge systems effectively and safely. Although simulations can offer realistic insights, it is still difficult to accurately map simulated and real-world behaviors.

8.2.10 Adaptive Systems

Robotics and automation experts use the term "adaptive systems" to describe machines that may change their behavior or operating parameters in response to environmental changes, job demands, or system circumstances. To constantly adapt and improve their performance, these systems make use of feedback mechanisms, machine learning (ML), and artificial intelligence (AI) techniques.

Here's how adaptive systems are applied in robotics and automation:

1. **Sensor Calibration and Compensation:** Adaptive systems can calibrate sensors automatically to take into account changes and drift, ensuring precise perception and data collecting.
2. **Dynamic Path Planning and Navigation:** Robots may modify their travel routes in real-time in response to shifting impediments, traffic, and environmental factors thanks to ML algorithms.
3. **Control Parameters Optimization:** Stability and optimal performance are guaranteed by adaptive control algorithms, which modify control parameters to take system dynamics fluctuations into account.
4. **Machine Learning Model Adaptation:** In order to retain accuracy in prediction and decision-making tasks, machine learning models can adapt to changes in data distribution or task requirements.

5. **Fault Detection and Recovery:** To ensure continued functioning, adaptive systems can identify problems or anomalies in real-time and switch to alternative modes or methods.
6. **Energy Efficiency Optimization:** By modifying operational parameters, such as power levels and duty cycles, in response to demand and environmental factors, adaptive algorithms reduce energy use.
7. **Process Optimization in Manufacturing:** Adaptive systems improve manufacturing procedures in industrial environments by modifying parameters to preserve quality and effectiveness.
8. **Human-Robot Collaboration:** Intelligent robots that can adapt to human preferences can collaborate with people more successfully and safely.
9. **Personalized Assistance:** Adaptive systems customize recommendations and help based on each user's preferences and needs in a variety of situations.
10. **Environmental Adaptation:** Robots with adaptive systems are able to change their behavior to adapt to various conditions, such as indoor versus outdoor settings.
11. **Context-Awareness:** Robots can respond to changing circumstances by using adaptive systems, which use contextual information to modify their behaviors.
12. **Noise and Interference Handling:** Adaptive algorithms improve accuracy and reliability by removing noise and interference from sensor data.
13. **Predictive Maintenance Adaptation:** Based on real-time data, adaptive predictive maintenance systems modify maintenance schedules to ensure prompt interventions.
14. **Learning and Self-Improvement:** Learning algorithms can be incorporated into adaptive systems to help them develop over time and become more accurate as they gain experience.

Robotic and automated systems benefit greatly from the flexibility, efficiency, and responsiveness that adaptive systems provide in response to dynamic situations and shifting requirements. Even while these systems have advantages, they also provide a number of obstacles, such as managing trade-offs between stability and adaptation, avoiding over-fitting, and guaranteeing effective handling of ambiguous or unexpected situations.

8.3 SUPERVISED AND UNSUPERVISED LEARNING

The two primary subcategories of Machine Learning (ML) approaches, supervised and unsupervised learning, have several applications in robotics and automation. They allow automated systems and robotics to learn from data and make wise decisions.

Here are some examples of how robotics and automation use supervised and unsupervised learning:

Supervised Learning: In supervised learning, a model is trained on labeled data with explicit input-output correlations. Based on these labeled samples, the model learns to generate predictions or classifications. In automation and robotics:

1. **Object Recognition:** Using supervised learning, models that can identify things from sensor data, such as pictures or point clouds taken by cameras or LiDAR sensors, are trained.
2. **Gesture Recognition:** Through supervised learning, robots may learn to identify and comprehend human gestures, enabling real human-robot interaction.
3. **Localization:** By connecting sensor data with ground truth positions, supervised learning helps train models for precise localization.

4. **Path Planning:** Robots can learn from labeled data to comprehend efficient and safe routes, avoiding hazards and dynamic changes.
5. **Control Systems:** Based on intended outcomes, supervised learning can be used to learn control rules for particular tasks, including robot arm movements.
6. **Defect Detection:** By comparing sensor data with known good samples, models developed with labeled data can locate flaws in manufactured goods.

Unsupervised Learning: Finding patterns or structures in data without labeled outputs is known as unsupervised learning. Exploration, grouping, and comprehension of underlying data relationships can all be accomplished using this method. In automation and robotics:

1. **Anomaly Detection:** Unsupervised learning aids in the identification of anomalies in sensor data that may point to malfunctions or irregularities in the apparatus.
2. **Cluster Analysis:** Robots can organize or sort activities by putting comparable items or data points together based on their shared traits.
3. **Reducing Dimensionality:** Unsupervised approaches make data representation simpler, which is helpful for quick processing and effective visualization.
4. **Reinforcement Learning Initialization:** Before fine-tuning models with reinforcement learning for particular tasks, unsupervised learning can be used to train them first.
5. **Exploratory Data Analysis:** Unsupervised learning facilitates the comprehension of intricate data distributions and correlations in sensor data.
6. **Behavior Profiling:** Through unsupervised learning, robots can discover the usual behaviours of various objects or entities, facilitating adaptive interaction.
7. **Adaptive Clustering:** By updating clusters and adjusting to new data patterns, robots may adapt to changing environments.

Systems may learn and adapt to a variety of tasks and settings with the help of both supervised and unsupervised learning techniques, which are useful tools in robotics and automation. The effectiveness of robotic and automated systems can be improved using either one or both strategies, depending on the application.

8.4 NEURAL NETWORKS AND DEEP LEARNING

Robotics and automation have made substantial use of the potent Machine Learning (ML) subsets of neural networks and deep learning. Through the use of these methods, robots and automated systems are able to learn patterns and representations from data and carry out difficult tasks. Here are some examples of how deep learning and neural networks are used in robotics and automation:

1. **Object Detection and Recognition:** Robots can recognize and interact with items in their surroundings thanks to Convolution Neural Networks (CNNs), which are excellent at identifying objects and patterns in images or sensor data.
2. **Semantic Segmentation:** Robots can comprehend scene layout and distinguish between various items and regions because to CNNs' pixel-wise classification.
3. **LiDAR and Sensor Data Processing:** Models for deep learning interpret information from LiDAR, radar, and other sensors to make precise maps, recognize objects, and help with navigation.

4. **Gesture and Pose Recognition:** Human-robot interaction is improved by neural networks' ability to recognize and comprehend human gestures, stances, and motions.
5. **Natural Language Processing (NLP):** Transformers and Recurrent Neural Networks (RNNs) process and produce natural language, enabling robots to comprehend and carry out spoken commands.
6. **Reinforcement Learning (RL):** When RL and neural networks are used together, robots can learn the best course of action through trial and error, enabling them to learn complicated tasks.
7. **Robot Control:** Robotic actuator control strategies can be learned using deep neural networks, allowing for precise and effective manipulation.
8. **Simulation and Training:** Before deploying robots in the real world, neural networks help simulate complicated physical interactions and train the robots in virtual environments.
9. **Autonomous Navigation:** In autonomous vehicles or drones, deep learning models evaluate sensor data to make judgments in real-time, enabling safe navigation.
10. **Human-Robot Collaboration:** Robots that use neural networks are better able to comprehend and anticipate human behavior, improving teamwork and assuring safety.
11. **Predictive Maintenance:** To predict equipment breakdowns, deep learning models analyze sensor data, enabling prompt maintenance actions.
12. **Adaptive Systems:** Neural networks enable robot behavior adaptation in response to changing circumstances, enabling robots to function in complex contexts.
13. **Quality Control and Inspection:** Defects, abnormalities, and quality problems in manufacturing processes are detected using deep learning.
14. **Exploration and Mapping:** Robots can explore and navigate uncharted places with the use of deep learning, which facilitates the analysis and interpretation of mapping data.

Robotics and automation have undergone a revolution thanks to neural networks and deep learning, which have made it possible for machines to tackle difficult jobs, make informed decisions, and adapt to a variety of situations. However, issues like dataset size, generalization, and interpretability continue to be crucial research areas that need to be addressed.

8.5 REINFORCEMENT LEARNING

A machine learning paradigm known as reinforcement learning (RL) teaches an agent how to interact with its environment in order to maximize cumulative rewards. Robotics and automation have made extensive use of RL to provide machines and automated systems the ability to learn difficult tasks by doing them incorrectly.

Here's how RL is applied in this context:

1. **Robotic Control:** RL is used to train robots to control their actuators, such as arms or wheels, to carry out tasks like picking up and placing objects or navigating through challenging surroundings.
2. **Grabbing and manipulating:** RL teaches robots the best manipulation and gripping methods for objects of various sizes, shapes, and characteristics.
3. **Path Planning and Navigation:** Robots can learn effective pathways and collision avoidance techniques in challenging situations thanks to RL-based navigation systems.

4. **Autonomous Vehicles:** Real-time learning (RL) is used to teach self-driving cars to make judgments based on current sensor data, traffic conditions, and safety considerations.
5. **Control of Drones:** RL helps drones learn to fly, navigate, and adjust to wind conditions, avoid obstacles, and optimize flight trajectories.
6. **Human-Robot Interaction:** By taking into account human intents and behavior, RL teaches robots how to interact with people safely and successfully.
7. **Skill Acquisition and Transfer:** RL enables robots to acquire a variety of skills in a simulation and transfer them to the real environment, speeding up learning and adaption.
8. **Resource Management:** RL makes manufacturing and industrial operations more efficient by allocating resources including time, energy, and materials.
9. **Logistics and storage:** RL directs robots to discover the best routes for managing storage and material handling jobs.
10. **Multi-Robot Collaboration:** RL makes it easier for several robots to work together and coordinate in order to accomplish common goals.
11. **Adaptive Systems:** RL-driven adaptive systems let robots learn and modify their behavior over time in response to alterations in the environment and task requirements.
12. **Task Prioritization:** RL teaches robots how to order tasks according to their significance and the resources at hand.
13. **Risk-Aware Decision-Making:** RL agents develop the ability to strike a balance between reward and risk, choosing actions that maximize long-term results while taking prospective dangers into account.
14. **Real-Time Learning and Adaptation:** Over iterations, RL agents can adapt to changing circumstances and learn from new experiences in real-time.

When there are no predetermined solutions or when the environment is complicated and dynamic, reinforcement learning is especially beneficial. The necessity for efficient exploration methodologies and issues with sample efficiency, however, are still active areas of research.

8.6 FOR TASKS LIKE OBJECT RECOGNITION AND PATH PLANNING

Let's examine more closely how supervised and unsupervised learning are especially used for tasks in robotics and automation like object detection and path planning:

Object Identification

1. **Supervised learning for object identification**

   Data collection: Engineers compile a dataset of pictures with labels identifying the things shown in the pictures.

   Model Training: A Convolutional Neural Network (CNN) is trained using the labeled dataset to discover the patterns and characteristics of various objects.

   Inference: When a new image is presented, the trained CNN can predict the presence of objects by classifying them into the learned categories.

2. **Unsupervised Object Recognition Learning:**

   Clustering: Unsupervised learning methods such as k-means can cluster related image attributes, assisting in the grouping of pictures of related objects.

Dimensionality Reduction: Principal Component Analysis (PCA) and other techniques can decrease the dimensionality of picture features while retaining crucial data, making it simpler to compare photos.

Path Planning

1. **Supervised Learning for Path Planning:**

   Data collection: Engineers offer labeled training data that comprises robot states, environments, and associated best routes.

   Model Training: With the help of this data, a supervised learning model is trained to associate the right paths with the states and situations of the robot and its surroundings.

   Inference: The robot's new pathways can be generated by the trained model using information about its environment and present condition.

2. **Unsupervised Learning for Path Planning:**

   Density Estimation: Unsupervised algorithms can calculate the density of prospective paths or obstacles in the surrounding area, which helps in identifying potential path regions.

   Self-Organization Maps (SOM): SOMs can map the environment and learn its structure, allowing robots to navigate across areas where they have previously observed surroundings with a similar structure.

3. **Hybrid Approaches:**

   Self-Supervised Learning for Object Recognition and Path Planning: When learning to recognize objects, the robot improves its comprehension of object structures by learning to forecast missing pieces of an image. The robot can predict its future states during path planning by using its current state and environment, allowing it to foresee difficulties and make plans accordingly.

   Transfer Learning: Models that have already been pre-trained for image identification on huge datasets can be tailored for particular item detection tasks. Similar to how custom path planning tasks can be added to pre-trained models for varied navigation conditions.

   Reinforcement Learning: By interacting with the environment and receiving incentives for wise decisions, reinforcement learning can assist robots in learning the best paths to follow.

The decision between supervised and unsupervised learning in both object identification and path planning depends on the availability of labeled data, the difficulty of the job, and the desired level of autonomy. In robotics and automation, these methods are frequently combined with other reinforcement learning or self-supervised learning techniques to get reliable results.

8.7 RL FOR TRAINING ROBOTS THROUGH TRIAL AND ERROR

A potent method for teaching robots to carry out tasks and make judgments in challenging, dynamic contexts is reinforcement learning (RL). By interacting with their surroundings and getting feedback in the form of rewards or penalties, RL enables robots to learn the best strategy.

Here's how RL is applied to training robots:

1. **Agent-Environment Interaction:** The robot engages with its surroundings as an agent. The physical space, objects, sensors, and other things that the robot can feel and influence are all included in the environment.

2. **State Representation:** The robot monitors its surroundings and records this data in a representation of its condition. This illustration effectively conveys important information about the current circumstance.
3. **Action Selection:** The robot chooses an action from a list of potential options based on the current state. Interactions and environmental changes result from these acts.
4. **Reward Signal:** The environment sends the robot a reward signal once it completes an action. The reward shows how effective the action was in attaining the goals of the assignment.
5. **Learning Policy:** The robot learns a policy that maximizes the cumulative reward it receives over time—a mapping from states to actions. Finding the policy that produces the maximum expected return is the objective.
6. **Exploration and Exploitation:** The robot strikes a balance between exploitation—using known activities that produce bigger rewards—and exploration—trying new actions to find better methods.
7. **Value Function:** The robot keeps track of a value function that calculates the anticipated total reward for each state. This aids in directing decision-making and the choice of action.
8. **Learning Algorithms:** RL algorithms, including Proximal Policy Optimization (PPO), Deep Q Networks (DQN), and Q-learning, direct the robot in discovering the best course of action based on incentives and experiences.
9. **Iterative Learning:** The robot performs tasks, earns incentives, modifies its policy in accordance with the learning algorithm, and repeats the process until its performance is improved.

Applications of Reinforcement Learning in Robotics

1. **Robotic Manipulation:** RL is used to teach robots how to manipulate items and carry out operations including picking, placing, and assembly.
2. **Autonomous Navigation:** RL teaches robots efficient and safe ways to navigate, dodging hazards and arriving at destinations.
3. **Dexterous Grasping:** Robots may acquire sophisticated and adaptive grasping techniques for items of various sizes and forms thanks to RL.
4. **Aerial and Underwater Robotics:** RL helps underwater robots and drones learn to swim, fly, and carry out tasks in difficult situations.
5. **Humanoid and Legged Robots:** RL teaches walking, running, and dynamic motions to humanoids and legged robots.
6. **Manufacturing and Industrial Automation:** In manufacturing environments, RL optimizes processes, control parameters, and resource management.

Robots may learn from their own experiences thanks to reinforcement learning, making it a useful method for instructing robots in tasks that are challenging to specify clearly. However, RL for robots is still undergoing research and development in areas like sample efficiency, safety during exploration, and generalization across many contexts.

8.8 ML FOR ROBOT PERCEPTION

Robot perception is greatly improved by machine learning, which enables robots to interpret and comprehend their environment from sensor data. Several important uses of machine learning for robot perception are listed below:

1. **Robots interaction**
   (*a*) Robots can interact with and manipulate items by using Convolutional Neural Networks (CNNs) and other deep learning models to recognize and detect objects in images and sensor data.
   (*b*) **Semantic segmentation:** Machine learning (ML) models assist robots in segmenting images into useful regions, differentiating between various object classes, or locating regions of interest.
   (*c*) **Pose Estimation:** Using ML algorithms, robots can precisely grip and operate things by determining their 3D poses.
   (*d*) **Recognition of Gestures and Emotions:** Machine learning models are able to decipher human gestures and emotions from visual and sensor data, increasing human-robot interaction.
2. **Scene Understanding:** ML-powered systems process sensor data to understand the overall scene layout, identifying objects and their spatial relationships.
3. **Depth Estimation:** Machine learning techniques estimate depth information from images, which is crucial for understanding 3D environments.
4. **LiDAR Data Analysis:** ML algorithms process LiDAR point clouds to detect objects, identify obstacles, and map surroundings for navigation.
5. **Visual SLAM (Simultaneous Localization and Mapping):** Machine learning aids in building accurate maps of environments and localizing robots within them using visual data.
6. **Human Detection and Tracking:** ML models detect and track humans in real time, enabling robots to navigate safely in environments with human presence.
7. **Voice and Speech Recognition:** ML-driven systems allow robots to understand and respond to spoken commands and interact with humans using natural language.
8. **Sensor Fusion:** ML techniques combine data from multiple sensors (cameras, LiDAR, etc.) to create a more comprehensive and accurate perception of the environment.
9. **Anomaly Detection:** Machine learning models identify anomalies in sensor data, helping robots detect unexpected events or faults.
10. **Adaptive Perception:** ML-based perception systems can adapt to changing lighting conditions, viewpoints, and environmental factors.
11. **Transfer Learning:** Pre-trained models can be fine-tuned on specific robot perception tasks, leveraging existing knowledge to improve performance with less data.

Robots can now sense their environment with a level of sophistication that was previously difficult to accomplish thanks to machine learning's capacity to analyze enormous volumes of sensor data and understand patterns. These abilities are essential for autonomous navigation, teamwork between humans and robots, and many other robotics applications.

8.9 ML FOR CONTROL

Robots may learn and adjust their behaviors, actions, and responses depending on data and experiences thanks to the use of machine learning to many elements of control in robotics. The following are some essential uses of machine learning for robotics control:

1. **Robot Arm Control:** ML algorithms help robots learn precise control of their arms and end effectors, enabling them to perform tasks like pick and place with high accuracy.

2. **Inverse Kinematics:** Machine learning models learn to compute the joint angles necessary to achieve a desired end effector position, simplifying complex manipulation tasks.
3. **Reinforcement Learning for Control:** RL algorithms guide robots in learning optimal control policies through trial and error, making them capable of complex tasks without explicit programming.
4. **Motor Control and Actuator Tuning:** ML techniques optimize control parameters of actuators to achieve desired responses and movements in various scenarios.
5. **Adaptive Control:** Machine learning models can adapt control strategies in real-time based on changing conditions, improving stability and performance.
6. **Trajectory Planning:** ML algorithms generate optimal trajectories that balance efficiency, safety, and task objectives, allowing robots to move smoothly and avoid obstacles.
7. **Learning from Demonstration:** Robots learn control strategies by observing human demonstrations, enabling them to imitate and perform tasks based on human actions.
8. **Path Following and Navigation:** Machine learning aids in learning navigation policies that guide robots to follow specific paths while avoiding collisions.
9. **Aerial Drone Control:** ML-based controllers help drones maintain stable flight, perform acrobatics, and adapt to different wind and weather conditions.
10. **Legged Robot Locomotion:** Machine learning algorithms assist legged robots in learning and adapting gaits and movements for different terrains and obstacles.
11. **Autonomous Vehicles:** ML models enable autonomous vehicles to learn driving behaviors, lane-keeping, and decision-making based on real-world data.
12. **Robotic Swarms and Collective Behavior:** Machine learning algorithms help coordinate the actions of multiple robots to achieve collective objectives, such as formation flying or collaborative assembly.
13. **Feedback Control:** ML techniques enhance feedback control systems by learning to predict system responses and optimize control gains.
14. **Energy-Efficient Control:** Machine learning optimizes control strategies to minimize energy consumption, extending robot operation time.

By allowing robots to adjust, learn, and optimize their behaviors in response to various surroundings and tasks, machine learning improves control in robotics. Robots become increasingly adaptable and capable of handling complicated and dynamic situations in a variety of applications because to this flexibility and adaptation.

8.10 ML FOR DECISION-MAKING

Robots can make well-informed decisions and take appropriate action based on data-driven insights thanks to the widespread usage of machine learning in robotics. Several uses of machine learning for robotics decision-making are listed below:

1. **Path Planning and Navigation:** Machine learning helps robots decide optimal paths by considering factors like obstacle avoidance, traffic conditions, and energy efficiency.
2. **Resource Allocation:** Machine learning algorithms optimize the allocation of resources such as time, energy, and materials in manufacturing and logistics tasks.
3. **Reinforcement Learning (RL):** RL enables robots to learn decision-making policies through trial and error, adapting their actions to maximize cumulative rewards over time.

4. **Optimal Control:** Machine learning assists in learning control policies that optimize system behavior by considering factors like stability, safety, and efficiency.
5. **Risk Assessment and Management:** Machine learning models analyze data to assess risks and make decisions that minimize potential negative outcomes.
6. **Human-Robot Collaboration:** Machine learning algorithms enable robots to anticipate human intentions and make decisions that enhance collaborative tasks.
7. **Dynamic Task Assignment:** Machine learning optimizes the assignment of tasks to different robots based on their capabilities, current workload, and objectives.
8. **Supply Chain Management:** Machine learning optimizes decisions in supply chain logistics, including inventory management, order fulfillment, and demand prediction.
9. **Quality Control and Inspection:** Machine learning models analyze data from manufacturing processes to make real-time decisions about product quality and defect detection.
10. **Emergency Response:** Robots can make decisions during disaster response, such as navigating hazardous environments or assisting in search and rescue operations.
11. **Trading and Finance:** Machine learning algorithms are used for algorithmic trading, portfolio management, and risk assessment in financial markets.
12. **Autonomous Vehicles:** Machine learning models help autonomous vehicles make real-time decisions about braking, acceleration, lane changes, and avoiding obstacles.
13. **Energy Management:** Machine learning optimizes energy consumption by making decisions on the operation of robotic systems to minimize energy usage.
14. **Adaptive Systems:** Machine learning enables robots to adapt their decisions based on changing conditions, enhancing their responsiveness and performance.

Robots' ability to handle complicated and uncertain contexts and make wise decisions that result in better results is improved through machine learning-based decision-making. However, it is still difficult to guarantee moral judgment, openness, and interpretability of machine learning models in crucial applications.

Numerical Problems with Solutions

Problem 1: *Linear Regression*: Suppose you have a dataset with the following values:

X = [1, 2, 3, 4, 5] and Y = [3, 5, 7, 9, 11]

Perform linear regression to find the equation of the line (y = mx + b) that best fits this data.

Solution: To find the line of best fit, we need to calculate the slope (m) and the y-intercept (*b*) using the formula:

$$m = \Sigma((X_i - \bar{X}) * (Y_i - \bar{Y}))/\Sigma((X_i - \bar{X})^2) b = \bar{Y} - m * \bar{X}$$

where $\bar{X}$ is the mean of X values and $\bar{Y}$ is the mean of Y values.

$$\bar{X} = (1 + 2 + 3 + 4 + 5)/5 = 3 \quad \bar{Y} = (3 + 5 + 7 + 9 + 11)/5 = 7$$

Calculate m:

$$\begin{aligned} m &= ((1-3) * (3-7) + (2-3) * (5-7) + (3-3) * (7-7) \\ &\quad + (4-3) * (9-7) + (5-3) * (11-7))/((1-3)^2 + (2-3)^2 \\ &\quad + (3-3)^2 + (4-3)^2 + (5-3)^2) \\ &= 2/10 = 0.2 \end{aligned}$$

Calculate b:

$$b = 7 - 0.2 * 3 = 6.4$$

The equation of the line is y = 0.2x + 6.4.

Problem 2: ***Principal Component Analysis (PCA):*** **Given a dataset with the following values:**

$$\mathbf{X = [2, 4, 6, 8, 10]\ Y = [3, 6, 9, 12, 15]}$$

Perform PCA to reduce the dimensionality of the dataset to one principal component.

Solution: To perform PCA, follow these steps:

1. Calculate the mean of each feature ($\bar{X}$ and $\bar{Y}$).
2. Compute the covariance matrix of the features.
3. Calculate the eigenvectors and eigenvalues of the covariance matrix.
4. Choose the eigenvector corresponding to the largest eigenvalue as the principal component.

For simplicity, let's calculate the first two steps:

$$\bar{X} = (2 + 4 + 6 + 8 + 10)/5 = 6\ \bar{Y} = (3 + 6 + 9 + 12 + 15)/5 = 9$$

Compute the covariance matrix:

$$\begin{aligned} \text{Cov}(X, Y) &= \Sigma((X_i - \bar{X}) * (Y_i - \bar{Y}))/(n - 1) \\ &= ((2 - 6) * (3 - 9) + (4 - 6) * (6 - 9) + (6 - 6) * (9 - 9) \\ &\quad + (8 - 6) * (12 - 9) + (10 - 6) * (15 - 9))/4 \\ &= 25/4 = 6.25 \end{aligned}$$

Since this is a simplified example, we'll skip the eigenvector and eigenvalue calculations. The eigenvector corresponding to the largest eigenvalue will provide the principal component.

Problem 3: ***K-Means Clustering*****: Given the following data points in a 2D space:**

Points: (2, 3), (3, 5), (4, 8), (7, 9), (10, 11), (12, 12)

Perform K-Means clustering with k = 2, starting with initial centroids at (3, 5) and (10, 11). Calculate the final cluster centroids.

Solution: Let's perform two iterations of K-Means clustering using the provided initial centroids.

Iteration 1: Assign points to clusters based on the nearest initial centroids:

(2, 3) → Cluster 1 (nearest to (3, 5))

(3, 5) → Cluster 1 (already assigned to this cluster)

(4, 8) → Cluster 1 (nearest to (3, 5))

(7, 9) → Cluster 2 (nearest to (10, 11))

(10, 11) → Cluster 2 (already assigned to this cluster)

(12, 12) → Cluster 2 (nearest to (10, 11))

Update centroids:

New centroid for Cluster 1: (2 + 4)/2, (3 + 8)/2 = (3, 5.5)

New centroid for Cluster 2: (7 + 10 + 12)/3, (9 + 11 + 12)/3 = (9.67, 10.67)

Iteration 2: Assign points to clusters based on the updated centroids:

(2, 3) → Cluster 1 (nearest to (3, 5))

(3, 5) → Cluster 1 (already assigned to this cluster)

(4, 8) → Cluster 1 (nearest to (3, 5))

(7, 9) → Cluster 2 (nearest to (9.67, 10.67))

(10, 11) → Cluster 2 (already assigned to this cluster)

(12, 12) → Cluster 2 (nearest to (9.67, 10.67))

Update centroids:

New centroid for Cluster 1: (2 + 4)/2, (3 + 8)/2 = (3, 5.5)

New centroid for Cluster 2: (7 + 10 + 12)/3, (9 + 11 + 12)/3 = (9.67, 10.67)

The final cluster centroids are approximately (3, 5.5) and (9.67, 10.67).

Problem 4: *Decision Tree*: Suppose you have a dataset with the following attributes and labels:

Outlook: Sunny, Overcast, Rainy Temperature: Hot, Mild, Cool Humidity: High, Normal Windy: True, False Play Tennis: Yes, No

Build a decision tree using the ID_3 algorithm to predict whether to play tennis based on the given attributes.

Solution: To build the decision tree using the ID_3 algorithm, follow these steps:

1. Calculate the entropy of the target variable (Play Tennis).
2. Calculate the information gain for each attribute and select the one with the highest information gain as the root node.
3. Repeat the process for the child nodes, considering the remaining attributes.

Given that the dataset has two classes (Yes and No), the entropy formula is:

$$\text{Entropy}(S) = -p(\text{Yes}) * \log2(p(\text{Yes})) - p(\text{No}) * \log2(p(\text{No}))$$

where p(Yes) is the proportion of instances with "Yes" label, and p(No) is the proportion of instances with "No" label.

Calculate the entropy of the entire dataset:

Total instances: 14 (9 Yes, 5 No) p(Yes) = 9/14 p(No) = 5/14 Entropy(S) = -p(Yes) * log2(p(Yes)) – p(No) * log2(p(No)) ≈ -0.649 * log2(0.649) – 0.357 * log2(0.357) ≈ 0.940

Next, calculate the information gain for each attribute by considering the entropy of subsets after splitting based on that attribute. The information gain formula is:

$$\text{Info Gain}(A) = \text{Entropy}(S) - \Sigma((|Sv|/|S|) * \text{Entropy}(Sv))$$

where A is the attribute, Sv is a subset after splitting based on attribute A, and |Sv| is the number of instances in subset Sv.

Calculate information gain for each attribute and select the one with the highest information gain to be the root node of the decision tree.

Continue this process to build the entire decision tree. Due to space limitations, I can't provide the full tree here, but this is the general approach.

Problem 5: *Support Vector Machine (SVM)*: Given the following dataset with two classes (positive and negative):

Positive Class: (1, 3), (2, 5), (3, 8) Negative Class: (6, 5), (7, 9), (8, 7)

Train a linear Support Vector Machine (SVM) classifier to separate the two classes using the optimal hyper plane.

Solution: To train a linear SVM classifier, we need to find the optimal hyper plane that maximizes the margin between the classes.

In this case, we have the following data points:

Positive Class: (1, 3), (2, 5), (3, 8) Negative Class: (6, 5), (7, 9), (8, 7)

Convert these points into feature vectors and labels:

$$X = [(1, 3), (2, 5), (3, 8), (6, 5), (7, 9), (8, 7)]$$
$$Y = [1, 1, 1, -1, -1, -1]$$

Train a linear SVM classifier using these data points. The goal is to find the hyper plane with the maximum margin that separates the positive and negative classes. The hyper plane can be represented as:

$$w * x + b = 0$$

where w is the weight vector perpendicular to the hyper plane and b is the bias.

The SVM optimization problem is:

Minimize: $0.5 * ||w||^2$. Subject to: $y_i * (w * x_i + b) >= 1$ for all data points (x_i, y_i)

The SVM optimization problem aims to maximize the margin while satisfying the constraint that all data points are correctly classified.

Once the SVM is trained, the weight vector w and bias b can be used to classify new data points.

Short Questions and Answers

Q.1. What is AI?

Ans. AI stands for Artificial Intelligence. It involves creating machines that can simulate human intelligence.

Q.2. What is ML?

Ans. ML stands for Machine Learning. It's a subset of AI that focuses on developing algorithms that allow computers to learn from data.

Q.3. What is the difference between AI and ML?

Ans. AI encompasses a broader scope of creating intelligent machines, while ML is a subset of AI that deals with algorithms learning from data.

Q.4. What are the types of Machine Learning?

Ans. Supervised Learning, Unsupervised Learning, Semi-Supervised Learning, Reinforcement Learning.

Q.5. What is supervised learning?

Ans. Supervised learning involves training a model with labeled data, where the input-output relationships are provided.

Q.6. What is unsupervised learning?

Ans. Unsupervised learning involves finding patterns and relationships in data without labeled outputs.

Q.7. What is deep learning?

Ans. Deep learning is a subset of ML that uses deep neural networks to model and understand complex patterns in data.

Q.8. What is a neural network?

Ans. A neural network is a computational model inspired by the human brain, consisting of interconnected nodes (neurons) that process and transmit information.

Q.9. What is overfitting?

Ans. Overfitting occurs when a model learns noise in the training data and performs poorly on new, unseen data.

Q.10. What is reinforcement learning?

Ans. Reinforcement learning is a paradigm where an agent learns to take actions in an environment to maximize cumulative rewards.

Q.11. What is a feature in ML?

Ans. A feature is an individual measurable property or characteristic used to describe data, often used as input for ML algorithms.

Q.12. What is a decision tree?

Ans. A decision tree is a tree-like model that makes decisions by recursively partitioning data into subsets based on feature values.

Q.13. What is clustering?

Ans. Clustering is an unsupervised learning technique that groups similar data points together based on certain criteria.

Q.14. What is NLP?

Ans. NLP stands for Natural Language Processing. It involves enabling computers to understand, interpret, and generate human language.

Q.15. What is computer vision?

Ans. Computer vision is a field of AI that enables computers to interpret and understand visual information from images and videos.

Q.16. What is regression?

Ans. Regression is a supervised learning technique used to predict continuous numerical values based on input features.

Q.17. What is transfer learning?

Ans. Transfer learning is the practice of using pre-trained models as a starting point and fine-tuning them for specific tasks with limited data.

Q.18. What is the bias-variance trade-off?

Ans. The bias-variance trade-off represents the balance between a model's ability to fit the training data and generalize to new, unseen data.

Q.19. What is a hyper parameter?

Ans. A hyper parameter is a parameter set before training a model that affects its performance and behavior, such as learning rate.

Q.20. What is data preprocessing?

Ans. Data preprocessing involves cleaning, transforming, and organizing data to prepare it for ML algorithms.

Q.21. What is a feature vector?

Ans. A feature vector is a numeric representation of the features that describe a data instance.

Q.22. What is anomaly detection?

Ans. Anomaly detection is the process of identifying rare and unusual patterns or instances in data.

Q.23. What is model evaluation?

Ans. Model evaluation involves assessing a model's performance on new, unseen data using metrics like accuracy, precision, and recall.

Q.24. What is a convolutional neural network (CNN)?

Ans. A CNN is a type of neural network designed for processing grid-like data, such as images, using convolutional layers.

Q.25. What is a recurrent neural network (RNN)?

Ans. An RNN is a type of neural network designed to work with sequences of data, where the network's state is updated based on previous inputs.

Q.26. What is a loss function?

Ans. A loss function measures the difference between predicted values and actual values, guiding the learning process in ML algorithms.

Q.27. What is bias in ML?

Ans. Bias refers to errors introduced by assuming a particular outcome due to a lack of data diversity or an overly simplistic model.

Q.28. What is variance in ML?

Ans. Variance refers to the model's sensitivity to fluctuations in training data, leading to poor generalization on new data.

Q.29. What is ensemble learning?

Ans. Ensemble learning combines multiple models to improve overall prediction accuracy and reduce overfitting.

Q.30. What is gradient descent?

Ans. Gradient descent is an optimization technique used to adjust model parameters iteratively to minimize the loss function.

Q.31. What is feature engineering?

Ans. Feature engineering involves selecting, transforming, and creating features to improve a model's performance.

Q.32. What is a support vector machine (SVM)?

Ans. An SVM is a classification algorithm that finds the hyperplane that best separates different classes in data.

Q.33. What is an auto encoder?

Ans. An auto encoder is a type of neural network used for unsupervised learning and dimensionality reduction.

Q.34. What is bias-variance decomposition?

Ans. The bias-variance decomposition breaks down a model's expected prediction error into bias, variance, and irreducible error components.

Q.35. What is the curse of dimensionality?

Ans. The curse of dimensionality refers to the challenges and limitations that arise when working with high-dimensional data.

Q.36. What is a feature extraction?

Ans. Feature extraction is the process of reducing the dimensionality of data while retaining its relevant information.

Q.37. What is regularization?

Ans. Regularization techniques control the complexity of models to prevent overfitting by adding penalty terms to the loss function.

Q.38. What is a confusion matrix?

Ans. A confusion matrix is a table that shows the performance of a classification model, including true positive, true negative, false positive, and false negative values.

Q.39. What is a recommendation system?

Ans. A recommendation system suggests items or content to users based on their preferences and behaviors.

Q.40. What is hyper parameter tuning?

Ans. Hyper parameter tuning involves finding the optimal values for hyperparameters to improve a model's performance.

Q.41. What is a generative adversarial network (GAN)?

Ans. A GAN is a type of neural network framework that generates new data by training a generator and a discriminator network against each other.

Q.42. What is k-fold cross-validation?

Ans. K-fold cross-validation divides the data into k subsets, trains the model on k–1 subsets, and validates on the remaining subset to assess its generalization performance.

Q.43. What is Bayes' theorem?

Ans. Bayes' theorem calculates the probability of a hypothesis being true based on evidence and prior beliefs.

Q.44. What is deep reinforcement learning?

Ans. Deep reinforcement learning combines deep learning and reinforcement learning to enable agents to learn complex tasks through trial and error.

Q.45. What is a word embedding?

Ans. A word embedding is a dense vector representation of a word in a language, capturing semantic relationships between words.

Q.46. What is a neural network activation function?

Ans. An activation function introduces non-linearity to a neural network by transforming input values to output values for each neuron.

Q.47. What is the Turing test?

Ans. The Turing test assesses a machine's ability to exhibit intelligent behavior indistinguishable from that of a human.

Q.48. What is anomaly detection?

Ans. Anomaly detection involves identifying data points or instances that deviate significantly from the norm, indicating potential anomalies or outliers.

Q.49. What is sentiment analysis?

Ans. Sentiment analysis is the process of determining the sentiment expressed in text, often categorized as positive, negative, or neutral.

Q.50. What is transfer learning in NLP?

Ans. Transfer learning in NLP involves using pre-trained language models as a starting point for specific tasks, saving time and resources.

Multiple Choice Questions and Answers

Q.1. What is the goal of Artificial Intelligence (AI)?

(*a*) Creating machines with human-like appearance
(*b*) Developing intelligent machines capable of mimicking human behaviors
(*c*) Creating machines that can replace human workers
(*d*) Creating machines with advanced mechanical capabilities

Q.2. Which is not a type of Machine Learning?

(*a*) Supervised Learning (*b*) Unsupervised Learning
(*c*) Reinforcement Learning (*d*) Predefined Learning

Q.3. What is the primary difference between supervised and unsupervised learning?

(*a*) Supervised learning uses labeled data, while unsupervised learning doesn't.
(*b*) Unsupervised learning involves human supervision, while supervised learning doesn't.
(*c*) Supervised learning requires less data.
(*d*) Unsupervised learning is only used for image data.

Q.4. What type of learning does Reinforcement Learning (RL) belong to?

(*a*) Unsupervised Learning (*b*) Semi-Supervised Learning
(*c*) Supervised Learning (*d*) None of the above

Q.5. Which ML technique is used for clustering similar data points together?

(*a*) Regression (*b*) Classification
(*c*) Clustering (*d*) Reinforcement Learning

Q.6. What is the purpose of a loss function in machine learning?

(*a*) To improve model complexity (*b*) To improve model accuracy
(*c*) To measure the difference between predicted and actual values
(*d*) To adjust hyperparameters

Q.7. Which type of neural network is often used for image recognition tasks?

(*a*) Recurrent Neural Network (RNN) (*b*) Convolutional Neural Network (CNN)
(*c*) Feedforward Neural Network (*d*) Multilayer Perceptron (MLP)

Q.8. Which technique is used to optimize model parameters by iteratively adjusting them to minimize the loss function?

(*a*) Gradient Descent (*b*) Clustering
(*c*) Reinforcement Learning (*d*) Principal Component Analysis (PCA)

Q.9. What is the main purpose of a validation set in model training?

(*a*) To test the model's performance on unseen data
(*b*) To train the model using labeled data
(*c*) To fine-tune the model's hyper parameters
(*d*) To evaluate the model's accuracy during training

Q.10. Which machine learning technique is used to predict continuous numerical values?

(*a*) Classification (*b*) Regression
(*c*) Clustering (*d*) Reinforcement Learning

Q.11. What is the primary goal of feature engineering?

(*a*) Creating more complex models (*b*) Reducing model complexity
(*c*) Reducing the dimensionality of data
(*d*) Creating informative and relevant features for better model performance

Q.12. What is the process of allowing a model to learn from its own mistakes and improve its performance over time called?

(*a*) Supervised Learning (*b*) Unsupervised Learning
(*c*) Reinforcement Learning (*d*) Transfer Learning

Q.13. What is the key idea behind transfer learning in machine learning?

(*a*) Training models from scratch for every new task
(*b*) Using pre-trained models as a starting point for new tasks
(*c*) Transferring data between different models
(*d*) Using labeled data for unsupervised learning

Q.14. Which type of learning is concerned with finding hidden patterns and relationships in data without labeled outputs?

(*a*) Supervised Learning (*b*) Unsupervised Learning
(*c*) Reinforcement Learning (*d*) Transfer Learning

Q.15. What is the term for the error that arises due to an overly complex model that fits the training data too closely?

(*a*) Bias (*b*) Variance (*c*) Underfitting (*d*) Overfitting

Q.16. Which term represents the combination of bias and variance in a model's prediction error?

(*a*) Irreducible Error (*b*) Total Error
(*c*) Generalization Error (*d*) Mean Squared Error

Q.17. Which algorithm finds the best hyperplane that separates data points into different classes?

(*a*) Support Vector Machine (SVM) (*b*) K-Means Clustering
(*c*) Naive Bayes (*d*) Principal Component Analysis (PCA)

Q.18. Which technique is used to combine multiple models to improve overall prediction accuracy and reduce overfitting?

(*a*) Regularization (*b*) Feature Engineering
(*c*) Ensemble Learning (*d*) Gradient Descent

Q.19. Which technique involves adjusting the learning rate to control the step size in gradient descent optimization?

(*a*) Regularization (*b*) Learning Rate Annealing
(*c*) Feature Scaling (*d*) Early Stopping

Q.20. What is the purpose of a confusion matrix in model evaluation?

(*a*) To measure the model's training time (*b*) To measure the model's prediction time
(*c*) To measure the model's accuracy on unseen data
(*d*) To assess the model's performance in classification tasks

Q.21. Which type of learning involves agents learning to perform tasks by interacting with an environment and receiving rewards?

(*a*) Unsupervised Learning (*b*) Supervised Learning
(*c*) Reinforcement Learning (*d*) Transfer Learning

Q.22. Which technique is used to automatically extract relevant features from raw data for improved model performance?

(*a*) Regularization (*b*) Dimensionality Reduction
(*c*) Clustering (*d*) Feature Scaling

Q.23. Which term describes a machine learning model's ability to generalize its predictions to new, unseen data?

(*a*) Adaptability (*b*) Transferability (*c*) Generalization (*d*) Robustness

Q.24. What does the term "cross-validation" refer to in machine learning?

(*a*) Training a model on multiple GPUs
(*b*) Validating a model's predictions on the training data
(*c*) Validating a model's performance on unseen data
(*d*) Splitting the data into multiple subsets for training and testing

Q.25. Which technique is used to prevent overfitting by adding penalty terms to the loss function?

(*a*) Regularization (*b*) Optimization
(*c*) Clustering (*d*) Ensemble Learning

Q.26. Which type of neural network is commonly used for processing sequences of data, such as text or time series?

(*a*) Convolutional Neural Network (CNN) (*b*) Recurrent Neural Network (RNN)
(*c*) Feedforward Neural Network (*d*) Radial Basis Function Network

Q.27. Which technique is used to transform raw text data into a numerical format suitable for machine learning algorithms?

(*a*) Dimensionality Reduction (*b*) Feature Scaling
(*c*) Word Embedding (*d*) Regularization

Q.28. What is the primary purpose of a neural network's activation function?

(*a*) To perform feature extraction (*b*) To introduce non-linearity
(*c*) To control the model's learning rate
(*d*) To determine the number of layers in the network

Q.29. Which type of learning is concerned with the process of assigning objects to predefined categories?

(*a*) Unsupervised Learning (*b*) Reinforcement Learning
(*c*) Supervised Learning (*d*) Semi-Supervised Learning

Q.30. Which term describes the process of using labeled data to train a model for making predictions?

(*a*) Training (*b*) Testing (*c*) Inference (*d*) Validation

Q.31. Which technique involves training a model to predict missing or hidden values in a dataset?

(*a*) Clustering (*b*) Classification
(*c*) Imputation (*d*) Dimensionality Reduction

Q.32. Which type of learning is concerned with improving a model's performance by iteratively adjusting its parameters?

(*a*) Reinforcement Learning (*b*) Transfer Learning
(*c*) Supervised Learning (*d*) Optimization Learning

Q.33. Which algorithm is commonly used for anomaly detection, identifying data points that deviate from the norm?

(*a*) K-Means Clustering (*b*) Support Vector Machine (SVM)
(*c*) Principal Component Analysis (PCA) (*d*) Isolation Forest

Q.34. Which term describes the process of identifying relevant patterns and relationships in data to aid decision-making?

(*a*) Data Collection (*b*) Data Manipulation
(*c*) Data Analysis (*d*) Data Preprocessing

Q.35. What is the term for a machine learning model that can make predictions about new, unseen data?

(*a*) Generalization Model (*b*) Predictive Model
(*c*) Inference Model (*d*) Validation Model

Q.36. Which technique involves using probability to make predictions or decisions based on uncertain data?

(*a*) Bayesian Inference (*b*) K-Means Clustering
(*c*) Principal Component Analysis (PCA) (*d*) Gradient Descent

Q.37. What does the term "bias" refer to in machine learning?

(*a*) The model's ability to generalize to new data

(*b*) The error introduced by overfitting the training data

(*c*) The error introduced by assuming overly simplistic relationships

(*d*) The model's ability to fit the training data closely

Q.38. What is the primary purpose of the "softmax" activation function in neural networks?

(*a*) To introduce non-linearity

(*b*) To transform output scores into probabilities

(*c*) To control the learning rate (*d*) To adjust the model's parameters

Q.39. Which type of neural network architecture is designed for image recognition tasks by using convolutional layers?

(*a*) Recurrent Neural Network (RNN) (*b*) Radial Basis Function Network

(*c*) Convolutional Neural Network (CNN) (*d*) Multilayer Perceptron (MLP)

Q.40. Which term refers to a technique that involves splitting data into training and testing sets to assess model performance?

(*a*) Validation Split (*b*) Cross-Validation (*c*) Data Splitting (*d*) Test Split

Q.41. Which technique involves using historical data to predict future values or trends?

(*a*) Clustering (*b*) Classification

(*c*) Regression (*d*) Dimensionality Reduction

Q.42. What is the purpose of a "dropout" layer in neural networks?

(*a*) To increase the learning rate (*b*) To reduce model complexity

(*c*) To prevent overfitting by randomly deactivating neurons

(*d*) To introduce additional layers

Q.43. Which technique involves training a model using both labeled and unlabeled data?

(*a*) Supervised Learning (*b*) Unsupervised Learning

(*c*) Semi-Supervised Learning (*d*) Reinforcement Learning

Q.44. Which term describes the ability of a model to perform well on data it has never seen before?

(*a*) Model Accuracy (*b*) Model Robustness

(*c*) Model Efficiency (*d*) Model Overfitting

Q.45. Which technique involves using pre-trained models as a starting point for new tasks with limited data?

(*a*) Reinforcement Learning (*b*) Transfer Learning

(*c*) Semi-Supervised Learning (*d*) Unsupervised Learning

Q.46. Which type of learning involves adjusting model parameters based on feedback from the environment?

(*a*) Reinforcement Learning (*b*) Supervised Learning

(*c*) Unsupervised Learning (*d*) Transfer Learning

Q.47. Which machine learning technique focuses on finding optimal paths for robots or agents to navigate through an environment?

(*a*) Reinforcement Learning (*b*) Clustering

(*c*) Regression (*d*) Path Planning

Q.48. What is the term for the process of converting raw data into a numerical format suitable for machine learning?

(*a*) Feature Extraction (*b*) Data Preprocessing

(*c*) Data Transformation (*d*) Data Normalization

Q.49. Which technique involves splitting data into subsets and training models on one subset while testing on another?

(*a*) Cross-Validation (*b*) Ensemble Learning

(*c*) Transfer Learning (*d*) Regularization

Q.50. Which type of neural network is commonly used for time series prediction and natural language processing tasks?

(*a*) Convolutional Neural Network (CNN) (*b*) Recurrent Neural Network (RNN)

(*c*) Feedforward Neural Network (*d*) Radial Basis Function Network

Key to Multiple Choice Questions

1. (*b*)	**2.** (*d*)	**3.** (*a*)	**4.** (*d*)	**5.** (*c*)	**6.** (*c*)	**7.** (*b*)	**8.** (*a*)	**9.** (*a*)
10. (*b*)	**11.** (*d*)	**12.** (*c*)	**13.** (*b*)	**14.** (*b*)	**15.** (*d*)	**16.** (*c*)	**17.** (*a*)	**18.** (*c*)
19. (*b*)	**20.** (*d*)	**21.** (*c*)	**22.** (*b*)	**23.** (*c*)	**24.** (*d*)	**25.** (*a*)	**26.** (*b*)	**27.** (*c*)
28. (*b*)	**29.** (*c*)	**30.** (*a*)	**31.** (*c*)	**32.** (*d*)	**33.** (*d*)	**34.** (*c*)	**35.** (*c*)	**36.** (*a*)
37. (*c*)	**38.** (*b*)	**39.** (*c*)	**40.** (*b*)	**41.** (*c*)	**42.** (*c*)	**43.** (*c*)	**44.** (*b*)	**45.** (*b*)
46. (*a*)	**47.** (*d*)	**48.** (*b*)	**49.** (*a*)	**50.** (*b*)				

Projects Works

1. **Object Recognition and Manipulation for Robotic Arm:** Develop a system that enables a robotic arm to identify and grasp objects from a cluttered environment. Use computer vision techniques for object recognition and reinforcement learning for improving manipulation skills.
2. **Autonomous Drone Navigation:** Build an autonomous drone that can navigate through indoor or outdoor environments while avoiding obstacles. Use techniques such as computer vision, SLAM (Simultaneous Localization and Mapping), and reinforcement learning to enhance its navigation capabilities.
3. **Automated Warehouse Management:** Create an AI-driven system for managing inventory in a warehouse. Use robotics to automate tasks like picking, packing, and sorting items, and integrate ML algorithms to optimize inventory levels and predict demand.
4. **Human-Robot Collaboration for Manufacturing:** Design a system where robots work collaboratively with humans in a manufacturing setting. Develop algorithms that allow robots to understand human intentions, adapt their actions, and assist in tasks like assembly or material handling.

5. **Predictive Maintenance for Industrial Machines:** Implement a predictive maintenance system using sensor data from industrial machines. Apply ML algorithms to predict when equipment is likely to fail, enabling maintenance teams to perform maintenance tasks before breakdowns occur.
6. **Gesture-Based Control of Robots:** Create a gesture recognition system that allows users to control robots using hand gestures. Utilize computer vision techniques to recognize different gestures and translate them into robot commands.
7. **Smart Agriculture with Autonomous Robots:** Develop autonomous robots for agricultural tasks like planting, irrigation, and harvesting. Implement AI algorithms to detect ripe fruits, assess crop health, and optimize planting patterns.
8. **Robotic Companion for Elderly and Disabled:** Design a robotic companion that can assist the elderly or disabled individuals in their daily tasks. Use AI to understand user preferences, provide reminders, and offer companionship.
9. **Drone-Based Inspection and Monitoring:** Create a drone-based system that performs inspections of infrastructure such as bridges, power lines, and pipelines. Implement image recognition and anomaly detection algorithms to identify potential issues.
10. **Robot Soccer Team with AI Strategies:** Build a team of small robots that can play soccer autonomously. Develop AI strategies for positioning, passing, and scoring goals, and compete against other robot teams in a simulated or physical environment.
11. **Robotic Waste Sorting:** Design a robot that can identify and sort different types of waste into appropriate categories (e.g., recyclables, compost, landfill). Use computer vision and ML to recognize and handle different materials.
12. **Autonomous Underwater Exploration:** Create an underwater robot that can autonomously explore and map underwater environments. Implement SLAM techniques specifically designed for underwater scenarios and integrate AI for decision-making.
13. **Automated Healthcare Delivery:** Develop a system where robots assist in delivering medication, meals, or other supplies to patients in hospitals or healthcare facilities. Use AI to navigate through the environment and optimize delivery routes.
14. **Robotic Cooking Assistant:** Build a robotic system that assists in cooking tasks, such as chopping vegetables, stirring dishes, and adjusting heat settings. Use AI to understand recipes, make ingredient substitutions, and learn user preferences.
15. **Autonomous Lawn Care Robot:** Design a robot that can autonomously mow lawns and maintain outdoor landscapes. Implement path planning algorithms and integrate ML to handle obstacles and varying terrain.

09

Graph Theory in Solving Real-World Problems Across Different Domains

9.1 INTRODUCTION

In mathematics, the study of graphs, which are made up of nodes (sometimes called vertices) and edges, is known as graph theory. Graphs are used to model and examine the connections, interactions, and relationships between different elements. Numerous domains, including computer science, social sciences, biology, transportation systems, and more, use graph theory in a variety of ways. A diverse and fascinating discipline, graph theory offers important insights into the composition and connections of various systems. Beyond mathematics, its practical applications can be used to address issues in a variety of fields.

9.2 BASIC CONCEPTS

Basic ideas in graph theory include the following:

(*a*) A node (vertex) is a basic building block of a graph that symbolizes a distinct entity or point.

(*b*) An edge is a link between two nodes that denotes a connection or interaction.

(*c*) A graph with edges that have a specified direction from one node (the source) to another node (the destination) is known as a directed graph (Digraph).

(*d*) **Undirected Graph:** A graph in which the connections between the nodes have no particular direction and are made in any order.

(*e*) **Weighted Graph:** A graph with edges given weights or values that reflect the expense, proximity, or significance of the link.

(*f*) **Degree of a Node:** The quantity of edges that an edge is incident with.

(*g*) **Path:** A collection of edges and nodes

(*h*) **Cycle:** A route that creates a closed loop and begins and finishes at the same node.

Types of Graphs

(*a*) Simple graphs lack self-loops (edges that connect one node to itself) and numerous edges that connect the same nodes.

(*b*) A graph that allows many edges to connect the same nodes is known as a multigraph.

(*c*) A directed graph with no cycles is known as a directed acyclic graph (DAG).

(*d*) **Complete Graph:** A straightforward graph with edges connecting every pair of unique nodes.

(*e*) **Bipartite Graph:** This type of graph has nodes that can be split into two sets, and all of its edges connect nodes in the two sets.

(*f*) **Tree:** A connected, undirected, cycle-free graph.

(*g*) **Forest:** A randomly arranged group of trees.

Graph Algorithms

Numerous algorithms for analyzing and resolving graph-related issues are included in graph theory. A few typical graph algorithms are as follows:

(*a*) **Breadth-First Search (BFS):** This traversal technique searches nodes one level at a time, beginning with a predetermined source node.

(*b*) **Depth-First Search (DFS):** A traversal method that investigates as much of each branch as possible before turning around.

(*c*) **Shortest Path Algorithms:** These include the Bellman-Ford algorithm and Dijkstra's method for determining the shortest path between two nodes.

(*d*) **Minimum Spanning Tree (MST) Algorithms:** These include Kruskal's method and Prim's algorithm for determining the minimum-weight spanning tree of a graph.

(*e*) **Topological Sorting:** A method for linearly sorting directed acyclic network nodes according to their dependencies.

(*f*) Cycle detection algorithms are frequently used to determine whether a system is in a deadlock by looking for cycles in a graph.

(*g*) Algorithms for coloring nodes in a graph so that no two neighboring nodes have the same color.

9.3 GRAPH REPRESENTATION AND ALGORITHMS

Graphs can be represented in a variety of ways depending on their properties and the particular algorithms being used. The adjacency matrix and the adjacency list are the two most used ways to represent graphs. Adjacency Matrix, first A 2D array with the dimensions V x V is known as an adjacency matrix, where V is the graph's node count. Matrix[i] [j] will be 1 if there is an edge connecting nodes i and j; otherwise, it will be 0. The matrix can store edge weights rather than 1s and 0s for weighted graphs.

Pros

- Simple and intuitive representation.
- Useful for dense graphs (when the number of edges is close to the maximum possible).

Cons

- Consumes more space for sparse graphs.
- Inefficient for some graph algorithms.

Adjacency List

An adjacency list is a set of linked lists in which each node has a list of nodes that are located nearby. The linked list's elements each stand for an edge.

Pros

- Efficient for sparse graphs, as it only stores information about existing edges.
- Requires less space compared to the adjacency matrix for sparse graphs.

Cons

- May consume more space for dense graphs.
- Slightly more complex representation compared to the adjacency matrix.

There are several fundamental graph algorithms used to analyze and manipulate graphs. Here are some key algorithms:

1. **Breadth-First Search (BFS):** BFS investigates nodes from a given source node level by level. It is helpful for both navigating a graph and determining the shortest path through one.
2. **Depth-First Search (DFS):** Before turning around, DFS investigates as far as it can along each branch. It's helpful for several graph-related tasks, including cycle detection, topological sorting, and connected component discovery.
3. **Dijkstra's Algorithm:** In a weighted network, Dijkstra's algorithm determines the shortest route between a single source node and every other node. For edge weights that are not negative, it works nicely.
4. **Bellman-Ford Algorithm:** Even when negative edge weights are present, Bellman-Ford determines the shortest route from a single source node to all other nodes in a weighted graph.
5. **Kruskal's Algorithm:** The Kruskal algorithm determines the graph's smallest spanning tree. Finding the smallest collection of edges connecting all nodes with the least amount of overall weight is useful.
6. **Prim's Algorithm:** The Kruskal algorithm determines the graph's smallest spanning tree. Finding the smallest collection of edges connecting all nodes with the least amount of overall weight is useful.
7. **Floyd-Warshall Algorithm:** In a graph, Floyd-Warshall determines the shortest paths for each pair. It manages edge weights with both positive and negative values.
8. **Topological Sorting:** A directed acyclic graph's nodes are arranged in a linear order via topological sorting based on their dependencies.
9. **Cycle Detection:** DFS-style algorithms are employed to find cycles in a graph. For many applications, including deadlock detection, cycle detection is essential.

These algorithms lay the groundwork for optimizing procedures in fields including route planning, network analysis, recommendation systems, and more.

9.4 CONNECTIVITY AND SHORTEST PATHS

In order to analyze and comprehend the relationships between nodes in a graph, it is essential to understand the fundamental notions of connectivity and shortest paths in graph theory.

Let's explore these concepts in more detail:

Connectivity

In graph theory, connectivity describes how closely a graph's nodes are related to one another. It aids in figuring out whether there are connections between nodes and whether the graph may be divided into separate entities. Important connecting ideas include:

1. **Connected Graph:** A graph is said to be connected if a path can be found connecting any two nodes. In other words, the graph doesn't contain any isolated nodes or clusters of nodes that are inaccessible to other nodes.
2. **Connected Component:** In a connected component, any two nodes are connected to one another through a path. If a graph is not totally connected, it can nevertheless have several connected parts.
3. **Strongly Connected Graph:** A graph is strongly connected in a directed graph if there is a directed path connecting any two nodes.
4. **Weakly linked Graph:** If the underlying undirected graph of a directed graph is linked, the graph is said to be weakly connected. In other words, all nodes are connected, regardless of edge directions.

Shortest Paths

The paths between nodes in a graph with the least overall edge weight are referred to as the shortest paths. In weighted graphs where edges have associated values, this idea is very crucial. Important shortest path algorithms are:

1. **Dijkstra's Algorithm:** In a weighted graph with non-negative edge weights, Dijkstra's algorithm determines the shortest route between a single source node and all other nodes. It is commonly used in network routing and navigation systems.
2. **Bellman-Ford Algorithm:** Even in the presence of negative edge weights, the Bellman-Ford algorithm locates the shortest route between a single source node and all other nodes in a weighted graph. Negative cycle graphs can also be handled by it.
3. **Floyd-Warshall Algorithm:** Regardless of negative edge weights, the Floyd-Warshall method determines the shortest pathways between any pair of nodes in a network. In dense graphs, where Dijkstra's approach could be less effective, it is advantageous.
4. **A* Algorithm:** The A* algorithm is a greedy search- and Dijkstra's algorithm-inspired heuristic-based algorithm. Video games and path finding software both frequently use it.
5. **Bidirectional Search:** In order to more quickly determine the shortest path, bidirectional search simultaneously investigates the graph from the source and goal nodes.

For example, finding the best routes, network analysis, resource allocation, and process optimization all require the understanding of connectedness and shortest paths. These ideas serve as the cornerstone for numerous algorithms used in network optimization, transportation planning, and navigation systems.

9.5 SPANNING TREES AND GRAPH TRAVERSAL

In graph theory, spanning trees and graph traversal are crucial ideas that aid in analysing the links and structure of a graph. Let's explore these ideas in greater depth:

Spanning Trees

A connected acyclic graph that forms a tree with all the nodes of the original graph is known as a spanning tree. While removing cycles, spanning trees maintain the graph's connectedness. Key details concerning spanning trees are as follows:

1. **Minimum Spanning Tree (MST):** A minimum spanning tree is one that, among all feasible spanning trees of a weighted graph, has the lowest total edge weight. In situations where edges have associated costs, such network design or transportation planning, minimum spanning trees are particularly helpful.
    - To identify the smallest spanning tree, algorithms like Kruskal's and Prim's algorithms are utilized.
2. **Applications of Spanning Trees:**
    - Creating cost-effective and efficient communication networks.
    - Establishing a low-cost connection between each node in a graph, like in cable arrangement.
    - Building effective electrical networks with the shortest possible wire lengths.

Graph Traversal

The process of traversing a graph entails going to every node and methodically following all of its edges. It's a crucial step in the analysis and information extraction process for graphs. There are two popular methods for traversing a graph:

1. **Breadth-First Search (BFS):**
    - Starting from a designated source node, BFS examines nodes one level at a time, maintaining the order of the nodes to be visited using a queue data structure.
    - BFS can be used to iteratively traverse a graph and discover the shortest path in an unweighted graph.
2. **Depth-First Search (DFS):**
    - DFS manages the traversal order using a stack (or recursion) and explores as far as feasible along each branch before backtracking.
    - DFS is employed for applications including path finding, topological sorting, and cycle identification.

Applications of Graph Traversal

Graph traversal techniques have a wide range of applications, including:

- In video games and navigational systems, path finding involves determining the shortest route between two nodes.
- **Maze Solving:** Finding a way out of a maze by navigating it.
- **Network analysis:** Examining the connections between online pages, social networks, and communication networks.
- Solving puzzles that involve nodes that are connected to one another.
- **Code Compilation:** During code compilation, an abstract syntax tree is built.
- **Circuit analysis:** The process of looking for connections and routes in electrical circuits.

Fundamental concepts in graph theory, such as spanning trees and graph traversal, have applications in a variety of fields, including network design and optimization, computer science, and social analysis.

9.6 NUMERICAL SOLUTIONS FOR KINEMATIC EQUATIONS IN ROBOTICS

Kinematics, as used in robotics, is the study of the motion of robotic systems with a particular emphasis on the locations, velocities, and accelerations of the robot's constituent parts (such as joints, links, and end effectors) without taking forces and torques into account. Understanding and managing robot movement requires the use of kinematic equations. Robot configurations (joint angles) and end effectors positions are calculated using numerical solutions for kinematic equations.

Here's a general outline of how numerical solutions are used for solving kinematic equations in robotics applications:

Forward Kinematics: By using the joint angles as input, forward kinematics determines the location and orientation of the robot's end effectors, which is frequently represented by a pose or transformation matrix. Forward kinematic equations are typically resolved using numerical techniques, such as the Denavit-Hartenberg (DH) parameters or the product-of-exponentials formulation.

Inverse Kinematics: The process of calculating the joint angles necessary to produce a particular intended end effector position and orientation is known as inverse kinematics. Inverse kinematics equations are frequently nonlinear, have a number of solutions, and can be difficult to analyse. Problems involving inverse kinematics are resolved using numerical techniques such as iterative algorithms.

Some common numerical approaches include:

Iterative Methods: To reduce the difference between the desired end effectors pose and the computed pose, joint angles are iteratively adjusted using techniques like the Newton-Raphson algorithm or the Jacobian-based iterative method.

Gradient Descent: By repeatedly altering joint angles in the direction of the error function's sharpest decline, gradient descent techniques reduce error.

Cyclic Coordinate Descent: With this iterative process, joint angles are changed one at a time while holding the others constant. It's especially helpful for robotic structures that are more basic.

Trajectory Generation: Robots frequently have to follow predetermined routes or trajectories. These trajectories can be produced numerically by interpolating between significant waypoints or locations along the intended path. Smooth trajectories that adhere to kinematic restrictions can be produced using techniques such as cubic splines or quintic splines.

Robot Simulation: In order to forecast how robots will move based on their kinematic equations, numerical solutions are frequently utilized in robot simulations. Before putting the idea into practice on actual robots, these simulations assist validate the design, test control tactics, and assess performance.

Robotics uses numerical solutions to solve difficult kinematic equations and provide robots the ability to move precisely and steadily. It's crucial to remember that numerically resolving kinematic equations may result in processing complexity and the requirement to take into account problems like singularities and workspace restrictions.

9.7 NUMERICAL SOLUTIONS FOR DYNAMIC EQUATIONS IN ROBOTICS

In robotics, numerical solutions to dynamic equations are essential for simulating and controlling robot motion. The relationship between forces, torques, and accelerations of a robot's parts (joints, linkages) is described by dynamic equations, which are crucial for comprehending the robot's behavior under various circumstances. The numerical solution of these equations enables precise robotic system simulation and control.

Here's an overview of how numerical solutions are used for dynamic equations in robotics:

1. ***Equations of Motion***: The effects of forces and torques on a robot's parts on its mobility are described by dynamic equations of motion. These equations, like the Newton-Euler equations or the Lagrange equations, are frequently constructed using physics and mechanical principles.
2. ***Numerical Integration***: On the basis of the dynamic equations, numerical integration techniques are utilized to approximate the state variables (positions, velocities, and accelerations) of the robot over time. The continuous differential equations are transformed using these techniques into discrete time steps that can be used for simulation or control. Typical approaches to numerical integration include:
   - **Euler Integration:** Straightforward and understandable, but due to its linear approximation, it may not be accurate for complicated systems.
   - **Runge-Kutta Methods:** These higher-order techniques improve accuracy by taking into account intermediate steps taken during each time period.
3. ***Simulation***: For modeling robot behavior in diverse contexts, numerical solutions are essential. Simulations can forecast how a robot will react to various inputs, external pressures, and environmental circumstances by integrating the dynamic equations across time.
4. ***Control***: Dynamic equations are used in robotic control to create control schemes that direct the behavior of the robot. Calculating control inputs (joint torques) is aided by numerical solutions while attempting to follow instructions, follow desired trajectories, or carry out precise activities.
5. ***Real-Time Control***: Numerical solutions are used to instantly solve dynamic equations in order to adapt to changing situations or react to sensor feedback for real-time control of robots. For effective real-time control, methods such as Model Predictive Control (MPC) or recursive algorithms (such as Recursive Newton-Euler) are used.
6. ***Multibody Dynamics Software***: Complex robotic systems can use specialized software, commonly referred to as multibody dynamics simulators, to numerically solve dynamic equations. Engineers and academics can simulate and examine robot behavior using these technologies in a variety of contexts. It's crucial to keep in mind that numerically solving dynamic equations can be computationally demanding, particularly for complicated robotic systems with numerous degrees of freedom. For precise and effective simulations and control of robotic systems, specialized software, effective algorithms, and advanced numerical techniques are frequently needed.

9.8 REPRESENTING AND SOLVING PROBLEMS IN MOTION PLANNING OF ROBOTICS

Motion planning is one of several robotics issues that can be resolved using graph theory. Motion planning is the process of selecting appropriate paths or trajectories for robots to follow from an initial configuration to a desired destination configuration, all the while avoiding obstructions and following rules. A well-organized framework for presenting, deciphering, and resolving motion planning issues is offered by graph theory. Graph theory is applied in this situation as follows:

1. ***Configuration Space (C-Space)***: All potential configurations (positions and orientations) that a robot could assume are represented in the configuration space. It might be multidimensional and intricate. This continuous space is discretized into a graph using graph theory, which is also known as a "roadmap" or "configuration space graph."
2. ***Nodes and Edges***: In the configuration space graph, nodes stand in for legitimate configurations, and edges link configurations that may be reached directly by a path devoid of collisions. Edges can describe possible trajectories as straight lines or curves.
3. ***Connectivity***: Any pair of nodes can travel along a plausible path thanks to the graph's connectedness. The robot might not be able to navigate to specific locations if the graph is not connected.
4. ***Search Algorithms***: The shortest or ideal path in the configuration space graph is found using graph search techniques like Dijkstra's algorithm, A* algorithm, and many iterations of these algorithms. These methods look at the nodes and edges of the network to find the least expensive path.
5. ***Sampling-Based Approaches***: The Probabilistic Roadmap (PRM) and Rapidly Exploring Random Trees (RRT) are two sample-based methods that employ random sampling to produce nodes in the configuration space and connect them to form a graph. These methods work especially well in high-dimensional and complicated settings.
6. ***Path Smoothing and Optimization***: Graph-based approaches can be used to smooth a path after it has been initially found in the graph or to optimize it for different criteria, such minimizing path length or averting sudden turns.
7. ***Dynamic Environments***: By updating the graph or recalculating pathways in real-time based on the shifting obstacle placements, graph-based techniques can be expanded to handle dynamic situations.
8. ***Multi-Robot Planning***: In multi-robot settings, where coordination and collision avoidance amongst numerous robots are taken into account, graph theory is also employed. Graph-based approaches aid in the discovery of paths that let robots accomplish their objectives without colliding.

Robotics engineers may successfully solve difficult motion planning problems and quickly produce collision-free pathways for robots in varied situations by using graph theory. Robots can navigate securely and effectively thanks to this method, which increases their ability to carry out duties in realistic environments.

9.9 REPRESENTING AND SOLVING TASK ALLOCATION IN ROBOTICS

Task allocation is one of several robotics issues that can be resolved using graph theory. To efficiently accomplish a group goal, specific jobs or activities are distributed across various robots or agents through task allocation. Task allocation issues can be modeled, examined, and solved using a structured framework that graph theory offers.

Here's how graph theory is used in this context:

1. ***Representation as a Graph***: Task allocation involves representing agents or robots as nodes in a graph and the tasks to be distributed as edges linking the nodes. This creates a bipartite network in which tasks and agents are represented by different sets of nodes in the graph.

2. ***Weighted Edges***: Weights can be applied to edges in the graph to indicate the expense, suitability, or compatibility of giving a task to a specific agent. These weights take into account things like distance, time, skill level, and the availability of resources.
3. ***Matching Algorithms***: Finding optimal or nearly optimal job assignments for agents can be done by using matching algorithms, which are frequently employed in graph theory. The Hungarian algorithm, which effectively resolves the assignment problem in bipartite graphs, is a popular algorithm.
4. ***Multi-Agent Coordination***: Graph theory can be applied to situations with numerous actors and tasks. It aids in task allocation optimization while taking into account limitations such few resources, agent coordination, and communication.
5. ***Task Constraints***: Task restrictions, such as dependencies or precedence relationships between tasks, can be incorporated into the allocation process using graph theory. As a result, assignments are assigned in a manner that respects these limitations.
6. ***Dynamic Task Allocation***: Graph-based approaches can respond to shifting circumstances by dynamically reassigning jobs as they become available or as they are finished. This is especially helpful in instances when the environment is unknown or changing.
7. ***Decentralized Approaches***: Decentralized job allocation techniques in multi-robot systems make use of graph-based algorithms to let robots negotiate and settle on work assignments on their own while still arriving at an overall optimal or nearly optimal solution.
8. ***Distributed Resource Allocation***: In situations where the optimal distribution of resources like bandwidth, electricity, or communication channels is required, graph theory can be used to share resources, not just jobs, among robots.

Robotics researchers and engineers may maximize resource utilization, improve task completion efficiency, and improve system performance in multi-agent and multi-task settings by using graph theory for task allocation.

Numerical Problems with Solutions

The following numerical problems and solutions demonstrate how graph theory can be applied to solve a wide range of real-world problems across different domains. Depending on the specific problem, various graph algorithms and techniques can be used to derive meaningful insights and optimize solutions.

Transportation and Networks

Problem 1. Find the shortest path from node A to node D in a weighted graph using Dijkstra's algorithm.

Solution: Perform Dijkstra's algorithm to find the shortest path: A -> B -> D with a total weight of 4.

Problem 2. Calculate the minimum spanning tree (MST) for a network with the following edge weights:

A-B : 2B-C : 3A-C : 4B-D : 1

Solution: The MST is A-B, B-D, and B-C, with a total weight of 6.

Problem 3. Determine if a given directed graph has a Hamiltonian path.

Solution: Use a Hamiltonian path algorithm to check for the existence of a Hamiltonian path.

Problem 4. Calculate the maximum flow in a flow network using Ford-Fulkerson algorithm.
Solution: Apply the Ford-Fulkerson algorithm to compute the maximum flow.

Social Network Analysis

Problem 5. Find the shortest path (degrees of separation) between two users in a social network.
Solution: Use BFS to find the shortest path between the two users.
Problem 6. Calculate the clustering coefficient of a social network graph.
Solution: Use the formula for clustering coefficient to compute it for the given graph.
Problem 7. Identify communities in a social network using community detection algorithms.
Solution: Apply community detection algorithms like Louvain or Girvan-Newman.

Circuit Design

Problem 8. Determine if a given graph is planar.
Solution: Apply Kuratowski's theorem or use planarity testing algorithms.
Problem 9. Find a Hamiltonian cycle in an electronic circuit layout.
Solution: Use algorithms like Held-Karp or brute force to find a Hamiltonian cycle.
Problem 10. Calculate the voltage across each component in a circuit using Kirchhoff's laws.
Solution: Apply Kirchhoff's laws to solve for voltages in the circuit.

Epidemiology

Problem 11. Model the spread of a disease using a susceptible-infected-recovered (SIR) model.
Solution: Simulate the SIR model to track the disease's progression.
Problem 12. Identify super-spreaders in a disease transmission network.
Solution: Analyze the network's structure and identify nodes with high degrees.
Problem 13. Calculate the effective reproduction number (R0) of a disease in a population.
Solution: Use mathematical models to estimate R0 based on disease parameters.

Recommendation Systems

Problem 14. Build a user-product recommendation graph based on user interactions.
Solution: Create a bipartite graph and incorporate user-product interactions.
Problem 15. Implement a collaborative filtering recommendation algorithm.
Solution: Use collaborative filtering techniques to make personalized recommendations.
Problem 16. Compute the Jaccard similarity between user preferences for two products.
Solution: Calculate the Jaccard similarity using the intersection and union of user sets.

Biology and Bioinformatics

Problem 17. Construct a phylogenetic tree from DNA sequence data.
Solution: Apply distance-based or character-based methods to build the tree.

Problem 18. Identify protein-protein interactions in a biological network.
Solution: Analyze the network's topology and identify interacting protein pairs.
Problem 19. Find the shortest sequence alignment between two DNA sequences.
Solution: Use dynamic programming algorithms like Needleman-Wunsch or Smith-Waterman.

Optimization and Logistics

Problem 20. Optimize the delivery routes for a fleet of trucks to minimize travel distance.
Solution: Apply the traveling salesman problem (TSP) algorithm to find optimal routes.
Problem 21. Schedule tasks on machines to minimize completion time (makespan).
Solution: Solve the job scheduling problem using algorithms like List Scheduling or Genetic Algorithms.

Web and Social Media

Problem 22. Analyze the connectivity of web pages in a web crawling graph.
Solution: Use algorithms like PageRank to measure page importance.
Problem 23. Find influential nodes in a social media network for targeted marketing.
Solution: Apply centrality measures like eigenvector centrality or betweenness centrality.

Finance and Risk Management

Problem 24. Model financial dependencies in a portfolio of assets using a correlation graph.
Solution: Construct a correlation graph and analyze its structure.
Problem 25. Calculate the risk exposure of a financial institution in a network of counterparties.
Solution: Use network analysis to measure systemic risk.

Energy and Power Grids

Problem 26. Determine the vulnerability of a power grid to node or edge failures.
Solution: Perform a vulnerability analysis on the grid's topology.
Problem 27. Optimize the placement of renewable energy sources in an electrical grid.
Solution: Apply optimization algorithms to determine the best locations.

Game Theory and Social Choice

Problem 28. Solve the stable matching problem for a set of participants and preferences.
Solution: Use algorithms like the Gale-Shapley algorithm to find stable matchings.
Problem 29. Determine the winner of an election using ranked preference voting.
Solution: Apply voting methods like Borda count or Ranked Pairs.

Computer Science and Algorithms

Problem 30. Solve the graph coloring problem for a map with regions and adjacent borders.
Solution: Apply graph coloring algorithms to color the map with the fewest colors.

Problem 31. Find strongly connected components in a directed graph.

Solution: Apply Kosaraju's algorithm or Tarjan's algorithm.

Problem 32. Detect cycles in a directed graph.

Solution: Use depth-first search (DFS) and track back edges.

Natural Language Processing

Problem 33. Represent a text corpus as a semantic graph for document similarity analysis.

Solution: Apply techniques like Latent Semantic Analysis (LSA) or Word Embeddings.

Problem 34. Find the shortest path between two words in a word association graph.

Solution: Use word embedding vectors and graph algorithms to find the path.

Environmental Science

Problem 35. Model the flow of pollutants in a river network.

Solution: Apply network flow algorithms to simulate pollutant dispersion.

Problem 36. Analyze the structure of ecological food webs.

Solution: Study food web properties like trophic levels and keystone species.

Supply Chain and Logistics

Problem 37. Optimize the supply chain for a company with multiple suppliers and distribution centers.

Solution: Apply network optimization techniques to minimize costs.

Problem 38. Determine the critical paths in a project management network.

Solution: Use the Critical Path Method (CPM) to identify critical tasks and project duration.

Social Sciences and Sociology

Problem 39. Analyze the spread of information or influence in a social network.

Solution: Use network diffusion models to simulate information propagation.

Problem 40. Study the collaboration network of researchers and identify influential authors.

Solution: Apply network analysis measures like degree centrality and co-authorship networks.

Healthcare and Medical Research

Problem 41. Analyze patient data to identify disease clusters in a healthcare network.

Solution: Apply clustering algorithms to group similar patients.

Problem 42. Model the spread of infectious diseases within a hospital network.

Solution: Use epidemiological models and network analysis to assess infection risk.

Environmental Engineering

Problem 43. Design a water distribution network to optimize water supply.

Solution: Apply network design and optimization techniques to minimize costs.

Problem 44. Analyze the transportation network for a waste management system.

Solution: Use graph algorithms to optimize waste collection routes.

Machine Learning and Deep Learning

Problem 45. Construct a convolutional neural network (CNN) architecture as a graph.

Solution: Design the CNN's architecture as a graph of layers and connections.

Problem 46. Represent a knowledge graph for a question-answering system.

Solution: Create a graph structure to represent relationships between concepts.

Physics and Materials Science

Problem 47. Model the crystal structure of a material using a lattice graph.

Solution: Represent the crystal lattice as a graph to study its properties.

Problem 48. Analyze the connectivity of particles in a granular material.

Solution: Use graph analysis to study the packing and stability of particles.

Robotics and Autonomous Systems

Problem 49. Plan the path of a robot through an obstacle-filled environment.

Solution: Apply graph-based path planning algorithms like A* or D*.

Problem 50. Model the sensor network of an autonomous vehicle for environmental monitoring.

Solution: Create a sensor network graph to optimize data collection and analysis.

Short Questions and Answers

Q.1. What is a graph in graph theory?

Ans. A graph is a mathematical structure comprising nodes (vertices) and edges that connect pairs of nodes.

Q.2. What is a directed graph?

Ans. A directed graph has edges with a specific direction, indicating one-way connections between nodes.

Q.3. What is an undirected graph?

Ans. An undirected graph has edges without a direction, representing symmetric relationships between nodes.

Q.4. What is a weighted graph?

Ans. A weighted graph assigns numerical values (weights) to edges, indicating costs, distances, or strengths of connections.

Q.5. What is a complete graph?

Ans. A complete graph is one where every pair of distinct nodes is connected by an edge.

Q.6. What is a cycle in a graph?

Ans. A cycle is a path that starts and ends at the same node, passing through a sequence of distinct nodes.

Q.7. What is a tree in graph theory?

Ans. A tree is a connected graph without any cycles, having exactly one path between any pair of nodes.

Q.8. What is a bipartite graph?

Ans. A bipartite graph has nodes divided into two disjoint sets, with edges connecting nodes from different sets.

Q.9. What is the degree of a node in a graph?

Ans. The degree of a node is the number of edges incident to it.

Q.10. What is a path in a graph?

Ans. A path is a sequence of nodes where each consecutive pair is connected by an edge.

Q.11. What is a spanning tree?

Ans. A spanning tree of a graph is a subgraph that covers all nodes while being a tree (connected and acyclic).

Q.12. What is a cycle basis?

Ans. A cycle basis is a set of cycles that can uniquely describe all other cycles in a graph.

Q.13. What is a planar graph?

Ans. A planar graph can be drawn on a plane without any edge intersections.

Q.14. What is Euler's formula for planar graphs?

Ans. For a connected planar graph with V vertices, E edges, and F faces: $V - E + F = 2$.

Q.15. What is graph connectivity?

Ans. Graph connectivity refers to how nodes are connected, whether there are paths between any pair of nodes.

Q.16. What is graph traversal?

Ans. Graph traversal is the process of visiting nodes and edges in a systematic manner to explore a graph's structure.

Q.17. What is breadth-first search (BFS)?

Ans. BFS is a graph traversal algorithm that explores nodes level by level from a starting node.

Q.18. What is depth-first search (DFS)?

Ans. DFS is a graph traversal algorithm that explores as far as possible along each branch before backtracking.

Q.19. What is Dijkstra's algorithm?

Ans. Dijkstra's algorithm finds the shortest path from a source node to all other nodes in a weighted graph.

Q.20. What is the Bellman-Ford algorithm?

Ans. The Bellman-Ford algorithm finds the shortest path from a source node to all other nodes, even with negative edge weights.

Q.21. What is the Floyd-Warshall algorithm?

Ans. The Floyd-Warshall algorithm computes all pairs shortest paths in a graph, handling both positive and negative edge weights.

Q.22. What is the Hamiltonian cycle?

Ans. A Hamiltonian cycle is a cycle that visits every node exactly once in a graph.

Q.23. What is the Traveling Salesman Problem (TSP)?

Ans. TSP is an optimization problem aiming to find the shortest possible route that visits a given set of cities and returns to the starting city.

Q.24. What is a graph coloring problem?

Ans. The graph coloring problem involves assigning colors to nodes in a way that no two adjacent nodes share the same color.

Q.25. What is the chromatic number of a graph?

Ans. The chromatic number of a graph is the minimum number of colors needed to color its nodes without adjacent nodes sharing the same color.

Q.26. What is the clique problem?

Ans. The clique problem involves finding the largest complete subgraph within a given graph.

Q.27. What is the independent set problem?

Ans. The independent set problem involves finding the largest subset of non-adjacent nodes in a graph.

Q.28. What is graph isomorphism?

Ans. Graph isomorphism is the concept of determining whether two graphs are structurally identical, even if node and edge labels differ.

Q.29. What is a matching in a graph?

Ans. A matching is a set of edges without common nodes.

Q.30. What is a maximum matching?

Ans. A maximum matching is a matching with the maximum number of edges in a graph.

Q.31. What is a bipartite matching?

Ans. A bipartite matching is a matching in a bipartite graph, where each node is adjacent to at most one edge.

Q.32. What is the cut-vertex of a graph?

Ans. A cut-vertex is a node whose removal disconnects the graph into two or more components.

Q.33. What is the cut-edge of a graph?

Ans. A cut-edge is an edge whose removal increases the number of components in the graph.

Q.34. What is the Eulerian path/circuit?

Ans. An Eulerian path/circuit is a path/cycle that visits all edges exactly once in a graph.

Q.35. What is the bridge in a graph?

Ans. A bridge is an edge whose removal increases the number of components in the graph.

Q.36. What is a spanning forest?

Ans. A spanning forest is a collection of disjoint spanning trees in a graph.

Q.37. What is a topological sort?

Ans. A topological sort is a linear ordering of nodes in a directed acyclic graph that respects edge directions.

Q.38. What is a directed acyclic graph (DAG)?

Ans. A directed acyclic graph is a graph without cycles and with directed edges.

Q.39. What is the chromatic polynomial?

Ans. The chromatic polynomial is a function that counts the number of ways to color the nodes of a graph with a given number of colors.

Q.40. What is the clique cover problem?

Ans. The clique cover problem involves finding a set of cliques that together cover all nodes in the graph.

Q.41. What is the girth of a graph?

Ans. The girth of a graph is the length of its shortest cycle.

Q.42. What is the flow network in graph theory?

Ans. A flow network is a directed graph with capacities on the edges, used to model the flow of resources.

Q.43. What is the max-flow min-cut theorem?

Ans. The max-flow min-cut theorem states that the maximum flow in a flow network is equal to the minimum cut capacity.

Q.44. What is the cycle space of a graph?

Ans. The cycle space is the vector space spanned by the cycles of a graph.

Q.45. What is the edge connectivity of a graph?

Ans. The edge connectivity of a graph is the minimum number of edges whose removal disconnects the graph.

Q.46. What is the vertex connectivity of a graph?

Ans. The vertex connectivity of a graph is the minimum number of nodes whose removal disconnects the graph.

Q.47. What is the line graph of a graph?

Ans. The line graph of a graph represents nodes as edges, where two nodes are connected if their corresponding edges share a common node.

Q.48. What is the Turán's theorem?

Ans. Turán's theorem provides an upper bound on the number of edges in an n-node graph that does not contain a certain subgraph.

Q.49. What is the graph automorphism?

Ans. A graph automorphism is an isomorphism from a graph to itself.

Q.50. What is the Tutte matrix?

Ans. The Tutte matrix is a matrix used to calculate the number of spanning trees in a graph.

Multiple Choice Questions and Answers

Transportation and Networks

Q.1. Which algorithm is commonly used to find the shortest path in a weighted graph?

(*a*) Depth-First Search (DFS)
(*b*) Breadth-First Search (BFS)
(*c*) Dijkstra's Algorithm
(*d*) Prim's Algorithm

Q.2. What is the purpose of finding the minimum spanning tree (MST) in a network?

(*a*) To find the longest path between two nodes
(*b*) To find the shortest path between two nodes
(*c*) To connect all nodes with the minimum total edge weight
(*d*) To disconnect all nodes

Q.3. In a directed graph, what does a Hamiltonian path refer to?

(*a*) A path that visits all nodes exactly once.
(*b*) A path that visits some nodes more than once.
(*c*) A path that only visits nodes with even degrees.
(*d*) A path that starts and ends at the same node.

Q.4. Which algorithm is used to find the maximum flow in a flow network?

(*a*) Dijkstra's Algorithm
(*b*) Kruskal's Algorithm
(*c*) Ford-Fulkerson Algorithm
(*d*) A* Algorithm

Q.5. In social network analysis, what is the "degree" of a node?

(*a*) The number of neighbors a node has.
(*b*) The number of communities the node belongs to.
(*c*) The number of likes or dislikes a user has given.
(*d*) The number of nodes in the entire network.

Circuit Design

Q.6. What is the primary objective when determining if a graph is planar?

(*a*) To find the shortest path between two nodes.
(*b*) To decide if the graph can be drawn without edge crossings.
(*c*) To find a Hamiltonian cycle.
(*d*) To calculate voltage across components.

Q.7. Which algorithm is used to find strongly connected components in a directed graph?

(*a*) Dijkstra's Algorithm
(*b*) Kruskal's Algorithm
(*c*) Tarjan's Algorithm
(*d*) Ford-Fulkerson Algorithm

Q.8. In Kirchhoff's laws for circuit analysis, what is the conservation of charge law based on?

(*a*) Conservation of voltage
(*b*) Conservation of current
(*c*) Conservation of power
(*d*) Conservation of resistance

Epidemiology

Q.9. What does the SIR model in epidemiology stand for?

(*a*) Susceptible, Infected, Resistant
(*b*) Sick, Inactive, Recovered
(*c*) Spread, Infection, Recovery
(*d*) Suspected, Recovered

Q.10. In epidemiology, what is a "super-spreader"?

(*a*) An individual with a strong immune system.
(*b*) An individual who is resistant to disease.
(*c*) An individual who infects a disproportionately large number of others.
(*d*) An individual who recovers quickly from a disease.

Q.11. What does R_0 represent in epidemiology?

(*a*) The number of people who have recovered from a disease.
(*b*) The number of people who are currently infected.
(*c*) The effective reproduction number of a disease.
(*d*) The initial reproduction number of a disease.

Recommendation Systems

Q.12. What is collaborative filtering in recommendation systems based on?

(*a*) User demographics
(*b*) User ratings and preferences
(*c*) Item popularity
(*d*) Item price

Q.13. Which similarity measure is often used in content-based recommendation systems?

(*a*) Jaccard similarity
(*b*) Pearson correlation coefficient
(*c*) Euclidean distance
(*d*) Cosine similarity

Q.14. In recommendation systems, what is the purpose of a recommender algorithm?

(*a*) To rank items by their price.
(*b*) To connect users with similar tastes.
(*c*) To maximize user privacy.
(*d*) To recommend items to users based on their preferences.

Biology and Bioinformatics

Q.15. What is a phylogenetic tree used for in biology?

(*a*) To study the relationships between species.
(*b*) To model the spread of diseases.
(*c*) To analyze protein-protein interactions.
(*d*) To design electrical circuits.

Q.16. Which graph-based data structure is often used to represent protein-protein interaction networks?

(*a*) Phylogenetic trees
(*b*) Bipartite graphs
(*c*) Knowledge graphs
(*d*) Social networks

Q.17. Which algorithm is used for finding the optimal sequence alignment between DNA sequences?

(*a*) Depth-First Search (DFS)
(*b*) Dijkstra's Algorithm
(*c*) Needleman-Wunsch Algorithm
(*d*) PageRank Algorithm

Optimization and Logistics

Q.18. Which problem does the Traveling Salesman Problem (TSP) aim to solve?

(*a*) Finding the shortest path between two nodes in a network.

(*b*) Finding the longest path between two nodes in a network.

(*c*) Finding the optimal route to visit a set of cities once and return to the starting city.

(*d*) Finding the optimal route to visit all cities multiple times.

Q.19. What is the primary goal of the Critical Path Method (CPM)?

(*a*) To find the shortest path in a network.

(*b*) To identify the most critical tasks in a project.

(*c*) To calculate voltage in an electrical circuit.

(*d*) To optimize delivery routes.

Web and Social Media

Q.20. What does PageRank measure in web search algorithms?

(*a*) The number of web pages on a website.

(*b*) The relevance of a web page to a search query.

(*c*) The popularity and importance of web pages in a network.

(*d*) The loading speed of a web page.

Q.21. Which centrality measure is used to identify influential nodes in a social network?

(*a*) Degree centrality (*b*) Closeness centrality

(*c*) Betweenness centrality (*d*) Eigenvector centrality

Finance and Risk Management

Q.22. How is systemic risk measured in a financial network?

(*a*) By analyzing the network's connectivity

(*b*) By calculating the network's maximum flow

(*c*) By counting the number of nodes (*d*) By measuring the total asset value

Q.23. Which of the following is a commonly used network model to represent financial dependencies?

(*a*) Bipartite graph (*b*) Planar graph

(*c*) Complete graph (*d*) Random graph

Energy and Power Grids

Q.24. What is the primary objective of optimizing the placement of renewable energy sources in an electrical grid?

(*a*) To maximize the cost of energy production. (*b*) To minimize energy consumption.

(*c*) To minimize environmental impact. (*d*) To maximize energy losses.

Q.25. What does the term "vulnerability" refer to in the context of power grids?

(*a*) The ability of a grid to withstand extreme weather conditions.
(*b*) The ease with which a grid can be repaired after a failure.
(*c*) The likelihood of node or edge failures causing widespread outages.
(*d*) The number of renewable energy sources in the grid.

Game Theory and Social Choice

Q.26. What problem does the Gale-Shapley algorithm solve?

(*a*) Finding the maximum flow in a network.
(*b*) Finding the minimum spanning tree in a graph.
(*c*) Finding stable matchings in a bipartite graph.
(*d*) Finding the optimal path in a maze.

Q.27. Which voting method is based on ranking preferences and eliminates candidates iteratively?

(*a*) Plurality voting
(*b*) Ranked Pairs
(*c*) Borda count
(*d*) Approval voting

Computer Science and Algorithms

Q.28. What is the primary objective of graph coloring algorithms?

(*a*) To find the shortest path between two nodes in a graph.
(*b*) To assign colors to nodes such that no adjacent nodes have the same color.
(*c*) To maximize the number of edges in a graph.
(*d*) To find the maximum flow in a network.

Q.29. Which algorithm is used to find the shortest path between two nodes in an unweighted graph?

(*a*) Kruskal's Algorithm
(*b*) Dijkstra's Algorithm
(*c*) Depth-First Search (DFS)
(*d*) Breadth-First Search (BFS)

Natural Language Processing

Q.30. What is Latent Semantic Analysis (LSA) commonly used for in natural language processing?

(*a*) Sentiment analysis
(*b*) Language translation
(*c*) Document clustering and retrieval
(*d*) Part-of-speech tagging

Q.31. Which similarity measure is often used in text document similarity analysis?

(*a*) Jaccard similarity
(*b*) Pearson correlation coefficient
(*c*) Euclidean distance
(*d*) Cosine similarity

Environmental Science

Q.32. What is the primary purpose of modeling pollutant dispersion in a river network?

(*a*) To calculate the velocity of the river water.
(*b*) To predict the direction of river flow.
(*c*) To study the spread of pollutants and assess environmental impact.
(*d*) To identify potential fishing spots.

Q.33. What are trophic levels in the context of ecological food webs?

(*a*) The levels of pollution in a water ecosystem.
(*b*) The levels of energy transfer in a food chain.
(*c*) The levels of soil fertility in a forest ecosystem.
(*d*) The levels of air quality in a region.

Supply Chain and Logistics

Q.34. What does the term "makespan" refer to in job scheduling?

(*a*) The total cost of scheduling jobs.
(*b*) The maximum time it takes to complete all jobs.
(*c*) The number of jobs in the schedule. (*d*) The priority of jobs.

Q.35. Which optimization problem involves finding the optimal allocation of resources to tasks?

(*a*) Job scheduling problem (*b*) Network flow problem
(*c*) Bipartite graph matching problem (*d*) Shortest path problem

Social Sciences and Sociology

Q.36. In social network analysis, what is the "centrality" of a node?

(*a*) How sociable the node is.
(*b*) The position or importance of the node in the network.
(*c*) The number of friends a node has. (*d*) The number of nodes in the network.

Q.37. What is network diffusion modeling used for in social sciences?

(*a*) Modeling the spread of diseases.
(*b*) Modeling the spread of information, innovations, or influence.
(*c*) Modeling the transportation of goods. (*d*) Modeling social interactions.

Healthcare and Medical Research

38. In healthcare, what does the term "clustering" refer to?

(*a*) The formation of groups of patients with similar characteristics.
(*b*) The spread of diseases in a population. (*c*) The organization of hospitals in a region.
(*d*) The allocation of medical resources.

Q.39. Which mathematical models are commonly used to simulate disease spread in a hospital setting?

(*a*) Financial models (*b*) Epidemiological models
(*c*) Transportation models (*d*) Weather models

Environmental Engineering

Q.40. What is the primary goal of optimizing the supply chain for a company?

(*a*) To maximize delivery times. (*b*) To minimize the number of suppliers.
(*c*) To minimize costs while ensuring timely delivery.
(*d*) To maximize product prices.

Q.41. Which type of network analysis is used to optimize waste collection routes?

(*a*) Social network analysis
(*b*) Transportation network analysis
(*c*) Electrical grid analysis
(*d*) Financial network analysis

Machine Learning and Deep Learning

Q.42. How is a convolutional neural network (CNN) architecture typically represented as a graph?

(*a*) As a directed acyclic graph (DAG) of layers and their connections.
(*b*) As a tree structure of nodes.
(*c*) As a fully connected graph.
(*d*) As a bipartite graph.

Q.43. What is a knowledge graph commonly used for in machine learning?

(*a*) Representing neural networks.
(*b*) Representing relationships between entities and concepts.
(*c*) Visualizing data
(*d*) Solving optimization problems.

Physics and Materials Science

Q.44. How is a crystal structure often represented using graph theory?

(*a*) As a directed graph with nodes representing atoms.
(*b*) As a weighted graph with nodes representing molecules.
(*c*) As a lattice graph with nodes representing lattice points.
(*d*) As a bipartite graph with nodes representing crystallographic data.

Q.45. What is the primary purpose of analyzing the connectivity of particles in a granular material?

(*a*) To determine the size of particles.
(*b*) To study the packing and stability of particles.
(*c*) To calculate the temperature of the material.
(*d*) To measure the electrical conductivity of the material.

Robotics and Autonomous Systems

Q.46. Which algorithm is commonly used to plan the path of a robot through an obstacle-filled environment?

(*a*) Depth-First Search (DFS)
(*b*) Dijkstra's Algorithm
(*c*) A* Algorithm
(*d*) PageRank Algorithm

Q.47. What is the primary purpose of modeling a sensor network for environmental monitoring?

(*a*) To predict weather patterns.
(*b*) To identify potential hazards in the environment.
(*c*) To optimize energy consumption.
(*d*) To optimize the placement of sensors for data collection.

Transportation and Networks

Q.48. In a transportation network, what does Dijkstra's algorithm primarily help to find?

(*a*) The longest path between two locations.

(*b*) The minimum spanning tree of the network.

(*c*) The shortest path between two locations. (*d*) The maximum flow in the network.

Social Network Analysis

Q.49. What does "centrality" measure in a social network?

(*a*) The number of users in the network.

(*b*) The importance or influence of a node in the network.

(*c*) The number of connections a node has.

(*d*) The degree of separation between two users.

Biology and Bioinformatics

Q.50. In bioinformatics, what is the purpose of constructing a phylogenetic tree?

(*a*) To represent the spread of diseases in a population.

(*b*) To analyze protein-protein interactions in a network.

(*c*) To study the evolutionary relationships between species.

(*d*) To optimize delivery routes in a supply chain.

Key to Multiple Choice Questions

1. (*c*)	**2.** (*c*)	**3.** (*a*)	**4.** (*c*)	**5.** (*a*)	**6.** (*b*)	**7.** (*c*)	**8.** (*b*)	**9.** (*a*)
10. (*c*)	**11.** (*d*)	**12.** (*b*)	**13.** (*d*)	**14.** (*d*)	**15.** (*a*)	**16.** (*b*)	**17.** (*c*)	**18.** (*c*)
19. (*b*)	**20.** (*c*)	**21.** (*d*)	**22.** (*a*)	**23.** (*a*)	**24.** (*c*)	**25.** (*c*)	**26.** (*c*)	**27.** (*b*)
28. (*b*)	**29.** (*d*)	**30.** (*c*)	**31.** (*d*)	**32.** (*c*)	**33.** (*b*)	**34.** (*b*)	**35.** (*a*)	**36.** (*b*)
37. (*b*)	**38.** (*a*)	**39.** (*b*)	**40.** (*c*)	**41.** (*b*)	**42.** (*a*)	**43.** (*b*)	**44.** (*c*)	**45.** (*b*)
46. (*c*)	**47.** (*d*)	**48.** (*c*)	**49.** (*b*)	**50.** (*c*)				

Projects Works

1. **Robot Path Planning Using Graph Algorithms:** Develop a project where a robot navigates through an environment using graph-based path planning algorithms like Dijkstra's, A*, or RRT. Implement these algorithms to find optimal paths while avoiding obstacles.
2. **Multi-Robot Coordination Using Graph Coloring:** Create a project that involves coordinating multiple robots to perform tasks in an environment. Use graph coloring algorithms to assign tasks to robots in a way that optimizes efficiency and minimizes conflicts.
3. **Warehouse Optimization with Graph Theory:** Design a project that optimizes the movement of robotic agents in a warehouse setting. Use graph-based algorithms to determine the most efficient routes for robots to pick and deliver items.

4. **Autonomous Drone Navigation:** Develop a project where drones navigate autonomously using graph-based methods. Explore how graphs can represent complex environments and help drones find collision-free paths.
5. **Network Topology Optimization:** Investigate how graph theory can be applied to optimize the communication network topology in a swarm of robots. Develop algorithms to ensure efficient data exchange and collaboration.
6. **Graph-Based Robotic Manipulation Planning:** Create a project that focuses on robotic manipulation tasks. Use graph algorithms to plan grasp poses, plan object manipulation sequences, or optimize the ordering of tasks in a manipulation sequence.
7. **Dynamic Task Allocation in Multi-Agent Systems:** Build a project that dynamically assigns tasks to multiple robots in real-time. Implement graph algorithms to allocate tasks based on changing conditions and priorities.
8. **Robot Coverage and Exploration:** Explore how graph-based algorithms can be applied to enable robots to efficiently cover and explore unknown environments. Investigate the use of spanning trees or graph traversal techniques.
9. **Robot Swarm Formation Control:** Develop a project where a swarm of robots forms and maintains specific formations using graph-based approaches. Implement algorithms to control the robots' positions and orientations.
10. **Graph Theory for Human-Robot Interaction:** Create a project that uses graph theory to model social interactions between robots and humans. Develop algorithms to optimize the robot's behavior to facilitate effective communication and collaboration.
11. **Sensor Network Localization:** Apply graph theory to solve the problem of localizing a network of sensors or robots in an unknown environment. Explore techniques like range-based or range-free localization.
12. **Optimal Sensor Placement:** Investigate how graph-based algorithms can be used to determine the optimal placement of sensors or cameras in an environment to achieve maximum coverage or observation quality.
13. **Graph-Based Fault Detection and Recovery:** Build a project that uses graph based algorithms to detect faults in a robotic system and devise recovery strategies. Model the system's components as nodes and use edge weights to represent dependencies.
14. **Energy-Efficient Robot Charging Strategy:** Design a project where robots autonomously decide when and where to recharge based on their battery levels and the distance to charging stations. Use graph-based algorithms to optimize the charging strategy.
15. **Graph-Based Robot Learning:** Develop a project that combines graph theory with machine learning to enable robots to learn and adapt to their environment. Explore techniques like reinforcement learning or graph neural networks.

10

Numerical Methods to obtain Practical Solutions by using Computational Algorithms and Techniques

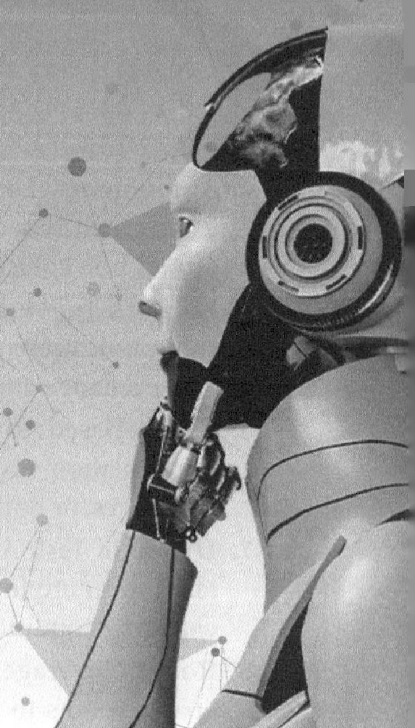

10.1 INTRODUCTION

The employment of numerical methods allows for the approximation of solutions to issues that analytical methods cannot completely resolve. When complicated equations or systems of equations are involved, these issues frequently occur in a variety of disciplines including engineering, physics, economics, and more. By utilizing computer algorithms and techniques, numerical approaches offer a mechanism to arrive at valuable and applicable solutions. Overall, numerical approaches are essential to modern research and engineering because they offer real-world answers to challenging issues that cannot be resolved analytically. They use the strength of computational resources to fill the gap between theoretical mathematics and practical applications.

10.2 KEY POINTS OF NUMERICAL METHODS

Here are some key points to understand about numerical methods:

1. **Approximation:** Numerical techniques entail estimating a problem's actual solution. Since there are often no closed-form solutions to mathematical problems, numerical methods offer a mechanism to find solutions that are close enough to be useful.
2. **Discretization:** A lot of issues require continuous mathematical objects, such functions or equations, which computers cannot directly interpret. The discretization of these continuous things into discrete values used in numerical methods makes them susceptible to computational manipulation.
3. **Algorithms:** Numerical approaches use algorithms, which are detailed instructions for carrying out calculations to produce the desired approximation. Programming languages are used to implement these algorithms, which are then carried out by computers.
4. **Errors:** Because of the limited precision of computers and the approximations used during computations, numerical methods create a variety of errors. These mistakes can be divided into truncation errors and round-off errors, which are caused by approximations, made when designing algorithms and restricted computer precision, respectively.

5. **Convergence:** A numerical method is considered to converge if, as the computational effort (iterations, steps, etc.) rises, the approximation it generates approaches the genuine answer. Convergence analysis aids in evaluating a numerical method's efficiency and dependability.
6. **Stability:** Stability describes a numerical method's capacity to deliver accurate results even in the presence of minute alterations or faults in the input data. A procedure that is unstable can produce wildly inaccurate findings.
7. **Problem Sorts:** Different sorts of issues can be solved using numerical methods, including as nonlinear equations, numerical integration (determining the region under a curve), differential equations, linear algebraic equations, optimization, interpolation, and more.
8. The Newton-Raphson method, the trapezoidal rule, Simpson's rule, the finite difference method, the Gaussian elimination method, and the finite difference method for solving differential equations are a few examples of popular numerical methods.
9. **Software and Tools:** Various numerical approaches can be implemented using a variety of software libraries, packages, and tools. For the implementation and use of numerical methods, programming languages like MATLAB, Python (with libraries like NumPy and SciPy), and Julia are frequently utilized.

10.3 NUMERICAL INTEGRATION

The process of numerical integration, often referred to as numerical quadrature, is used to make an approximation of a function's definite integral across a certain range of values. The area under a curve between two specified bounds is represented by the definite integral. When the integral cannot be computed analytically using conventional techniques or when the function is only known from sampled data points, numerical integration is especially helpful.

There are numerous numerical techniques for approximate integrals, each with advantages and disadvantages. Here are a few typical approaches:

1. **Trapezoidal Rule:** With this technique, a trapezium is used to roughly represent the area under the curve between two nearby data points. The sum of the areas of these trapezoids is used to approximate the integral. Although the trapezoidal rule is very straightforward, it gets more precise as the number of intervals (subdivisions) rises.
2. **Simpson's Rule:** By approximating portions of the curve with quadratic polynomials, Simpson's technique offers a more precise approximation. Each pair of neighboring subintervals is used to fit a quadratic polynomial. The interval is divided into an even number of subintervals. The areas under these polynomials are then weighted to approximate the integral.
3. **Midpoint Rule:** By breaking the interval into smaller segments, this technique can roughly estimate the integral. It presumes that each subinterval's function is roughly constant.
4. **Numerical Integration Libraries:** Numerous programming languages include libraries or functions that can be used to conduct numerical integration. The SciPy library in Python, for instance, has a number of integration methods, including quad for general integration, trapz for the trapezoidal rule, and simps for Simpson's rule.

The number of subintervals selected will affect how accurate these methods are. In general, increasing the number of subintervals increases computational work while simultaneously improving approximation accuracy. The needed degree of accuracy and the properties of the function being integrated frequently influence the technique of choice and the number of subintervals.

Here's a simplified example of using the trapezoidal rule in Python:

import numpy as np from scipy.integrate import trapz # Define the function to be integrated def f(x): return x**2 # Define the interval [a, b] a = 0 b = 2 # Generate x values within the interval x_vals = np.linspace(a, b, num=5) # Calculate corresponding y values y_vals = f(x_vals) # Use the trapezoidal rule to approximate the integral approx_integral = trapz(y_vals, x=x_vals) print("Approximate Integral:", approx_integral)

Always keep in mind that the method you use relies on the particular issue at hand as well as the desired level of precision. More sophisticated numerical integration techniques could be required when working with more complicated functions or when accuracy requirements are higher.

10.4 NUMERICAL DIFFERENTIATION

Utilizing discrete data points, numerical differentiation is a method for approximating a function's derivative at a given location. The pace at which a function's values vary in relation to its independent variable is indicated by the function's derivative. When working with data that is only available at specified locations or when there is no analytical expression for the derivative, numerical differentiation is helpful.

There are a few different methods commonly used for numerical differentiation:

1. **Forward Difference:** In this method, the derivative at a point is approximated using the slope between the function values at the current point and a nearby point slightly ahead. Mathematically, the forward difference approximation is given by:

   $$f'(x) \approx (f(x + h) - f(x))/h$$

   where h is a small step size.
2. **Backward Difference:** Similar to the forward difference method, the backward difference approximates the derivative using the slope between the current point and a nearby point slightly behind. The formula is:

   $$f'(x) \approx (f(x) - f(x - h))/h$$
3. **Central Difference:** This method provides a more accurate approximation by using the slope between two nearby points, one slightly ahead and one slightly behind. The formula for central difference is:

   $$f'(x) \approx (f(x + h) - f(x - h))/(2 * h)$$

   The central difference method typically provides better accuracy than the forward or backward difference methods.
4. **Five-Point Stencil:** This is a higher-order method that uses more data points to achieve greater accuracy. It combines the function values at the point of interest and two points on each side. The formula involves a weighted combination of these points.

   The accuracy of numerical differentiation techniques increases as the step size h decreases. However, round-off mistakes might cause numerical instability when utilising a very tiny h. It's crucial to select a suitable value for h based on the size of the issue and the required degree of accuracy.

   Here is a condensed illustration of how to use Python's central difference approach for numerical differentiation:

   def f(x): return x**2 def numerical_derivative(f, x, h): return (f(x + h) – f(x – h))/(2 * h) x_0 = 1.0 # Point at which derivative is approximated h = 0.01 # Step size approx_derivative = numerical_ derivative(f, x_0, h) print("Approximate Derivative:", approx_derivative) .Keep in mind that numerical differentiation methods are subject to errors due to the discretization of data and the choice of step size. Additionally, noisy or irregularly spaced data can further complicate the process of numerical differentiation.

10.5 ROOT-FINDING METHODS

The procedures used to locate the roots (or solutions) of equations numerically are referred to as "root-finding methods," "root approximation methods," or "numerical root-finding techniques." A value at which an equation evaluates to zero is known as its root. These techniques are especially helpful when it is difficult or impossible to identify the roots analytically.

Here are some common root-finding methods:

1. **Bisection Method:** The Intermediate Value Theorem is the foundation of this approach. It operates by periodically reducing a region where a root resides. The procedure establishes which of the smaller subintervals of the interval holds the root. Repeat this procedure until a reliable root is identified.
2. **Newton-Raphson Method (Newton's Method):** The tangent line is used in this iterative approach to roughly determine a function's root. The approach uses the following formula to iteratively revise the estimate starting with an initial guess:

   x_{n+1} = x_n – f(x_n)/f'(x_n)

   where x_n is the current estimate and f'(x_n) is the derivative of the function at that point. This method is efficient for well-behaved functions but can diverge or get stuck in certain cases.
3. **Secant Method:** Similar to the Newton-Raphson method, the secant method approximates the root using a linear interpolation of two points. However, it doesn't require the computation of derivatives. The formula for iteration is:

   x_{n+1} = x_n – f(x_n) * (x_n – x_{n–1})/(f(x_n) – f(x_{n–1}))
4. **Fixed-Point Iteration:** By using this technique, the equation is changed into an equivalent form in which the root represents a fixed point in a function. The method then uses this altered function to iteratively update the guess. To ensure convergence, the converting function must be carefully chosen for the procedure.
5. **Regula Falsi (False Position) Method:** The bisection method and linear interpolation are combined in this approach. It keeps the two points that bracket the root and uses the secant line's slope to approximately determine the root. In order for the function values at the two places to have opposing signs, the algorithm chooses the following estimate accordingly.
6. **Brent's Method:** The bisection method, secant method, and inverse quadratic interpolation are all combined in Brent's approach. In order to give both quick convergence and resilience, it seeks to integrate the advantages of these methods.

The characteristics of the function and the desired level of accuracy must be taken into consideration while selecting the best root-finding technique. For some kinds of functions or educated estimates, several techniques might perform better. The behavior of convergence, possible divergence, and the existence of many roots are additional crucial considerations.

Here's a simplified example of using the bisection method in Python to find a root of a function:

def f(x): return x**2 – 4 def bisection_method(f, a, b, tol, max_iterations): if f(*a*) * f(*b*) >= 0: return None # Root is not guaranteed to exist in the interval for _ in range(max_iterations): c = (a + b)/2 if abs(f(*c*))

< tol: return c # Found a root within the specified tolerance elif f(c) * f(a) < 0: b = c else: a = c return None # Maximum iterations reached without finding a root a = 0.0 b = 3.0 tolerance = 1e–6 max_iterations = 100 root = bisection_method(f, a, b, tolerance, max_iterations) if root is not None: print("Approximate Root:", root) else: print("No root found in the specified interval.")

When there are no analytical solutions to an equation or when it is difficult to discover the answers directly, root-finding techniques can be very useful.

10.6 OPTIMIZATION ALGORITHMS

An objective function is supplied, and optimization algorithms are computer methods used to determine the optimal solution—either the maximum or minimum—within a defined domain. Numerous disciplines, including mathematics, engineering, economics, machine learning, and more, frequently deal with optimization issues. These algorithms aid in the discovery of variable values that satisfy certain requirements and produce the best results.

There are two main categories of optimization problems:

1. **Unconstrained Optimization:** Without any restrictions on the variables, the objective function must be optimized in this kind of issue. The objective is to determine the function's global maximum or minimum.
2. **Constrained Optimization:** The optimization in this case is subject to some variable limitations. Finding the best values that adhere to the constraints and the optimization goal is the goal.

There are various optimization techniques, and the choice of one depends on the parameters of the objective function, the constraints, and the nature of the issue at hand.

Here are some commonly used optimization algorithms:

1. **Gradient Descent:** a well-liked approach for optimization without constraints. It incrementally modifies the variables in the direction of the objective function's steepest descent. For big datasets and various gradient-based techniques for differentiable functions, there exist variations such as stochastic gradient descent.
2. **Newton's Method:** Using both the gradient and the second derivative (Hessian) of the objective function is another approach to unconstrained optimization. Although it can reach convergence more quickly than gradient descent, it requires that the Hessian matrix be positive definite and computable.
3. **Conjugate Gradient:** A method that iteratively solves huge systems of linear equations and can be applied to unrestricted optimization. To get the minimum of the objective function along specific conjugate directions, the gradient and a conjugate direction are combined.
4. **Simplex Method:** a form of limited optimization that is frequently used to solve linear programming issues. In order to find the best answer, it uses the vertices of a simplex, a higher-dimensional generalization of a triangle or tetrahedron.
5. **Interior Point Methods:** Problems involving linear and nonlinear programming can be solved using these strategies. They use iteratively modifying the variables to move through the interior of the viable region.
6. **Genetic Algorithms:** An evolutionary algorithm class that draws inspiration from natural selection. In order to find the best solution, these algorithms generate and evolve a population of candidate answers.

7. **Particle Swarm Optimization (PSO):** A population-based optimization method motivated by the cooperative nature of fish or birds. To locate the ideal solution, particles navigate the solution space.
8. **Simulated Annealing:** This algorithm, which takes its cue from the metallurgical annealing procedure, accepts poor solutions with a probability that diminishes over time, enabling escape from local maxima.
9. **Bayesian Optimization:** A strategy that actively chooses points in the search space to sample while using a probabilistic model to estimate the objective function, with the goal of locating the best answer with the fewest possible evaluations.

The parameters of the issue, the dimensionality of the variable space, the smoothness of the objective function, and the presence of constraints all affect the optimization strategy that is chosen. A combination of different algorithms or heuristics may be necessary for some situations in order to achieve successful optimization.

10.7 NUMERICAL SOLUTIONS FOR KINEMATIC EQUATIONS

In robotics, numerical solutions are essential for resolving kinematic equations. Kinematics is the study of how things move without taking the forces into account. Kinematic equations are used in robotics to explain the connection between joint angles and locations of robotic manipulators as well as the pose of the end-effector in a certain configuration. Finding these linkages frequently calls for numerical solutions, particularly for complex robot geometries and non-analytical answers.

Here are some key applications of numerical solutions for kinematic equations in robotics:

1. **Forward Kinematics:** Forward kinematics is used to establish the position and orientation of the end-effector in the workspace of the robot given the joint angles or joint displacements. Through the use of Denavit-Hartenberg parameters or alternative representations, the transformation matrix from the robot's base to its end-effector must be calculated. To solve for the transformation matrix, numerical techniques like the matrix exponentiation approach or iterative techniques are frequently utilized.
2. **Inverse Kinematics:** To determine the joint angles or joint displacements necessary to produce the desired end-effector position and orientation, inverse kinematics is applied. Generally speaking, there is no one-of-a-kind analytical solution to this problem, particularly for complicated robots with several degrees of freedom. The required joint values are iteratively solved using numerical techniques such as the Newton-Raphson method, Jacobian-based methods, and optimization algorithms.
3. **Workspace Analysis:** To establish a robotic manipulator's attainable workspace, numerical solutions are applied. This entails determining the corresponding end-effector positions while altering the joint angles within predetermined ranges. The workspace of the robot can be numerically represented by discrediting the joint space and assessing the forward kinematics.
4. **Path Planning:** Path planning algorithms use numerical solutions to establish workable routes for a robot's end-effector to walk along while avoiding impediments. To create smooth and collision-free routes, methods such as numerical optimization, spline interpolation, and numerical integration are used.
5. **Trajectory Generation:** Numerical techniques are used to construct trajectories that take joint restrictions, velocity constraints, and smooth motion into account for applications requiring

coordinated motion of robotic arms. For trajectory interpolation, techniques like quintic and cubic splines are employed.

6. **Singularity Analysis:** In the workspace of a robot, singular configurations can be found where particular joint configurations result in poorly defined or problematic behavior. Numerical analysis aids in avoiding the problems that high joint velocities or redundancy, which are frequently associated with singularities.
7. **Sensor Calibration:** Sensors attached on robotic arms, such as cameras and depth sensors, are calibrated using numerical techniques. To identify the transformation between the sensor and robot coordinate frames, this calibration requires the solution of nonlinear equations.

Overall, due to the complexity of many robotic systems and the dearth of analytical solutions for some issues, numerical solutions are crucial for solving kinematic equations in robotics. These techniques make it possible for robot manipulators to be accurately positioned, motion planned, and analyzed for use in a variety of applications, from production and automation to research and exploration.

10.8 ROBOTICS APPLICATIONS AS NUMERICAL SOLUTIONS FOR DYNAMIC EQUATIONS

Dynamic equations in robotics, which describe the motion of robotic systems while taking into account the forces and torques involved, are solved in large part through the use of numerical solutions. The interactions between the joints, linkages, and environment of the robot are taken into account via dynamic equations. Understanding the behavior of the robot, simulating its motion, and precisely regulating its movements all depend on the solution of dynamic equations.

Here are some key applications of numerical solutions for dynamic equations in robotics:

1. **Robot Simulation:** To mimic the dynamic behavior of robots in diverse settings, numerical solutions are applied. Simulation software can shed light on how a robot's joints and linkages react to applied forces, torques, and external perturbations by solving the dynamic equations over time. This aids in foretelling the robot's motion and evaluating its performance prior to application in the actual world.
2. **Control Algorithms:** For creating control algorithms that regulate a robot's movements, dynamic equations are essential. The right control inputs (torques or forces) are determined via numerical solutions of dynamic equations by controllers, such as proportional-integral-derivative (PID) controllers or more sophisticated techniques like model-based controllers, to achieve desired trajectories and stability.
3. **Trajectory Tracking:** To properly track intended trajectories, numerical solutions are used. Dynamic equations are iteratively solved in trajectory tracking to modify control inputs and reduce the discrepancy between the desired and actual trajectories. This guarantees that the robot will precisely follow the designated path.
4. **Impact Analysis:** Dynamic equations are used to examine the impact forces, contact forces, and their influence on the robot's motion when it comes to activities involving collision or contact with the environment. The robot's behavior throughout these interactions can be predicted via numerical simulations.
5. **Robot Dynamics Identification:** The dynamic parameters of a robot, such as its mass distribution and inertial characteristics, are determined using numerical solutions. The best-fit parameters are found using a technique known as system identification that compares the robot's actual motion with simulated motion while utilizing different parameter values.

6. **Energy Consumption Analysis:** Dynamic equations aid in determining how much energy a robot uses while moving. Researchers and engineers can evaluate the effectiveness of various control systems and optimize the robot's movements to reduce energy consumption by numerically solving the equations.
7. **Virtual Prototyping:** In virtual prototyping, complicated robotic systems are modeled and analyzed using numerical solutions. Prior to actual construction, this enables engineers to test out various designs, elements, and configurations to optimize the robot's performance.
8. **Dynamic Motion Planning:** Dynamic equations are used to create motion plans that take into consideration the robot's dynamic capabilities and limits in situations when the robot's motion involves quick manoeuvres or intricate interactions.

Numerical integration methods, such as the Euler method, Runge-Kutta methods, or more sophisticated methods designed for solving differential algebraic equations (DAEs), are frequently used to solve dynamic equations numerically. These methods make it possible to simulate and analyze a robot's behavior over time while taking into account the intricate interactions between forces, torques, and motion.

Numerical Problems with Solutions

Root Finding (Bisection Method)

Problem 1. Find the root of the equation $f(x) = x^3 - 2x - 5$ in the interval [1, 3] using the bisection method.

Solution: Implement the bisection method to find the root, which is approximately 2.0946.

Newton-Raphson Method

Problem 2. Use the Newton-Raphson method to find the square root of 5 with an initial guess of $x_0 = 2$.

Solution: Iterate using the formula $x_1 = (x_0 + 5/x_0)/2$ until convergence. The square root of 5 is approximately 2.2361.

Interpolation (Lagrange Polynomials)

Problem 3. Given a set of data points (x, y), interpolate a polynomial using Lagrange interpolation and find the value at x = 4.

Solution: Calculate the Lagrange polynomial and evaluate it at x = 4.

Numerical Integration (Trapezoidal Rule)

Problem 4. Approximate the integral of $f(x) = x^2$ from 0 to 2 using the trapezoidal rule with n = 4 subintervals.

Solution: Apply the trapezoidal rule formula to calculate the integral.

Solving Linear Systems (Gaussian Elimination)

Problem 5. Solve the system of linear equations: $2x + y - z = 8$,$- 3x - y + 2z = -11$ and $- 2x + y + 2z = -3$

Solution: Use Gaussian elimination to find the values of x, y, and z.

Eigenvalues and Eigenvectors (Power Method)

Problem 6. Find the largest eigenvalue and corresponding eigenvector of a 3×3 matrix using the power iteration method.

Solution: Apply the power iteration method to find the largest eigenvalue and eigenvector.

Ordinary Differential Equations (Euler's Method)

Problem 7. Approximate the solution of the differential equation $dy/dx = x^2 - y$ with $y(0) = 1$ using Euler's method with a step size of 0.2 from $x = 0$ to $x = 1$.

Solution: Apply Euler's method iteratively to approximate the solution.

Numerical Differentiation (Central Difference Method)

Problem 8. Estimate the derivative of $f(x) = \sin(x)$ at $x = \pi/4$ using the central difference method with a step size of $h = 0.1$.

Solution: Calculate the derivative using the central difference formula.

Curve Fitting (Least Squares Method)

Problem 9. Fit a straight line ($y = mx + b$) to a set of data points using the least squares method.

Solution: Find the values of m and b that minimize the sum of squared residuals.

Monte Carlo Integration

Problem 10. Approximate the value of π by using Monte Carlo integration to estimate the area of a quarter-circle.

Solution: Generate random points and calculate the ratio of points inside the quarter-circle to total points.

Gauss-Seidel Method (Iterative Solver)

Problem 11. Solve the system of linear equations using the Gauss-Seidel method: $4x + y - z = 7$, $2x + 7y + 2z = -5$ and $3x + 4y - 8z = 6$.

Solution: Perform iterations until convergence to find the values of x, y, and z.

Simpson's Rule (Numerical Integration)

Problem 12. Approximate the integral of $f(x) = 3x^3 - 2x^2 + x + 1$ from 0 to 2 using Simpson's rule with $n = 4$ subintervals.

Solution: Apply Simpson's rule formula to calculate the integral.

Newton-Cotes Quadrature (Composite Trapezoidal Rule)

Problem 13. Compute the integral of $f(x) = e^x$ from 0 to 2 using the composite trapezoidal rule with $n = 6$ subintervals.

Solution: Divide the interval into subintervals and apply the trapezoidal rule to each subinterval.

LU Decomposition (Matrix Factorization)

Problem 14. Use LU decomposition to solve the following system of linear equations: $3x + 2y - z = 1$, $x - y + 4z = -2$ and $2x + 3y + 2z = 5$.
Solution: Perform LU decomposition and forward/backward substitution to find x, y, and z.

Runge-Kutta Method (Ordinary Differential Equations)

Problem 15. Solve the initial value problem for the first-order ODE $dy/dx = x^2 - y$ with $y(0) = 1$ using the fourth-order Runge-Kutta method over the interval [0, 1].
Solution: Apply the fourth-order Runge-Kutta method iteratively to approximate the solution.

Interpolation (Newton's Divided Difference)

Problem: Interpolate a polynomial using Newton's divided difference method for the following data points: (0, 1), (1, 2), (2, 4).
Solution: Calculate the divided difference coefficients and construct the interpolating polynomial.

Eigenvalues and Eigenvectors (QR Algorithm)

Problem 17. Find the eigenvalues and eigenvectors of a 3×3 matrix using the QR algorithm.
Solution: Implement the QR algorithm to compute eigenvalues and eigenvectors.

Finite Difference Method (Partial Differential Equations)

Problem 18. Solve the 2D heat equation $\partial u/\partial t = \partial^2 u/\partial x^2 + \partial^2 u/\partial y^2$ subject to boundary conditions using the finite difference method.
Solution: Discretize the equation and solve iteratively to obtain the temperature distribution.

Monte Carlo Simulation (Integration)

Problem 19. Use Monte Carlo simulation to estimate the value of the integral $\int(0 \text{ to } 1)e^{-x^2}dx$.

Solution: Generate random points and compute the integral using the Monte Carlo method.

Singular Value Decomposition (SVD)

Problem 20. Compute the singular value decomposition of a given matrix A.
Solution: Perform SVD to factorize the matrix into its singular values and left/right singular vectors.

Root Finding (Bisection Method)

Problem 21. Find the root of $f(x) = x^3 - 2x - 5$ in the interval [1, 3] using the bisection method.
Solution: Implement the bisection method and iterate until convergence to find the root.

Newton-Raphson Method

Problem 22. Use the Newton-Raphson method to find a root of the equation $f(x) = e^x - 2x^2 + 3$, starting with $x_0 = 1$.

Solution: Apply the Newton-Raphson formula iteratively to approximate the root.

Interpolation (Lagrange Polynomials)

Problem 23. Given data points (1, 4), (2, 7), and (3, 10), interpolate a polynomial using Lagrange interpolation and find its value at x = 2.5.

Solution: Calculate the Lagrange polynomial and evaluate it at x = 2.5.

Numerical Integration (Simpson's Rule)

Problem 24. Approximate the integral of $f(x) = 2x^3 - 5x^2 + 4x - 1$ from 0 to 2 using Simpson's rule with n = 4 subintervals.

Solution: Apply Simpson's rule formula to estimate the integral.

Linear Systems (Gaussian Elimination)

Problem 25. Solve the system of linear equations using Gaussian elimination: $2x + 3y - z = 1$, $4x + y + 2z = 2$, $3x + 2y + 3z = 3$.

Solution: Perform Gaussian elimination to find the values of x, y, and z.

Eigenvalues and Eigenvectors (Power Method)

Problem 26. Find the largest eigenvalue and corresponding eigenvector of a 3 × 3 matrix using the power iteration method.

Solution: Apply the power iteration method iteratively to find the largest eigenvalue and eigenvector.

Ordinary Differential Equations (Euler's Method)

Problem 27. Use Euler's method to approximate the solution of the initial value problem $y' = 2x - y$, $y(0) = 1$ on the interval [0, 1] with a step size of 0.1.

Solution: Apply Euler's method to approximate the solution at various points.

Numerical Differentiation (Finite Differences)

Problem 28. Estimate the derivative of $f(x) = \cos(x)$ at $x = \pi/4$ using the central difference method with h = 0.1.

Solution: Calculate the derivative using the central difference formula.

Curve Fitting (Least Squares Method)

Problem 29. Fit a linear regression model to a dataset with points (1, 2), (2, 3), (3, 5), and (4, 7).

Solution: Find the best-fit line using the least squares method.

Monte Carlo Simulation

Problem 30. Use Monte Carlo simulation to estimate the value of π by randomly sampling points inside a unit circle.

Solution: Generate random points and calculate the ratio of points inside the circle to total points.

Short Questions and Answers

Q.1. What are numerical methods?

Ans. Numerical methods are mathematical techniques used to approximate solutions to problems that cannot be solved exactly using analytical methods.

Q.2. Why are numerical methods important?

Ans. Numerical methods provide practical solutions to complex problems in various fields by using computational algorithms.

Q.3. What is interpolation?

Ans. Interpolation is the process of estimating values between known data points.

Q.4. What is extrapolation?

Ans. Extrapolation involves estimating values beyond the range of known data points.

Q.5. Name a common interpolation method.

Ans. Polynomial interpolation using methods like Lagrange or Newton interpolation.

Q.6. What is the Newton-Raphson method used for?

Ans. It's used to find approximations to the roots of real-valued functions.

Q.7. What's the main idea behind the bisection method?

Ans. The method narrows down the interval where a root exists by repeatedly halving it.

Q.8. What is numerical integration?

Ans. Numerical integration approximates the definite integral of a function using discrete data points.

Q.9. Name a method for approximating integrals using trapezoids.

Ans. Trapezoidal rule.

Q.10. What's the basic principle of Simpson's rule?

Ans. It uses quadratic polynomials to approximate sections of the curve and provides a more accurate integration result.

Q.11. What's the purpose of Gaussian elimination?

Ans. Gaussian elimination is used to solve systems of linear equations by transforming them into row-echelon form.

Q.12. What does "convergence" mean in numerical methods?

Ans. Convergence refers to the property of a numerical method to approach the true solution as the number of iterations or steps increases.

Q.13. What is a "round-off error"?

Ans. Round-off error occurs due to limited precision in representing numbers in a computer.

Q.14. What is a "truncation error"?

Ans. Truncation error results from approximating a mathematical process, such as differentiation or integration, with a finite algorithm.

Q.15. What is a "stable" numerical method?

Ans. A stable method produces accurate results even in the presence of small perturbations or errors in the input data.

Q.16. What is the main goal of root-finding methods?

Ans. Root-finding methods aim to find the solutions (roots) of equations that make them equal to zero.

Q.17. Name an iterative method for solving systems of linear equations.

Ans. Jacobi method or Gauss-Seidel method.

Q.18. Why Newton-Raphson method is considered iterative?

Ans. It iteratively refines the approximation to the root in each step.

Q.19. What does "Jacobian" refer to in optimization methods?

Ans. The Jacobian is a matrix of partial derivatives used in optimization algorithms.

Q.20. What's the primary application of the simplex method?

Ans. Solving linear programming problems.

Q.21. What are "local optima" and "global optima"?

Ans. Local optima are solutions that are the best in their immediate neighborhood, while global optima are the best solutions in the entire search space.

Q.22. What's the basic idea behind genetic algorithms?

Ans. Genetic algorithms use principles from natural selection to find optimal solutions through evolution.

Q.23. What's the primary use of simulated annealing?

Ans. Simulated annealing is used for optimization by allowing "bad" moves in the search process to avoid getting stuck in local optima.

Q.24. What's the purpose of Bayesian optimization?

Ans. Bayesian optimization is used to optimize expensive, black-box functions by choosing evaluation points based on probabilistic models.

Q.25. In robotics, why are numerical solutions important for kinematic equations?

Ans. Numerical solutions help determine the relationship between joint angles and positions of robotic manipulators, especially for complex geometries.

Q.26. What's the difference between forward and inverse kinematics?

Ans. Forward kinematics calculates the end-effector's position given joint angles, while inverse kinematics finds the required joint angles to achieve a desired end-effector position.

Q.27. Why is numerical simulation important in robotics?

Ans. Numerical simulation helps predict a robot's behavior in various scenarios, aiding in design, analysis, and control development.

Q.28. How are numerical methods used in trajectory tracking?

Ans. Numerical methods are used to iteratively adjust control inputs to minimize the error between desired and actual trajectories.

Q.29. What's the significance of dynamic equations in robotics?

Ans. Dynamic equations describe a robot's motion while accounting for forces and torques, crucial for control and simulation.

Q.30. What is the primary role of numerical solutions in control algorithms?

Ans. Numerical solutions help determine control inputs to regulate a robot's motion and achieve desired behaviors.

Q.31. How do numerical methods contribute to sensor calibration in robotics?

Ans. Numerical methods are used to solve nonlinear equations to determine transformations between sensor and robot frames.

Q.32. What's the importance of solving dynamic equations in robot simulation?

Ans. Dynamic equations allow simulation software to predict how robots will respond to forces, torques, and perturbations.

Q.33. What is the "Jacobain" in robotics?

Ans. The Jacobian matrix represents the relationship between joint velocities and end-effector velocities, critical for motion planning and control.

Q.34. Why might dynamic motion planning require numerical solutions?

Ans. Dynamic motion planning considers the robot's capabilities and constraints, necessitating numerical methods for generating feasible trajectories.

Q.35. How are numerical methods used in trajectory generation for robots?

Ans. Numerical methods, like spline interpolation, help generate smooth and collision-free paths for coordinated robot motion.

Q.36. Why is solving dynamic equations important in virtual prototyping?

Ans. Dynamic equations allow engineers to simulate and analyze complex robot systems before physical construction.

Q.37. What's the role of numerical integration in solving differential equations?

Ans. Numerical integration approximates the solution of differential equations by discretizing the time domain.

Q.38. Why might using a small step size in numerical integration be problematic?

Ans. Small step sizes can lead to increased computation time and accumulated numerical errors.

Q.39. What's the advantage of using higher-order numerical integration methods?

Ans. Higher-order methods provide more accurate approximations by considering additional terms in the expansion.

Q.40. How do numerical methods help solve initial value problems?

Ans. Numerical methods iteratively estimate solutions by using the current state to predict the next state of a system.

Q.41. What is the central difference method used for?

Ans. The central difference method is used to approximate derivatives by evaluating the function at points on both sides of the desired derivative point.

Q.42. Why is numerical differentiation important in sensor data processing?

Ans. Numerical differentiation helps estimate rates of change or velocities from sampled sensor data.

Q.43. How do numerical methods aid in solving optimization problems?

Ans. Numerical methods optimize functions by finding points where the derivative or gradient is zero.

Q.44. What's the purpose of gradient descent in optimization?

Ans. Gradient descent is an iterative optimization method that updates parameters to minimize a cost function.

Q.45. What is the basic idea behind the simplex method for linear programming?

Ans. The simplex method moves along the edges of a feasible region to find the optimal solution.

Q.46. Why are numerical solutions important in trajectory planning for robotic arms?

Ans. Numerical solutions enable trajectory planning algorithms to generate smooth and accurate paths while considering joint constraints.

Q.47. How do numerical methods contribute to inverse kinematics in robotics?

Ans. Numerical methods iteratively adjust joint angles to find solutions that match desired end-effector positions.

Q.48. What's the role of numerical integration in calculating robot joint torques?

Ans. Numerical integration helps compute joint torques by integrating the dynamics equations over time.

Q.49. What is the relationship between numerical solutions and stability in control systems?

Ans. Numerical methods should be chosen carefully to ensure stable and accurate control system behavior.

Q.50. How can numerical methods help improve the efficiency of robot motion planning?

Ans. Numerical methods provide algorithms that efficiently search for optimal motion paths while considering constraints and objectives.

Multiple Choice Questions and Answers

Q.1. Numerical methods are used to:

(*a*) Find exact solutions to all mathematical problems

(*b*) Approximate solutions to complex problems

(*c*) Replace analytical methods entirely

(*d*) Only solve linear equations

Q.2. Which error arises due to limited computer precision?

(*a*) Truncation error (*b*) Round-off error (*c*) Absolute error (*d*) Relative error

Q.3. What is the main advantage of numerical methods over analytical methods?

(*a*) Numerical methods are always more accurate

(*b*) Numerical methods can solve any mathematical problem

(*c*) Numerical methods can handle complex problems without exact solutions
(*d*) Numerical methods are faster than analytical methods

Q.4. Which numerical method involves subdividing an interval into smaller intervals and approximating the integral using trapezoids?
(*a*) Simpson's rule (*b*) Forward difference
(*c*) Trapezoidal rule (*d*) Bisection method

Q.5. Which numerical method is used to find roots of equations?
(*a*) Newton-Raphson method (*b*) Simpson's rule
(*c*) Gaussian elimination (*d*) Trapezoidal rule

Q.6. What type of error arises due to approximating a mathematical process with a finite algorithm?
(*a*) Round-off error (*b*) Truncation error
(*c*) Absolute error (*d*) Convergence error

Q.7. Which method combines the strengths of the bisection method and linear interpolation?
(*a*) Regula Falsi method (*b*) Secant method
(*c*) Gradient descent (*d*) Newton-Raphson method

Q.8. Which method uses the tangent line to approximate the root of a function iteratively?
(*a*) Bisection method (*b*) Regula Falsi method
(*c*) Secant method (*d*) Newton-Raphson method

Q.9. Which numerical method is used for solving systems of linear equations by transforming them into row-echelon form?
(*a*) Gaussian elimination (*b*) Gradient descent
(*c*) Trapezoidal rule (*d*) Simpson's rule

Q.10. What is the goal of root-finding methods?
(*a*) To find optimal solutions (*b*) To find the maximum value of a function
(*c*) To find the minimum value of a function
(*d*) To find solutions that make a function equal to zero

Q.11. Which optimization algorithm is inspired by the social behavior of birds or fish?
(*a*) Genetic algorithm (*b*) Simulated annealing
(*c*) Particle Swarm Optimization (PSO) (*d*) Gradient descent

Q.12. Which method involves iteratively adjusting control inputs to minimize the error between desired and actual trajectories?
(*a*) Trajectory generation (*b*) Path planning
(*c*) Trajectory tracking (*d*) Dynamic analysis

Q.13. Which method combines aspects of the bisection method and linear interpolation to find roots?
(*a*) Regula Falsi method (*b*) Newton-Raphson method
(*c*) Secant method (*d*) Gradient descent

Q.14. Which type of error arises due to the finite precision of computers in numerical methods?

(*a*) Round-off error (*b*) Truncation error (*c*) Absolute error (*d*) Relative error

Q.15. What is the primary application of the simplex method?

(*a*) Root-finding (*b*) Optimization
(*c*) Numerical integration (*d*) Trajectory tracking

Q.16. Which numerical method is often used for finding optimal solutions through evolution-inspired processes?

(*a*) Bisection method (*b*) Regula Falsi method
(*c*) Genetic algorithm (*d*) Gradient descent

Q.17. Which method uses quadratic polynomials to approximate sections of a curve for integration?

(*a*) Simpson's rule (*b*) Trapezoidal rule
(*c*) Bisection method (*d*) Newton-Raphson method

Q.18. What is the purpose of gradient descent in optimization?

(*a*) To maximize the function value (*b*) To minimize the function value
(*c*) To solve systems of linear equations (*d*) To interpolate data points

Q.19. Which method involves iteratively updating the control inputs to minimize the error between desired and actual trajectories?

(*a*) Forward difference (*b*) Backward difference
(*c*) Simpson's rule (*d*) Trajectory tracking

Q.20. Which algorithm approximates the solution of a mathematical process using discrete data points?

(*a*) Trajectory generation (*b*) Numerical integration
(*c*) Path planning (*d*) Genetic algorithm

Q.21. Which method uses a weighted combination of function values at different points for numerical integration?

(*a*) Trapezoidal rule (*b*) Simpson's rule
(*c*) Bisection method (*d*) Regula Falsi method

Q.22. What's the primary goal of numerical differentiation?

(*a*) Approximating integrals (*b*) Solving linear equations
(*c*) Finding the minimum of a function (*d*) Approximating derivatives

Q.23. Which algorithm uses principles from natural selection for optimization?

(*a*) Simulated annealing (*b*) Genetic algorithm
(*c*) Gradient descent (*d*) Bisection method

Q.24. What's the main purpose of the central difference method?

(*a*) Numerical differentiation (*b*) Numerical integration
(*c*) Root-finding (*d*) Optimization

Q.25. Which method is commonly used for approximating integrals using the area of trapezoids?

(*a*) Newton-Raphson method
(*b*) Simpson's rule
(*c*) Bisection method
(*d*) Trapezoidal rule

Q.26. What is the central concept in the bisection method?

(*a*) Quadratic interpolation
(*b*) Linear interpolation
(*c*) Iterative halving of intervals
(*d*) Secant lines

Q.27. Which optimization algorithm aims to mimic the annealing process in metallurgy?

(*a*) Genetic algorithm
(*b*) Simulated annealing
(*c*) Gradient descent
(*d*) Particle Swarm Optimization (PSO)

Q.28. Which numerical method involves solving a system of equations by eliminating variables through row operations?

(*a*) Gradient descent
(*b*) Gaussian elimination
(*c*) Trapezoidal rule
(*d*) Genetic algorithm

Q.29. What type of error arises due to approximating a mathematical process with a finite algorithm?

(*a*) Round-off error
(*b*) Truncation error
(*c*) Absolute error
(*d*) Convergence error

Q.30. What is the primary application of the Newton-Raphson method?

(*a*) Numerical integration
(*b*) Root-finding
(*c*) Optimization
(*d*) Trajectory tracking

Q.31. Which numerical method involves calculating a transformation matrix from joint angles to end-effector position?

(*a*) Forward kinematics
(*b*) Inverse kinematics
(*c*) Genetic algorithm
(*d*) Gradient descent

Q.32. Which method involves using a series of linear equations to approximate derivatives?

(*a*) Newton-Raphson method
(*b*) Gradient descent
(*c*) Central difference method
(*d*) Regula Falsi method

Q.33. What's the primary application of the secant method?

(*a*) Numerical differentiation
(*b*) Numerical integration
(*c*) Root-finding
(*d*) Optimization

Q.34. Which numerical method uses a combination of the forward and backward difference approximations?

(*a*) Central difference method
(*b*) Secant method
(*c*) Trapezoidal rule
(*d*) Genetic algorithm

Q.35. What is the main idea behind the Gauss-Seidel method for solving systems of linear equations?

(*a*) It uses row operations to eliminate variables.

(*b*) It updates variables one at a time using the most recent values.
(*c*) It finds the root of a function through iterative refinement.
(*d*) It approximates integrals using quadratic polynomials.

Q.36. Which method is used to estimate solutions to equations that cannot be solved analytically?
(*a*) Numerical integration (*b*) Numerical differentiation
(*c*) Numerical methods (*d*) Analytical methods

Q.37. Which method uses a weighted combination of function values at different points for numerical integration?
(*a*) Trapezoidal rule (*b*) Simpson's rule
(*c*) Bisection method (*d*) Regula Falsi method

Q.38. What is the primary purpose of the secant method?
(*a*) To solve systems of linear equations (*b*) To estimate derivatives
(*c*) To approximate integrals (*d*) To find roots of equations

Q.39. What type of error arises due to the finite precision of computers in numerical methods?
(*a*) Round-off error (*b*) Truncation error
(*c*) Absolute error (*d*) Relative error

Q.40. What's the main goal of root-finding methods?
(*a*) To maximize the function value (*b*) To solve systems of linear equations
(*c*) To find the minimum value of a function
(*d*) To find solutions that make a function equal to zero

Q.41. Which optimization algorithm is inspired by the social behavior of birds or fish?
(*a*) Genetic algorithm (*b*) Simulated annealing
(*c*) Particle Swarm Optimization (PSO) (*d*) Gradient descent

Q.42. Which method involves iteratively adjusting control inputs to minimize the error between desired and actual trajectories?
(*a*) Trajectory generation (*b*) Path planning
(*c*) Trajectory tracking (*d*) Dynamic analysis

Q.43. Which method combines aspects of the bisection method and linear interpolation to find roots?
(*a*) Regula Falsi method (*b*) Newton-Raphson method
(*c*) Secant method (*d*) Gradient descent

Q.44. Which type of error arises due to the finite precision of computers in numerical methods?
(*a*) Round-off error (*b*) Truncation error (*c*) Absolute error (*d*) Relative error

Q.45. What is the primary application of the simplex method?
(*a*) Root-finding (*b*) Optimization
(*c*) Numerical integration (*d*) Trajectory tracking

Q.46. Which numerical method is often used for finding optimal solutions through evolution-inspired processes?

(*a*) Bisection method (*b*) Regula Falsi method
(*c*) Genetic algorithm (*d*) Gradient descent

Q.47. Which method uses quadratic polynomials to approximate sections of a curve for integration?

(*a*) Simpson's rule (*b*) Trapezoidal rule
(*c*) Bisection method (*d*) Newton-Raphson method

Q.48. What is the purpose of gradient descent in optimization?

(*a*) To maximize the function value (*b*) To minimize the function value
(*c*) To solve systems of linear equations (*d*) To interpolate data points

Q.49. Which method involves iteratively updating the control inputs to minimize the error between desired and actual trajectories?

(*a*) Forward difference (*b*) Backward difference
(*c*) Simpson's rule (*d*) Trajectory tracking

Q.50. Which algorithm approximates the solution of a mathematical process using discrete data points?

(*a*) Trajectory generation (*b*) Numerical integration
(*c*) Path planning (*d*) Genetic algorithm

Key to Multiple Choice Questions

1. (*b*)	**2.** (*b*)	**3.** (*c*)	**4.** (*c*)	**5.** (*a*)	**6.** (*b*)	**7.** (*a*)	**8.** (*d*)	**9.** (*a*)
10. (*d*)	**11.** (*c*)	**12.** (*c*)	**13.** (*a*)	**14.** (*a*)	**15.** (*b*)	**16.** (*c*)	**17.** (*a*)	**18.** (*b*)
19. (*d*)	**20.** (*b*)	**21.** (*b*)	**22.** (*d*)	**23.** (*b*)	**24.** (*a*)	**25.** (*d*)	**26.** (*c*)	**27.** (*b*)
28. (*b*)	**29.** (*b*)	**30.** (*b*)	**31.** (*a*)	**32.** (*c*)	**33.** (*c*)	**34.** (*a*)	**35.** (*b*)	**36.** (*c*)
37. (*b*)	**38.** (*d*)	**39.** (*a*)	**40.** (*d*)	**41.** (*c*)	**42.** (*c*)	**43.** (*a*)	**44.** (*a*)	**45.** (*b*)
46. (*c*)	**47.** (*a*)	**48.** (*b*)	**49.** (*d*)	**50.** (*b*)				

Projects Works

1. **Inverse Kinematics Solver for Robotic Arm:** Develop a numerical solver for the inverse kinematics problem of a robotic arm with multiple degrees of freedom. Implement different numerical methods to find joint angles that correspond to a desired end-effector position.
2. **Trajectory Planning and Optimization:** Create a program that generates smooth and collision-free trajectories for a robot's end-effector using numerical optimization techniques. Consider joint limits and dynamic constraints during the trajectory planning process.
3. **Dynamic Simulation of Robotic Systems:** Build a dynamic simulator that uses numerical integration methods to model the motion of robotic systems under the influence of forces and torques. Visualize the simulated motion and analyze the system's behavior.

4. **Solving Path Planning Problems:** Develop a program that employs numerical optimization to find optimal paths for a robot to navigate through an environment with obstacles. Consider real-world constraints such as non-holonomic motion and varying terrain.
5. **Optimal Gripper Design:** Use numerical methods to optimize the design of a robotic gripper for a specific task. Implement an optimization algorithm to determine the gripper's parameters that maximize its performance in terms of grasp stability, force exertion, or object manipulation.
6. **Optimal Control of Robotic Manipulators:** Create a control system that uses numerical optimization to find control inputs that achieve desired trajectories while minimizing energy consumption or joint torques for a robotic arm.
7. **Dynamic Path Planning for Mobile Robots:** Develop a dynamic path planner for mobile robots that considers dynamic obstacles and environmental changes. Use numerical methods to adapt the robot's trajectory in real-time to avoid collisions.
8. **Sensor Fusion and Calibration:** Use numerical techniques to calibrate multiple sensors on a robotic platform. Implement algorithms that fuse data from different sensors to provide accurate information about the robot's environment.
9. **Efficient Motion Planning for Redundant Robots:** Explore numerical methods to efficiently solve the inverse kinematics problem for redundant robots. Investigate how redundancy can be utilized to optimize the robot's motion for specific tasks.
10. **Optimization of Robotic Assembly Sequences:** Create an optimization framework that uses numerical methods to find the optimal assembly sequence for a complex robotic assembly task. Consider factors like part compatibility, tool access, and task efficiency.
11. **Robot Motion Control for Cooperative Tasks:** Develop a control algorithm for a team of robots to cooperatively manipulate an object. Utilize numerical methods to coordinate the robots' motions and maintain stability during the task.
12. **Human-Robot Interaction Planning:** Design a planning system that generates trajectories for a robot that interacts with humans in a safe and socially acceptable manner. Use numerical methods to ensure collision avoidance and natural movement.
13. **Real-time Path Adaptation for Unmanned Vehicles:** Create an algorithm for unmanned ground or aerial vehicles that uses numerical methods to adapt their paths in real-time based on changing environmental conditions or dynamic obstacles.
14. **Multi-Robot Coordination and Task Allocation:** Develop an optimization-based approach using numerical methods to allocate tasks and coordinate the movements of multiple robots to achieve a common goal, such as exploration or surveillance.
15. **Predictive Maintenance for Robotic Systems:** Build a system that uses numerical techniques to predict when maintenance is needed for robotic systems based on sensor data, wear and tear models, and historical usage patterns.

11

Signal Processing for analyzing Vast Data Generated in Digital World

11.1 INTRODUCTION

In the study area of signal processing, signals are manipulated, analysed, and interpreted. Information is represented through signals, which can be in the form of audio, video, pictures, sensor readings, and more. Signal processing seeks to improve the quality of signals and prepare them for a variety of applications by extracting meaningful information from them. It is essential to many different industries, including telecommunications, radar, sonar, audio and picture processing, and more. A vast variety of mathematical tools, methods, and approaches are used in signal processing. It is an interdisciplinary field having connections to computer science, engineering, physics, and mathematics. Our capacity to analyse and interpret the enormous amounts of data has improved as a result of the important technological improvements brought about by the developments in signal processing.

11.2 KEY ASPECTS OF SIGNAL PROCESSING

Here are some key concepts and aspects of signal processing:

1. **Signal kinds:** Signals can be divided into many sorts according to their features. Continuous-time, discrete-time, analogue, and digital signals are examples of common signal types.
2. **Signal Representation:** Signals are typically represented mathematically, either through graphs, equations, or value sequences.
3. **Sampling and Quantization:** Analogue signals are sampled in digital signal processing to create discrete-time signals. When a continuous signal is quantized, a finite set of discrete values are created.
4. **Signal Transformations:** Signal transformations, such as the Fourier transform and the Laplace transform, enable the analysis of signals in various domains (such as the complex plane or frequency plane) to expose its frequency components and behavior.
5. **Filtering:** The removal, augmentation, or modification of particular frequency components is the process of filtering, which modifies a signal. It is employed for feature extraction, signal amplification, and noise reduction.

6. Convolution is a procedure that combines two signals to create a third signal that depicts how one signal alters the other.
7. **Modulation and Demodulation:** While demodulation separates the information signal from the modulated carrier, modulation entails altering the parameters of a carrier signal in accordance with the information signal.
8. **Digital Signal Processing (DSP):** DSP is the use of methods and techniques to process digital signals on hardware or digital computers.
9. **Image and Video Processing:** The manipulation and analysis of visual data are the subjects of image and video processing. Techniques include object recognition, compression, filtering, and image enhancement.
10. **Audio Processing:** For purposes such as audio compression, voice recognition, and music production, audio processing entails the analysis and modification of audio signals.
11. Applications Signal processing is used in many different industries, including radar and sonar systems, medical imaging (MRI, CT scans), voice recognition, image and video compression, telecommunications (data transmission), and more.
12. **Machine Learning with Signal Processing:** To extract patterns, classify data, and generate predictions from signals, machine learning algorithms are frequently used with signal processing techniques.
13. **Real-Time Processing:** A number of applications, including control systems, voice communication, and live video streaming, call for the processing of signals in real-time.

11.3 FOURIER ANALYSIS

A complicated waveform or signal is broken down into a collection of simpler sinusoidal functions using the Fourier analysis, a fundamental mathematical tool used in both signal processing and mathematics. It bears the name of Jean-Baptiste Joseph Fourier, who first proposed the idea back in the 1800s. Numerous applications in signal processing, communications, physics, engineering, and other fields benefit from the use of Fourier analysis to examine and comprehend the frequency components contained in signals.

11.4 KEY CONCEPTS IN FOURIER ANALYSIS

Key concepts in Fourier analysis include:

1. **Fourier Series:** A periodic signal is represented by the Fourier series as the accumulation of sinusoidal components. It divides a periodic signal into harmonics, which are integer multiples of the fundamental frequency (also known as the fundamental or the first harmonic).
2. **Continuous Fourier Transform:** By examining signals across the entire real line, the continuous Fourier transform extends the idea of Fourier series to non-periodic events. It converts a signal's frequency components and their magnitudes from the time domain to the frequency domain.
3. **Discrete Fourier Transform (DFT):** The DFT is used to analyze discrete, sampled signals. It is a finite variant of the continuous Fourier transform. It converts an N-sample sequence into a complex number sequence.
4. **The Fast Fourier Transform (FFT):** It is a useful approach for determining the DFT of a sequence. As a result of the huge reduction in computing complexity, it is appropriate for use in large-scale and real-time applications.

5. **Frequency Domain Analysis:** Fourier analysis enables frequency domain signal analysis, which is necessary for activities like noise filtering, spotting dominating frequencies, and comprehending a signal's properties.
6. **Power Spectral Density (PSD):** The PSD is a measurement of how evenly a signal's power is distributed across its range of frequencies. It sheds light on how a signal's energy is distributed throughout its range of frequencies.
7. **Inverse Fourier Transform:** This technique returns a signal to the time domain from the frequency domain. After examining a signal's frequency components, it is used to reconstruct the signal.

11.4.1 Fourier Series

A periodic function can be represented mathematically using the Fourier series as the sum of sine and cosine functions with various frequencies. It is a fundamental idea in several disciplines, including mathematics, physics, engineering, and signal processing. For a periodic function f(x) with period T, the general form of a Fourier series is as follows:

$$f(x) = a_0 + \int_{n=1}^{\infty} a_n \cos\frac{2\pi nx}{T} + b_n \sin\frac{2\pi nx}{T}$$

where,

a_0 is the average value of the function over one period.

a_n and b_n are the Fourier coefficients, which determine the amplitude of the cosine and sine terms at frequency n/T, respectively.

n is the positive integer representing the harmonic order or frequency.

T is the period of the function.

The Fourier coefficients a_n and b_n are calculated using the following formulas:

$$a_n = \frac{2}{T}\int_0^T f(x)\cos\frac{2\pi nx}{T}dx$$

The steps to determine the Fourier coefficients and create the Fourier series representation are broken down as follows:

1. Find a_0:

$$a_0 = \frac{1}{T}\int_0^T f(x)dx$$

2. Calculate a_n and b_n for each positive integer n:

$$a_n = \int_0^T f(x)\cos\int\frac{2\pi nx}{T}dx$$

$$b_n = \frac{2}{T}\int_0^T f(x)\sin\int\frac{2\pi nx}{T}dx$$

3. Once you have the values of a_0, a_n and b_n for various n, you can construct the Fourier series representation of the function f(x).

Not all functions can be accurately described by a Fourier series, and the convergence features of Fourier series depend on the properties of the function being represented. The Gibbs phenomenon, in which the Fourier series approximation overshoots around the discontinuities, may be present in functions having discontinuities or sharp corners.

Practical applications frequently involve dealing with finite Fourier series, where a finite number of terms in the series are taken into account rather than an infinite sum. Assuming the function is suitable for Fourier analysis, the closer the approximation to the original function, the more terms you include.

11.4.2 Continuous Fourier Transform

A mathematical technique called the Continuous Fourier Transform is used to examine the frequency components of a continuous, time-domain signal. A function of time (or space) is converted into a function of frequency. The following integral, designated as F(), defines the Continuous Fourier Transform of the function f(t):

$$F(\omega) = \int_{-\infty}^{\infty} f(t)\cdot e^{-i\omega t} dt$$

Where,

$f(t)$ is the time-domain signal you want to transform.

$F(\omega)$ is the frequency-domain representation of the signal.

ω is the angular frequency in radians per unit time ($\omega=2\pi f$, where f is the frequency in Hz).

e is the base of the natural logarithm, and i is the imaginary unit.

The inverse Continuous Fourier Transform allows to recover the original signal from its frequency-domain representation:

$$f(t) = \frac{1}{2\pi}\int_{-\infty}^{\infty} F(\omega)\cdot e^{i\omega t} d\omega$$

1. The duality between the time domain and the frequency domain is established by the pair of equations above.
2. Not all functions have a Continuous Fourier Transform, which is an important distinction to make. The transform cannot exist or perform properly unless the function satisfies specific requirements.
3. **Linearity:** The linear combination of the individual transforms and a linear combination of functions yields the same transform.
4. **Time Shift:** A phase shift in the frequency domain is produced by a continual time-domain signal shift:

$$\int\{f(t-\tau)\} = F(\omega)\cdot e^{-i\omega\tau}$$

5. **Frequency Shift:** Shifting the signal's frequency by ω_0 corresponds to a phase rotation in the time domain:

$$\mathcal{F}\{e^{iw_0 t} f(t)\} = F(\omega - \omega_0)$$

6. **Scaling in Time:** Stretching or compressing the signal in time corresponds to a scaling in frequency:

$$\mathcal{F}\{f(at)\} = \frac{1}{|a|}F\int\frac{\omega}{a}$$

7. **Convolution:** The transform of the convolution of two signals is the product of their individual transforms:

$$\int\{f \cdot g\} = F(\omega) \cdot G(\omega)$$

8. **Parseval's Theorem:** The energy (or power) of a signal in the time domain is equal to the integral of the squared magnitude of its transform in the frequency domain:

$$\int_{-\infty}^{\infty} |f(t)|^2 \, dt = \frac{1}{2\pi}\int_{-\infty}^{\infty} |F(\omega)|^2 \, d\omega$$

The Continuous Fourier Transform is a potent mathematical technique used to examine the frequency content of continuous signals in a number of disciplines, including signal processing, communication theory, optics, and more.

11.4.3 Discrete Fourier Transform

A mathematical method called the Discrete Fourier Transform (DFT) is used to examine the frequency components of a discrete, sampled signal. It converts a sequence of N complex numbers (which represent discrete samples of a signal) into an other sequence of N complex numbers (which represents the frequency-domain representation of the signal). For the analysis and manipulation of digital signals, the DFT is extensively employed in the signal processing, communication, and a variety of other domains.

Given a sequence of N complex numbers $x[n]$, where $n = 0, 1, 2, \ldots, N-1$, the DFT $X[k]$ is defined as:

$$X[k] = \int_{n=0}^{N-1} x[n] \cdot e^{-i\frac{2\pi}{N}kn}$$

where,

$X[k]$ is the frequency-domain representation of the signal at frequency index k.

$x[n]$ is the time-domain signal value at sample index n.

N is the total number of samples.

i is the imaginary unit.

$k = 0, 1, 2, \ldots, N-1$ represents the frequency bin index.

The inverse Discrete Fourier Transform (IDFT) allows to recover the original sequence $x[n]$ from its frequency-domain representation:

$$x[n] = \frac{1}{N}\int_{k=0}^{N-1} X[k] \cdot e^{i\frac{2\pi}{N}kn}$$

The pair of equations above establish the duality between the time domain and the frequency domain in the discrete case.

Key Properties of the DFT

1. **Linearity:** The DFT is linear, much like the Continuous Fourier Transform. The linear combination of the individual DFTs and a linear combination of sequences yields the same DFT.
2. **Periodicity:** With a period of N, the DFT displays periodicity in the frequency domain. This is known as the "circular convolution property," and it's crucial to take into account when working with signals of various lengths.

3. **Shift Theorem:** A phase shift in the frequency domain is produced when the time-domain signal is shifted by m samples:

$$\int\{x[n-m]\} = X[k]\cdot e^{-i\frac{2\pi}{N}km}$$

4. **Scaling in Time:** Stretching or compressing the signal in time corresponds to a scaling in frequency:

$$\equiv\{x[an]\} = \frac{1}{|a|}X\equiv\frac{k}{a}$$

5. **Convolution Theorem:** The DFT of the convolution of two sequences is the element-wise product of their individual DFTs:

$$\int\{x[n]\cdot y[n]\} = X[k]\cdot Y[k]$$

Computing Complexity: The DFT can be calculated with a great degree of efficiency using the Fast Fourier Transform (FFT) algorithm. The FFT algorithm drastically reduces the number of operations by making use of the symmetry and periodicity aspects of the DFT.

Digital signal processing applications range from audio and picture processing to communication systems and scientific computing, and the DFT and the FFT's effective computation are fundamental ideas in this field.

11.4.4 Fast Fourier Transform

The Discrete Fourier Transform (DFT) and its inverse (IDFT) can be calculated quickly using the Fast Fourier Transform (FFT) technique. It considerably lessens the difficulty of calculation. The FFT algorithm is frequently employed in a wide range of applications, including image processing, data analysis, and digital signal processing.

Cooley-Tukey FFT Algorithm: One of the most well-known algorithms for quickly computing the DFT while utilizing divide-and-conquer strategies and taking advantage of the symmetry and periodicity aspects of the DFT is the Cooley-Tukey FFT algorithm.

The algorithm can be explained recursively as follows:

Base Case: The DFT is just the input sequence's value if N, the length of the sequence, is 1.

- Divide the input sequence into even-indexed and odd-indexed subsequences in the second iterative step.
- Recursively calculate each subsequence's DFT.
- Compute the DFT for the full sequence by combining the outcomes of the two sub problems.

Mathematically, this can be represented as follows:

Given an input sequence *x*[*n*] of length *N*, and assuming *N* is a power of 2 for simplicity, the FFT algorithm computes the DFT *X*[*k*] using the following equation:

$$X[k] = \int_{n=0}^{N-1} x[n]\cdot e^{-i\frac{2\pi}{N}kn}$$

$$= \int_{n=0}^{N/2-1} x[2n]\cdot e^{-i\frac{2\pi}{N/2}kn} + e^{-i\frac{2\pi}{N}kn}\int_{n=0}^{N/2-1} x[2n+1]\cdot e^{-i\frac{2\pi}{N/2}kn}$$

The FFT algorithm significantly reduces the number of computations by exploiting the symmetry properties of the DFT. The key insight is that the DFT of a sequence of length *N* can be computed by combining the DFTs of two sequences of length *N*/2.

Efficiency and Applications: The FFT algorithm has a computational complexity making it much faster than the direct computation of the DFT.

Applications for the FFT include:

- **Signal processing:** Examining and filtering signals across a number of domains.
- Modulation, demodulation, and channel equalization in communication systems.
- Image processing techniques include convolution and fourier analysis.
- **Spectral analysis:** Taking data and extracting its frequency components.
- **Numerical simulations:** Spectral techniques for solving partial differential equations.

Overall, the Fast Fourier Transform technique has revolutionized several fields by making it possible to compute the Discrete Fourier Transform in a practical and efficient manner, which is crucial for processing and analyzing digital signals and data.

11.4.5 Frequency Domain Analysis

A mathematical method called frequency domain analysis is used to examine signals and systems in terms of their frequency content. It entails employing methods like the Fourier Transform or the Discrete Fourier Transform (DFT) to convert signals and functions from the time domain to the frequency domain. This enables us to comprehend how various frequency components affect how the signal or system behaves.

In this section, we give a succinct review of the mathematical ideas related to frequency domain analysis.

1. **Continuous Fourier Transform (CFT):** The Continuous Fourier Transform is used to transform a continuous-time signal *f*(*t*) into its frequency-domain representation

$$F(\omega) = \int_{-\infty}^{\infty} f(t)\cdot e^{-i\omega t}dt$$

Here, ω represents angular frequency in radians per unit time.

The inverse Continuous Fourier Transform allows you to recover the original signal from its frequency-domain representation:

$$f(t) = \frac{1}{2\pi}\int_{-\infty}^{\infty} F(\omega)\cdot e^{i\omega t}d\omega$$

2. **Discrete Fourier Transform (DFT):** The Discrete Fourier Transform is used to transform a sequence of discrete samples $x[n]$ into its frequency-domain representation

$$X[k] = \int_{n=0}^{N-1} x[n]\cdot e^{-i\frac{2\pi}{N}kn}$$

Where *N* is the number of samples in the sequence, and *k* represents frequency bin index.

The inverse DFT allows you to recover the original sequence from its frequency-domain representation:

$$x[n] = \frac{1}{N}\int_{k=0}^{N-1} X[k]\cdot e^{i\frac{2\pi}{N}kn}$$

3. **Frequency Response of Systems:** In frequency domain analysis, systems are often described by their frequency response. The frequency response of a linear time-invariant system relates the output to the input in the frequency domain. The frequency response can be obtained by applying the system's transfer function to the input signal in the frequency domain.
4. **Convolution in Frequency Domain:** Convolution in the time domain corresponds to multiplication in the frequency domain, and vice versa. Mathematically, the convolution theorem states that the Fourier Transform of a convolution of two functions is the product of their individual Fourier Transforms:

$$\int \{f \cdot g\} = F(\omega) \cdot G(\omega)$$

The frequency components of a signal and their interactions with systems are revealed through frequency domain analysis. Understanding filtering procedures, signal modulations, system stability, and other phenomena involving the interaction of many frequencies benefit greatly from this knowledge. In many cases, the analysis can be made simpler and a greater knowledge of complicated systems and signals can be obtained by shifting from the time domain to the frequency domain.

11.4.6 Power Spectral Density

The Power Spectral Density (PSD) is a fundamental concept in signal processing and spectral analysis. It describes how the power of a signal is distributed across different frequency components. The PSD is widely used to analyze the frequency content of stationary stochastic (random) processes. It provides crucial information about the signal's energy distribution in the frequency domain.

The PSD is usually denoted by $S(\omega)$ or $S(f)$, where:

$S(\omega)$ represents the PSD as a function of angular frequency ω, typically in radians per unit time.

$S(f)$ represents the PSD as a function of regular frequency f, typically in Hertz.

The relationship between $S(\omega)$ and $S(f)$ is given by the following equation:

$$S(f) = \frac{1}{2\pi} S(\omega)$$

The PSD is defined as the limit of the average power of a signal over an infinitesimally small frequency range as the range approaches zero. Mathematically, the PSD of a continuous-time signal $x(t)$ is given by:

$$S(\omega) = \lim_{\Delta\omega \to 0} \frac{1}{\Delta\omega} \int_{-\frac{\Delta\omega}{2}}^{\frac{\Delta\omega}{2}} |X(\omega + \delta\omega)|^2 \, d\delta\omega$$

where, $X(\omega)$ is the continuous Fourier Transform of the signal $x(t)$.

For discrete-time signals, the PSD is defined based on the Discrete Fourier Transform (DFT) and the signal's autocorrelation function. If $x[n]$ is a discrete-time signal, the PSD is given by:

$$S(\omega) = \frac{1}{N} \left| \int_{n=0}^{N-1} x[n] e^{-i\omega n} \right|^2$$

where,

N is the number of samples in the sequence.

The sum represents the DFT of the signal.

Properties of the Power Spectral Density

1. **Non-Negativity:** The PSD is always non-negative: $\geq S(\omega) \geq 0$.
2. **Total Power:** The total power of the signal $x(t)$ is equal to the integral of its PSD over all frequencies:

$$P_x = \int_{-\infty}^{\infty} S(\omega)d\omega$$

3. **Parseval's Theorem:** The total power of the signal $x(t)$ is also equal to the integral of the signal's auto-correlation function:

$$P_x = \frac{1}{2\pi}\int_{-\infty}^{\infty} R(\tau)d\tau$$

where $R(\tau)$ is the auto-correlation function of the signal $x(t)$.

The Power Spectral Density is a helpful tool for many applications, including signal processing, communication systems, noise analysis, and the study of random processes. It offers crucial insights into the frequency characteristics of a signal.

11.4.7 Inverse Fourier Transform

A mathematical process called the Inverse Fourier Transform, also referred to as the Inverse Fourier Series or Inverse Fourier Integral, enables you to reconstruct a time-domain signal from its frequency-domain representation. The Inverse Fourier Transform does the opposite of what the Fourier Transform does, which is to transform a signal from the time domain to the frequency domain. It is a crucial tool in both mathematics and signal processing.

The Inverse Fourier Transform of a frequency-domain representation $F(\omega)$ is denoted as $f(t)$ and is defined as:

$$f(t) = \frac{1}{2\pi}\int_{-\infty}^{\infty} F(\omega)\cdot e^{i\omega t}d\omega$$

where,

$F(\omega)$ is the frequency-domain representation of the signal.

$f(t)$ is the time-domain signal you want to recover.

ω is the angular frequency in radians per unit time.

i is the imaginary unit.

This equation means that to obtain the time-domain signal, you integrate the product of the frequency-domain representation $F(\omega)$ and a complex exponential $e^{i\omega t}$ with respect to the frequency ω over the entire range of frequencies.

It is crucial to remember that the aforementioned integral, given that the signal and its Fourier Transform satisfy specific requirements that let this integral to exist, is a definite integral that spans from negative infinity to positive infinity. This prerequisite is comparable to the requirements for the existence of the Fourier Transform.

Practical Points to Consider

1. The frequency-domain representation F() may frequently be conjugate symmetric, which makes it easier to calculate the inverse Fourier Transform. You just need to integrate over positive frequencies in such circumstances.

2. **Convolution Theorem:** Since the Inverse Fourier Transform and the Fourier Transform are closely connected, these operations are covered by the Convolution Theorem. According to the Convolution Theorem, the convolution of each signal's individual time-domain representation is equal to the inverse Fourier Transform of the product of two frequency-domain representations.
3. **Numerical Approximations:** When working with discrete frequency-domain samples produced using methods like the Discrete Fourier Transform (DFT), the Inverse Fourier Transform may be numerically approximated in practise.

Signal processing, communication, image reconstruction, and other areas all depend on the Inverse Fourier Transform, a potent mathematical tool that enables you to transfer signals from the frequency domain back to the time domain.

11.5 APPLICATIONS OF FOURIER ANALYSIS

Fourier analysis has wide-ranging applications:

(*a*) Fourier analysis is used in telecommunications to facilitate the transmission of signals over various frequencies through modulation and demodulation procedures.

(*b*) Fourier analysis is employed in audio processing to transform audio inputs into representations in the frequency domain for operations like equalization and compression.

(*c*) The Fourier transform is employed in image processing for picture enhancing, compression, and filtering.

(*d*) Fourier analysis is a technique used in physics to examine the frequency components of various physical phenomena, including electromagnetic signals and sound waves.

(*e*) Fourier analysis aids in understanding the frequency response of systems and locating resonances in the investigation of vibrations and control systems.

Overall, Fourier analysis is a potent tool for figuring out the frequency components that make up complicated signals, allowing us to analyze, work with, and interpret signals in a variety of contexts and applications.

11.6 FILTERS AND CONVOLUTION IN SIGNAL PROCESSING

Fundamental ideas in the subject of signal processing, which deals with the analysis, modification, and interpretation of signals, include filters and convolution. Signals can take on many different shapes, including acoustic, visual, or time-series data. In processes like noise reduction, feature extraction, and signal augmentation, filters and convolution are crucial.

Filters

In signal processing, a filter is a mathematical operation that is applied to a signal to extract or change specific signal properties. Filters can be applied to a signal to remove undesirable noise, amplify certain frequencies, or extract important information. There are two broad groups into which filters fall:

FIR (Finite Impulse Response) Filters: These filters only take into account a limited number of previous input values when determining the output because they feature a finite-length impulse response. FIR filters frequently use convolutional techniques and have linear phase characteristics.

A particular kind of digital filter used in signal processing is called a Finite Impulse Response (FIR) filter. It only takes into account a limited number of previous input values when calculating the output since it has a finite-length impulse response. Due to its linear phase properties, or ability to maintain the structure of the signal without introducing phase distortion, FIR filters are well-liked.

The mathematical representation of an FIR filter involves specifying its impulse response coefficients. Let's say we have an FIR filter with 'N' taps (coefficients), denoted as h[n] for n = 0, 1, ..., N−1. The filter's output 'y[n]' at time 'n' is obtained by convolving the input signal 'x[n]' with the filter coefficients 'h[n]':

$$y[n] = \Sigma[\, h[k] * x[n-k] \,], \text{ for } k = 0 \text{ to } N-1$$

This equation represents the basic operation of an FIR filter, where 'x[n]' is the input signal, 'y[n]' is the output signal, and 'h[n]' are the filter coefficients. The output 'y[n]' is computed as a weighted sum of the past 'N' input samples, each multiplied by the corresponding filter coefficient.

Stability (BIBO stability), linear phase response, and the capacity to be constructed to satisfy certain frequency response specifications are only a few of the favorable characteristics of FIR filters. The windowing method, the Parks-McClellan algorithm (also known as the Remez exchange algorithm), or the frequency-sampling method are common methods for computing the filter coefficients 'h[n]' that satisfy the given specifications after the desired frequency response (magnitude and phase characteristics) has been specified.

The coefficients 'h[n]' that determine the frequency response of a FIR filter can be computed using the Discrete Fourier Transform (DFT) or other frequency-domain methods. FIR filters can be made to have low-pass, high-pass, band-pass, and notch frequency characteristics, among others.

Due to its simplicity, design flexibility, and predictable behavior, FIR filters are generally utilized in a wide range of signal processing applications, including audio processing, picture processing, and communication systems. Comprehension means how FIR filters function on input signals to produce desired filtering effects requires a comprehension of its mathematics, notably the convolution operation with the filter coefficients.

IIR (Infinite Impulse Response) Filters: These filters can be designed with feedback loops since they have an infinite-length impulse response. IIR filters may have nonlinear phase characteristics yet may be more computationally efficient.

Low-pass, high-pass, band pass, and notch filters are examples of common filter types, each of which is intended to modify a certain frequency range of the signal.

An additional kind of digital filter used in signal processing is an infinite impulse response (IIR) filter. IIR filters can contain feedback loops in their design because, unlike FIR filters, they have an infinite-length impulse response. As a result, IIR filters are able to produce certain frequency response characteristics with less filter taps than FIR filters. However, if IIR filters are not properly built, they may be more prone to stability problems.

The main distinction between IIR and FIR filters is that IIR filters are recursive, which results in an endless impulse response. An IIR filter's outputs are computed at time 'n' depending on both the current input, 'x[n]' and past outputs and inputs, forming a feedback loop.

Following is a representation of the generic difference equation for causal linear IIR filters of order 'N':

$$y[n] = b[0] * x[n] + b[1] * x[n-1] + \ldots + b[N] * x[n-N] - a[1] * y[n-1] - a[2] * y[n-2] - . - a[M] * y[n-M]$$

In this equation:

'x[n]' is the input signal at time 'n'.

'y[n]' is the output signal at time 'n'.

'b[i]' are the feed forward coefficients of the filter (coefficients for the input samples).

'a[i]' are the feedback coefficients of the filter (coefficients for the output samples).

The filter order is 'N' for the input coefficients and 'M' for the output coefficients.

The 'b[i]' and 'a[i]' coefficients are chosen depending on the filter's desired frequency response properties. Specifying the desired frequency response (for example, low-pass, high-pass, or band pass) is a common step in the design of IIR filters. The filter coefficients are then calculated using methods such as the bilinear transform, Butterworth, Chebyshev, or elliptic filter design methods.

IIR filters can be more effective than FIR filters in terms of the number of coefficients needed to obtain a particular frequency response, but because of their recursive nature, they can be more difficult to build and analyze. To avoid unstable filter behavior, ensuring stability is an essential component of IIR filter design. Understanding the recursive computation of the output based on the current and previous input and output samples, as defined in the difference equation, is part of the mathematics of IIR filters. Another crucial idea is the transfer function of an IIR filter, which explains the relationship between the input and output signals in the frequency domain.

IIR filters are significant in signal processing because of their effective construction and capacity to provide particular frequency response characteristics, but they necessitate careful design and stability consideration. To obtain the necessary filtering effects, the accompanying mathematics uses the difference equation, the transfer function, and other filter design strategies.

Convolution

In signal processing, convolution is a mathematical procedure that combines two signals to create a third signal. Convolution is frequently used to apply a filter to a signal in the context of filters. In order to create the output signal, the convolution operation entails sliding the filter signal over the input signal, multiplying the values at each point where the two signals overlap, and summing the results.

Mathematically, if 'f' is the input signal and 'h' is the filter, the convolution operation (denoted by '*') is defined as:

$$(f * h)(t) = \int f(\tau) \cdot h(t-\tau)]d\tau \text{ (integration over all possible } \tau \text{ values)}$$

In discrete time, the convolution operation for discrete signals is a sum:

$$(f * h)[n] = \sum[f[m] \cdot h[n-m]] \text{ (summation over all possible m values)}$$

Convolutional neural networks (CNNs), which are employed in deep learning for tasks like image identification, are a useful tool for a variety of signal processing tasks, including filtering, edge detection in images, and convolution. Signal processing uses the basic mathematical operation of convolution, among other applications. In order to show how one function "affects" the other as it slides or overlaps over its domain, it involves combining two functions to create a third function. Convolution is particularly crucial for filtering and analyzing signals in the context of signal processing.

In continuous time, the convolution of two continuous functions 'f(t)' and 'g(t)' is defined as follows:

$$(f * g)(t) = \int[f(\tau) \cdot g(t-\tau)]d\tau \text{, integrated over all possible } \tau \text{ values}$$

In discrete time (which is often used in digital signal processing), the convolution of two discrete sequences 'x[n]' and 'h[n]' is given by:

$$(x * h)[n] = \sum[x[k] \cdot h[n-k]] \text{, summed over all possible k values}$$

Here's how convolution works:

1. **Sliding Window:** Assume that you want to apply a "filter" or "kernel" to one of the functions (let's say, "h(t)" or "h[n]") to another function (let's say, "f(t)" or "x[n]"). The domain of the second function is where the filter begins.
2. **Overlap and Multiply:** You multiply the values of the first function ('f(t)' or 'x[n]') that overlaps with the filter at each point of the filter with the values of the second function ('h(t)' or 'h[n]').
3. **Summarization:** To determine the value of the convolved function at that place, add all of these products together after multiplying. Every location of the filter goes through this procedure once, producing the convolved function as a whole.

Convolution is a powerful operation with various applications in signal processing, including:

Filtering: Signals are subjected to convolutional filters. For instance, you can apply a low-pass, high-pass, or any other filter to remove or enhance particular frequency components using a convolution process.

Signal analysis: Operations like correlation, which aid in identifying patterns or similarities between signals, employ convolution.

Image Processing: Through the use of convolutional kernels, convolution is utilized in image processing to perform tasks like edge recognition, blurring, and feature extraction.

Signal Transformation: Using methods like the Fourier Transform, convolution is used to convert signals between domains, such as from the time domain to the frequency domain.

Understanding the principles and uses of convolution, a fundamental idea in signal processing, is crucial for efficiently processing and analyzing information. In conclusion, filters and convolution are fundamental ideas in signal processing that make it possible to manipulate and analyze signals for a variety of purposes.

11.7 SAMPLING AND RECONSTRUCTION

In signal processing, sampling and reconstruction are fundamental ideas, especially when working with continuous-time signals in a digital setting. They are essential for sampling analogue signals into digital representations, then reconstructing the digital representations back into analogue signals. Applications like digital communication, audio processing, and image processing all depend on this process.

1. **Sampling**: By choosing a finite number of values from the continuous signal at predetermined intervals, sampling is the process of turning a continuous-time signal into a discrete-time signal. The outcome is a discrete value sequence that can be thought of as a sampled representation of the original signal. The sampling rate (also known as sampling frequency) and sampling interval (the amount of time between consecutive samples) are the two most important parameters in sampling. According to the Nyquist-Shannon sampling theorem, the sampling rate must be at least twice the maximum frequency present in the continuous signal to prevent aliasing (misinterpretation of high-frequency components as lower frequencies).

   Mathematically, if 'x(t)' is the continuous-time signal, 'x[n]' is the sampled sequence, 'T' is the sampling interval (inverse of the sampling rate, T = 1/fs), and 'n' is the discrete sample index, the relationship is:

$$x[n] = x(nT)$$

2. **Reconstruction**: The process of reconstructing involves transforming a discrete-time signal (a signal that has been sampled) back into a continuous-time signal. This is essential when the digital signal needs to be transformed back into analogue for output or additional continuous domain processing.

   In order to estimate the values of the continuous signal between the discrete samples, interpolation techniques are often used throughout the reconstruction process. Linear interpolation, sinc interpolation (ideal interpolation based on the sinc function), polynomial interpolation, and more sophisticated approaches like spline interpolation are examples of common interpolation techniques. The accuracy of the analog-to-digital conversion is directly tied to the quality of the reconstruction, which ensures that the continuous signal from the original can be retrieved from the discrete samples properly.

   The relationship between the sampled signal and the continuous reconstructed signal can be expressed as:

   $$x\_r(t) = \sum[x[n]\cdot \text{sinc}((t-nT)/T)], \text{ summed over all sample indices 'n'}$$

   Here, 'sinc' is the sinc function, and the summation accounts for the contribution of each sample to the reconstructed signal.

Understanding the fundamentals of sampling and reconstruction is essential to avoiding problems like aliasing, information loss, and distortion when converting between analogue and digital representations of signals. Sampling and reconstruction are key components of digital signal processing.

11.8 SENSOR DATA PROCESSING AS A CRITICAL ASPECT OF ROBOTICS APPLICATIONS

Processing sensor data is a key component of robotics applications that enables robots to properly perceive and interact with their environment. Multiple sensors collect data about the robot's surroundings, and processing this data enables the robot to decide what to do and how to do it.

Here are some key robotics applications that heavily rely on sensor data processing:

Localization and Mapping

Robots can navigate and create maps of their surroundings using sensors like LiDAR (Light Detection and Ranging), cameras, and inertial measurement units (IMUs). Robots are able to detect their position in real-time and produce precise maps with the use of sensor data processing techniques like SLAM (Simultaneous Localization and Mapping).

Fundamental robotics applications like localization and mapping are essential for helping robots navigate, comprehend their surroundings, and carry out jobs efficiently. Let's explore these two related ideas in more detail:

1. **Localization**: The process of figuring out a robot's position (pose) inside a specific environment is referred to as localization. For robots to function freely and navigate from one place to another, this is necessary. Robot localization involves a number of techniques:

   Odometry: By monitoring wheel rotations and assuming no slippage, this technique determines the position of a robot. It is suitable for short-term navigation but less accurate for long-term use because it tends to accumulate errors over time.

Global Positioning System (GPS): GPS can deliver quite precise positional data outside. However, it has drawbacks, such as decreased accuracy in canyons in cities or interior settings.

Visual Landmarks: Robots can locate themselves using visual cues like recognizable landmarks. This strategy is frequently applied in conjunction with other techniques.

Simultaneous Localization and Mapping (SLAM): SLAM is a technique in which a robot simultaneously localizes itself within a map of its environment while creating a map of that environment. This approach is frequently employed in robotics, particularly when great accuracy is required when there is no GPS coverage.

2. **Mapping**: Making a map of the robot's surroundings entails simulating that environment. A map gives the robot a spatial grasp of potential barriers, notable locations, and other important aspects. In order to navigate safely and effectively, accurate maps are essential.

   Occupancy Grid Maps: These are grids that depict the environment in 2D or 3D, with each cell indicating whether it is occupied by a space-filling object or an impediment.

   Maps that emphasize specific environmental elements, such as lines, points, or landmarks. Feature-based mapping and visual-based localization are frequently combined.

   Topological maps: These maps provide a high-level representation ideal for some types of navigation by depicting the environment as a collection of connected regions or nodes.

   Semantic Maps: Semantic maps include information on the kinds of items or regions in the environment in addition to geometric data. Making wiser decisions may benefit from this.

   For many robotics applications, such as autonomous cars, mobile robots, drones, and even some industrial automation tasks, localization and mapping are crucial. Robots are able to navigate challenging areas, avoid hazards, and successfully fulfill their given tasks by fusing precise localization with thorough mapping.

Obstacle Detection and Avoidance

To identify obstructions in the robot's path, sensors like ultrasonic sensors, infrared sensors, and 3D cameras are used. For the robot to design safe paths and prevent collisions, data processing algorithms analyze this data. Robots must be able to recognize and avoid obstacles in order to move safely and effectively across challenging settings. Both autonomous and semi-autonomous robots must have this capacity in order to function properly and avoid colliding with objects or injuring themselves or others.

Here's an overview of obstacle detection and avoidance in robotics:

1. **Obstacle Detection**:

   Robots may identify barriers using a variety of sensors. Ultrasonic, infrared, LiDAR (light detection and ranging), radar, 3D cameras, and depth sensors are examples of common sensors. These sensors offer data on the existence, proximity, and occasionally even the geometry of adjacent barriers.

   Data Fusion: To improve the precision and dependability of obstacle recognition, many robots employ data fusion techniques to merge input from various sensors. Individual sensor limitations may be overcome by sensor fusion.

   Computer vision: Robots may recognize barriers using computer vision methods based on visual data thanks to cameras and depth sensors. It is feasible to travel around items thanks to object recognition algorithms' ability to recognize and categorise them.

2. **Obstacle Avoidance:**

   Route Planning: Robots employ route planning algorithms to compute safe and effective paths to travel around detected obstacles after they have been identified. Finding a collision-free path to the specified location is the objective.

   Reactive Control: Some robots use reactive control techniques that let them react instantly to moving impediments or unanticipated changes in their surroundings. Systems that are reactive can quickly adjust to prevent impending collisions.

   Machine Learning: Robots can be taught to learn obstacle avoidance behaviors using machine learning techniques like reinforcement learning or neural networks. These techniques can develop over time and can adjust to various surroundings.

3. **Applications:**

   Mobile Robots: In order to safely traverse and carry out duties like material handling or cleaning, robots functioning in settings like warehouses, factories, hospitals, or homes must avoid impediments.

   Autonomous Vehicles: To ensure safe navigation on roads or in the air, self-driving cars and drones use obstacle detection and avoidance.

   Robots employed in search and rescue operations must go across disaster-stricken areas while avoiding debris and other obstructions.

   Agricultural Robotics: To explore fields and carry out duties like planting, spraying, or harvesting, agricultural robots need to be able to recognize obstacles.

   Robots must be able to identify and avoid obstacles in order to navigate real-world situations safely and independently, which will allow them to complete a variety of jobs quickly and without assistance from humans.

Object Recognition and Manipulation

Robots having vision sensors, including cameras or depth sensors, are able to identify and control objects in their surroundings. The identification of objects, location determination, and action planning are all accomplished using image processing and computer vision techniques.

Robots can interact with the physical world, comprehend the items in their surroundings, and carry out tasks that involve handling or manipulating objects thanks to the key robotics applications of object identification and manipulation. For a wide range of robotic systems across numerous industries, these characteristics are essential. An overview of robotic object manipulation and recognition is provided below:

1. **Object Recognition:**

   Computer Vision: Robots can take pictures or videos of their surroundings using cameras and vision sensors. To identify objects based on visual input, computer vision techniques such as image processing and machine learning algorithms are used.

   Robots can be taught to categories objects into particular groups or to recognize particular objects based on their appearance, shape, color, or texture.

   Object Localization: In addition to object recognition, robots are also capable of localizing items in the environment, which is crucial for manipulation jobs.

2. **Object Manipulation:**

   Pick and area: Robots using grippers, robotic arms, or end-effectors can take items out of one area and put them in another. In manufacturing, logistics, and material handling operations, this is frequently employed.

Assembly: By manipulating individual pieces, aligning them, and firmly joining them, robots may construct components. This is useful in the electronics and auto production industries.

Dexterous Manipulation: Some robots can execute difficult and delicate manipulation jobs, such precise assembly, picking up delicate things, or performing surgical procedures, thanks to their highly dexterous hands.

Gripping Techniques: Successful object manipulation requires the development of efficient gripping techniques. Robots must decide how to grasp an object most effectively based on its shape, size, weight, and other characteristics.

3. **Applications**:

   Manufacturing: Tasks like product assembly, picking and arranging things on conveyor belts, and quality control through visual inspection are all done by robots in factories.

   Logistics and Warehousing: Robots can efficiently load and unload goods as well as sort, package, and organize things in warehouses.

   Healthcare: Surgical robots can manipulate objects precisely, enabling highly precise less invasive operations.

   Agriculture: Robots are employed to perform activities that require precision object manipulation, such as harvesting crops or pruning plants.

   Retail: Robots can help with customer service, stocking stores, and inventory management.

   Robots' capabilities are increased via object identification and manipulation, which enables them to carry out activities that were difficult or impossible for machines before. These programs improve automation, boost productivity, and significantly affect many different industries.

Environmental Monitoring

Robots can be utilized in disaster response, agriculture, and other environmental monitoring jobs. To monitor and react to changes in the environment, sensor data processing is done on data such as temperature, humidity, gas levels, and more. A crucial robotics application is environmental monitoring, which involves utilizing robots and sensors to collect information about the outside world, the built environment, and numerous factors that have an impact on ecosystems, climate, pollution levels, and other things. This knowledge is crucial for sustaining the sustainability and health of our world as well as for scientific research, resource management, and catastrophe response.

Here are some key aspects of environmental monitoring as a robotics application:

1. **Data Collection**:

   Sensor Deployment: Robots can be placed in hazardous or isolated areas to gather real-time data using a variety of sensors, including weather sensors, air quality monitors, water quality analyzers, and radiation detectors.

   Autonomous Operation: Robots that have been designed to do so can collect data for long periods of time without human interaction. For ongoing environmental condition monitoring, this is especially helpful.

2. **Ecosystem Monitoring**:

   Biodiversity Studies: Robots can be used to observe animal movements, keep tabs on wildlife populations, and evaluate the state of ecosystems. For conservation efforts and an understanding of how climate change affects biodiversity, this information is essential.

Habitat Assessment: Robots can traverse difficult terrain to map habitats, investigate ecosystems, and keep track of changes in land use and vegetation.

3. **Climate and Weather Monitoring:**

   Gathering Climate Data: Robots are capable of gathering information about the climate, including temperature, humidity, atmospheric pressure, wind speed, and precipitation. Understanding the effects of climate change and doing climate research are both dependent on this information.

   Extreme Weather Response: Robots can be used to assess damage, monitor conditions, and support disaster response activities in disaster-affected areas.

4. **Pollution Detection and Monitoring:**

   Air and Water Quality: Robots with sensors can keep an eye on pollutants in the air and water to help pinpoint their sources, evaluate the success of pollution management strategies, and safeguard the general public's health.

   Industrial Monitoring: Robots can monitor emissions from industrial sites to make sure that environmental rules are being followed.

5. **Natural Resource Management:**

   Agricultural Monitoring: Robots can monitor soil conditions, assess crop health, and optimise irrigation, resulting in more productive and sustainable agriculture.

   In forestry, robots can be used to monitor forest health, spot wildfires, and determine how deforestation is affecting the environment.

6. **Ocean and Marine Monitoring:**

   Oceanographic Research: Robots may explore and gather data from the depths of the oceans, investigating marine life, seafloor geology, and ocean currents. Examples of these robots are autonomous underwater vehicles (AUVs) and remotely operated vehicles (ROVs).

   Robots that monitor the environment help us better understand our planet, manage resources wisely, respond to environmental emergencies, and advance scientific knowledge. These programs contribute to maintaining a healthy environment for both present and future generations.

Human-Robot Interaction

Robots can communicate with people using sensors like cameras and microphones. To comprehend and react to human input, data processing techniques including speech recognition, natural language processing, and facial recognition are used. The goal of human-robot interaction (HRI), a broad and rapidly developing discipline in robotics, is to make it possible for humans and robots to interact and collaborate effectively. In order to build robots that can collaborate with people in a variety of settings, comprehend human language, and help with jobs that call for human supervision or engagement. Here are some essential components of robotics applications involving human-robot interaction:

1. **Natural Language Processing:**

   Speech Recognition: Robots are capable of comprehending spoken language, enabling users to instruct, query, or converse with the robot.

   Text-to-Speech: Robots have the ability to translate written text into spoken language, making it simpler for the robot to convey information to people.

2. **Gesture and Expression Recognition**:

   Vision-Based Interaction: Robots can recognize human gestures, facial expressions, and body language using cameras and computer vision algorithms, allowing for more casual kinds of communication.

   Emotion Detection: By analyzing facial expressions, robots can determine the emotions of humans and modify their behavior or replies accordingly.

3. **Collaborative Tasks**:

   Cooperative Work: Humans and robots can work together to complete cooperative activities like assembly, construction, or logistics.

   Assistance: By doing activities like fetching goods, turning on lights, or opening doors, robots can assist persons with physical limitations or restricted mobility.

4. **Social Interaction**:

   Social Robots: Some robots are made to interact socially with people, whether it is as a companion, a source of amusement, or a helper in a medical context.

   Personalization: Robots can modify their behavior in accordance with user preferences, enhancing and personalizing interactions.

5. **Safety and Trust:**

   Safe Interactions: It's crucial to make sure that humans and robots can coexist peacefully, especially in public areas. Systems for interacting with humans and robots must be able to recognize and steer clear of potential dangers.

   Trust: Establishing trust between people and robots requires the design of predictable robot behavior and clear intention-communication.

6. **Education and Training**:

   Robotic Learning: Humans may learn from and be trained by robots in a variety of tasks including education, healthcare, and industrial work.

   Skill Transfer: Humans can learn new skills more quickly thanks to robots' ability to guide and demonstrate tasks.

7. **Entertainment and Assistance**:

   Entertainment Robots: Robots can entertain people, play games, or participate in creative endeavors with people.

   Personal Assistants: Robots can manage schedules, offer information, and carry out activities to help people in their daily lives.

   Making robots more approachable, practical, and incorporated into our society depends critically on human-robot interaction. We may anticipate ever more sophisticated and seamless interactions between humans and robots as the field develops, which will result in cutting-edge applications in numerous industries.

Autonomous Vehicles

Sensor data processing plays a significant role in autonomous vehicles and drones. Radar, LiDAR, cameras, GPS, and other sensors gather data about the environment around the car, and sophisticated algorithms evaluate this information to decide whether to fly or drive. Autonomous vehicles are a well-known and revolutionary robotics application, sometimes known as self-driving cars or autonomous cars.

By allowing cars to drive and run without direct human input, they have the potential to revolutionize transportation.

Here's an overview of autonomous vehicles as a robotics application:

1. **Sensing and Perception:**

   Sensor Array: Autonomous vehicles come with a variety of sensors, such as GPS, LiDAR, cameras, radar, and ultrasonic sensors. These sensors collect information about the area around the car, such as the locations of other cars, people walking along the road, traffic signs, and road conditions.

   Data Fusion: To generate a thorough picture of the surroundings and enable real-time decision-making, the vehicle's onboard computer analyses data from various sensors.

2. **Localization and Mapping:**

   HD Maps: High-definition maps are utilized to display specific details about the road system, lane markings, traffic signs, and pertinent landmarks. These maps are used by autonomous cars for accurate localization and navigation.

   Simultaneous Localization and Mapping, or SLAM, is a technique used by some autonomous vehicles to simultaneously localize themselves inside and generate and update maps of their surroundings.

3. **Path Planning and Control**:

   Path Planning: The vehicle's control system determines a safe and effective path to the destination based on sensor data, maps, and navigational objectives, taking into account numerous elements like traffic, obstructions, and road conditions.

   Trajectory Generation: Using acceleration, braking, steering, and lane changes, the vehicle creates a trajectory that specifies how it should proceed over time.

4. **Decision-Making**:

   Machine Learning: To make complex decisions in dynamic and unpredictable surroundings, autonomous cars frequently use machine learning algorithms. These algorithms provide the car the ability to take lessons from the past and adjust to new circumstances.

   Safety and Ethics: To protect the safety of its occupants and other road users, the vehicle must make moral judgments, such as avoiding crashes, and follow traffic laws.

4. **Applications**:

   Personal Transportation: By eliminating the need for human drivers, autonomous cars can provide comfortable and secure means of personal transportation.

   Ride-Sharing and Mobility Services: Autonomous cars can be utilized for platforms that provide ride-sharing and mobility-as-a-service, providing a reliable and reasonably priced mode of transportation.

   Logistics and Delivery: Last-mile delivery using autonomous vehicles can optimize logistics and lower delivery costs.

   Public Transportation: By offering more adaptable and accessible options, autonomous shuttles and buses can improve existing public transportation networks.

Autonomous vehicles have the potential to increase road safety, reduce traffic congestion, improve fuel efficiency, and provide greater mobility for individuals who are unable to drive. However, their widespread adoption also presents technical, legal, ethical, and societal challenges that need to be addressed for a successful integration into our transportation systems.

Healthcare Robotics

To deliver precise and secure interactions with the human body, robotic equipment used in surgery or patient care rely on sensor data. In medical robots, data processing ensures precise motions and feedback. Healthcare robotics is an exciting and quickly developing subject that uses robots technology to advance medical procedures, increase patient care, and streamline administrative processes in healthcare facilities. Healthcare robots have the potential to improve accuracy, efficacy, and accessibility to medical treatments.

Here are some key applications of robotics in healthcare:

1. **Surgical Robotics**:

   Robotic Surgery: Using the da Vinci Surgical System, surgical robots can help surgeons execute minimally invasive treatments with greater accuracy, dexterity, and visualization. With the help of this technology, patients may experience less trauma, heal more quickly, and have better results.

   Robots can be used in orthopedic surgery to assist with joint replacement procedures, ensuring precise implant placement and maximizing patient mobility.

   Neurosurgery: Using stable hands and real-time image input, robots can help surgeons perform delicate brain and spinal surgeries.

2. **Rehabilitation Robotics**:

   Physical Therapy: Robots can help therapists give patients recovering from accidents or operations focused and reliable rehabilitation exercises.

   Exoskeletons and Prosthetic Limbs: Robotics technology is utilized to create wearable exoskeletons and advanced prosthetic limbs, restoring freedom and movement to those who have lost limbs or have mobility issues.

3. **Medical Imaging and Diagnosis**:

   Robot-Assisted Imaging: Robots can place MRI and CT scanners precisely for the best imaging quality with the least amount of radiation exposure.

   Image-Guided Procedures: Robots can help with biopsies, tumour ablation, and other image-guided interventions by helping to precisely target certain locations.

4. **Telemedicine and Remote Care**:

   Robotic Telepresence: Robots can function as telepresence tools, enabling medical practitioners to engage with patients remotely, have consultations, and continuously check vital signs.

   Home Healthcare: For patients who are elderly or suffer from chronic diseases, robots can help with chores like medication reminders, fall detection, and basic health monitoring.

5. **Laboratory and Drug Discovery**:

   Laboratory Automation: Robots can do repetitive operations in labs like sample processing, drug screening, and high-throughput analysis, resulting in quicker and more precise results.

   Drug Dispensing: Robots can help chemists dispense prescriptions in chemist's and hospitals precisely.

6. **Assistive Robotics**:

   Elderly Care: Robots can assist the elderly with everyday duties, health monitoring, and fall prevention.

Special Needs Care: By assisting with mobility, communication, and social contact, robots can support people with impairments.

7. **Infection Control**:

   Robotic Disinfection: By autonomously cleaning and disinfecting hospital rooms and other high-risk areas, robots with UV lights or disinfectant sprayers help stop the spread of illnesses.

   Healthcare robots has a lot of potential to enhance patient outcomes, save healthcare costs, and address the scarcity of healthcare workers. However, to ensure the efficient and secure incorporation of robots technology into healthcare practices, it also necessitates careful consideration of ethical, privacy, and safety considerations, as well as regulatory permissions.

Industrial Automation

Robots in manufacturing use sensors to carry out operations including quality assurance, material handling, and assembly. Processing sensor data makes operations more efficient and aids in process optimization. Manufacturing and production processes are being revolutionized by industrial automation, which is a well-known and significant application of robotics. Automation and robotics technology can be integrated into industries to boost productivity, cut costs, improve product quality, and promote workplace safety.

Here are key aspects of industrial automation as a robotics application:

1. **Robotic Manufacturing**:

   Assembly: Robots are capable of performing exacting and repetitive assembly operations like screw insertion, component joining, and component attachment. This results in more rapid production and uniform quality.

   Welding: To carry out precise and effective welding operations, robotic welding systems are frequently utilized in sectors like automotive, aerospace, and construction.

   Painting: High-precision robots can apply paint or coatings, assuring uniform coverage and cutting waste.

2. **Material Handling**:

   Palletizing: By automatically loading and unloading things onto pallets, robots help improve logistics at distribution centers and warehouses.

   Packaging: Robots can package goods, sort goods, and set them up for shipping, which speeds up order fulfillment and uses less labor from humans.

3. **Pick and Place**:

   Bin Picking: Robots are capable of picking objects out of bins or other containers, a job that previously required human dexterity and vision.

   Sorting: Robots may arrange objects according to predetermined standards like size, weight, or color.

4. **Quality Control**:

   Inspection: Robots with sensors and vision systems can check products for flaws and ensure that only high-quality goods are released onto the market.

   Testing: To ensure that products satisfy predetermined performance requirements, robots can test their functionality.

5. **CNC Machining:**

   Computer Numerical Control (CNC): CNC machines, such as lathes, mills, and routers, can be operated by robots to create complex and accurate parts for a variety of industries.

6. **Industrial Cleaning and Maintenance**:

   Robotic cleaning may ensure effective operations and increase the lifespan of machines by cleaning and maintaining industrial equipment.

   Routine Maintenance: Routine maintenance chores can be carried out by robots, minimizing downtime and increasing the overall efficacy of the equipment.

7. **Collaborative Robots (Cobots)**:

   Human-Robot Collaboration: Collaborative robots are made to work alongside humans, increasing their strengths and helping with activities that call for human-machine cooperation.

8. **Data-Driven Automation**:

   Data Analytics: Automation systems can gather and analyze data from a variety of sensors and processes, enabling businesses to plan ahead and perform predictive maintenance.

   Manufacturing, logistics, and the construction industries are just a few of the industries that have been transformed by industrial automation. Industries can increase production, increase flexibility, and boost overall competitiveness by using robotic technologies and automation systems.

   Overall, processing sensor data enables robots to perceive and adapt to their surroundings, interact with people, and carry out a variety of tasks independently. It is a crucial part of modern robotics.

11.9 IMAGE AND SPEECH RECOGNITION IN ROBOTICS APPLICATION

Robots can observe and interact with their surroundings, comprehend human orders, and carry out jobs efficiently thanks to image and speech recognition technologies.

Let's explore how image and speech recognition are used in robotics:

1. **Image Recognition:**

   Object Detection: Robots with cameras and computer vision technology can recognize items around them. For activities like picking up and arranging objects, navigation, and interacting with the environment, this ability is necessary.

   Semantic Segmentation: Image recognition enables robots to distinguish between various regions and objects by classifying each pixel, which enables them to comprehend the layout of a scene.

   Visual Localization: Robots locate themselves on a map or in their environment by using picture recognition. For precise navigation and path planning, this is essential.

   Quality Control: To ensure high-quality production, picture recognition can be used in manufacturing to check products for flaws.

   Facial Recognition: Based on facial traits, robots can identify and communicate with humans, enabling customized interactions and security applications.

2. **Speech Recognition:**

   Voice Commands: Human speech is understood by robots, enabling natural human-robot communication. This is especially helpful in situations with constrained physical interfaces.

Natural Language Processing (NLP): Speech recognition is utilised to extract meaning from spoken words in NLP, allowing for more complex interactions and context awareness.

Healthcare: Healthcare robotics can employ speech recognition to transcribe patient notes, transcribe medical dictation, and respond to orders from medical personnel.

Assistive Robotics: Robots that are made to help the elderly or people with impairments can carry out vocal commands to do things like turn on lights, make calls, or remind people of things.

3. **Integration:**

Multimodal Perception: By combining speech and visual recognition, robots can gain a deeper comprehension of their surroundings and interactions. For instance, a robot may understand spoken instructions about things while also recognizing them.

Human-Robot Interaction: The effectiveness of human-robot interactions is improved when voice recognition is combined with picture recognition and other sensory inputs. Robots are capable of deciphering both verbal and nonverbal clues, facilitating more effective and natural conversation.

Speech and picture recognition are essential for enhancing the autonomy, adaptability, and user-friendliness of robots. These technologies serve as the building blocks for applications that call for robots to comprehend and engage with their environment, making them useful tools in a variety of fields such as manufacturing, healthcare, customer service, and more.

NUMERICAL PROBLEMS WITH SOLUTIONS

Problem 1. Given a continuous-time signal $x(t) = \cos(2\pi * 1000 * t)$, where t is in seconds, what is the frequency of this signal in hertz (Hz)?

Solution: The frequency of a sinusoidal signal is given by the coefficient of t in the argument of the cosine function. In this case, the frequency is 1000 Hz.

Problem 2. A discrete-time signal $x[n] = [2, -1, 3, 0, -2]$. Find the length of this signal.

Solution: The length of a discrete-time signal is simply the number of elements it contains. In this case, the length is 5.

Problem 3. Given a continuous-time signal $x(t) = \sin(2\pi * 500 * t)$, where t is in milliseconds (ms), what is the frequency of this signal in kilohertz (kHz)?

Solution: First, convert the frequency from Hz to kHz: 500 Hz = 0.5 kHz.

Problem 4. Find the amplitude of the discrete-time signal $x[n] = 5 * \cos(\pi/4 * n)$.

Solution: The amplitude of a sinusoidal signal is the coefficient of the cosine (or sine) function. In this case, the amplitude is 5.

Problem 5. Calculate the energy of the continuous-time signal $x(t) = 2 * \cos(2\pi * 1000 * t)$, where t is in seconds, over the interval [0, 1] seconds.

Solution: The energy of a continuous-time signal x(t) over an interval [a, b] is given by the integral of the square of the signal magnitude:

$$E = \int[a,b]\, |x(t)|^2\, dt = \int[0,1](2 * \cos(2\pi * 1000 * t))^2\, dt$$

$$= \int[0,1]\, 4 * \cos^2(2\pi * 1000 * t)\, dt = 2$$

So, the energy of the signal over the interval [0, 1] seconds is 2.

Problem 6. Find the average power of the discrete-time signal x[n] = [1, 2, 3, 4, 5] over the entire sequence.

Solution: The average power of a discrete-time signal x[n] is given by the sum of the squares of the signal values divided by the number of samples:

$$P\_avg = (1^2 + 2^2 + 3^2 + 4^2 + 5^2)/5 = 15$$

So, the average power of the signal is 15.

Problem 7. Given a continuous-time signal $x(t) = e^{(-t)} * u(t)$, where t is in seconds, find the value of x(1).

Solution: The signal is a decaying exponential multiplied by the unit step function u(t). The unit step function is 0 for t < 0 and 1 for t >= 0. Therefore, when t = 1, $x(t) = e^{(-1)}$.

Problem 8. Find the convolution of the discrete-time signals x[n] = [1, 2, 3] and h[n] = [1, 1, 1].

Solution: Convolution in discrete-time is calculated by summing the element-wise products of the two signals, where one signal is flipped and shifted over the other. The convolution of x[n] and h[n] is:

$$y[n] = [1, 3, 6, 5, 3]$$

Problem 9. Calculate the Discrete Fourier Transform (DFT) of the discrete-time signal x[n] = [1, 2, 3, 4].

Solution: The DFT of a signal x[n] of length N is given by the formula:

$$X[k] = \Sigma \text{ from } n = 0 \text{ to } N - 1 \ (x[n] * e^{(-j * 2\pi * k * n/N)})$$

Applying this formula, we get:

$$X[0] = 10 \ X[1] = -2 + 2j \ X[2] = -2 \ X[3] = -2 - 2j$$

Problem 10. Find the Inverse Discrete Fourier Transform (IDFT) of the sequence X[k] = [3, –1 + j, –1, –1 – j].

Solution: The IDFT of a sequence X[k] of length N is given by the formula:

$$x[n] = (1/N) * \Sigma \text{ from } k=0 \text{ to } N-1 \ (X[k] * e^{(j * 2\pi * k * n/N)})$$

Applying this formula, we get:

$$x[0] = 0.5 \ x[1] = 1.5 \ x[2] = 2.5 \ x[3] = 3.5$$

Problem 11. Find the autocorrelation of the discrete-time signal x[n] = [1, 2, 3, 4].

Solution: The autocorrelation of a discrete-time signal x[n] is calculated by shifting the signal and multiplying it by its time-reversed version. The autocorrelation of x[n] is:

$$Rxx[0] = 30 \ Rxx[1] = 20 \ Rxx[2] = 10 \ Rxx[3] = 20 \ Rxx[4] = 30$$

Problem 12. Calculate the cross-correlation between the discrete-time signals x[n] = [1, 2, 3, 4] and y[n] = [2, 1, –1, 0].

Solution: The cross-correlation of two discrete-time signals x[n] and y[n] is calculated similarly to the autocorrelation, but one of the signals is time-flipped before shifting it over the other signal. The cross-correlation between x[n] and y[n] is:

$$Rxy[0] = 4 \ Rxy[1] = 7 \ Rxy[2] = 7 \ Rxy[3] = 4$$

Problem 13. Given the continuous-time signal x(t) = 2 * cos(2π * 1000 * t), where t is in milliseconds (ms), sample this signal at a rate of 4 kHz for 2 ms.

Solution: To sample the continuous-time signal, we use the sampling rate to determine the interval between samples. In this case, the interval is 1/4 kHz = 0.25 ms. We need to sample for 2 ms, so the number of samples is 2 ms/0.25 ms/sample = 8 samples. The samples of the signal at t = 0 ms, 0.25 ms, 0.5 ms, ..., 1.75 ms are [2, 0, −2, 0, 2, 0, −2, 0].

Problem 14. Given a discrete-time signal x[n] = [2, 4, 6, 8, 10], find the discrete-time derivative of this signal.

Solution: The discrete-time derivative of a signal x[n] is calculated by taking the difference between consecutive samples. The derivative of x[n] is:

$$x'[n] = [2, 2, 2, 2]$$

Problem 15. Calculate the Discrete Cosine Transform (DCT) of the sequence x[n] = [1, 2, 3, 4].

Solution: The DCT of a sequence x[n] is given by the formula:

$$X[k] = 2/N * \Sigma \text{ from } n=0 \text{ to } N-1\ (x[n] * \cos(\pi * k * (2n+1)/(2N)))$$

Applying this formula, we get:

$$X[0] = 5\ X[1] = 0\ X[2] = 0\ X[3] = 0$$

Problem 16. Given a discrete-time signal x[n] = [1, 2, 3, 4], calculate the energy and power of this signal.

Solution: The energy of a discrete-time signal x[n] is calculated by summing the squares of the signal values, and the power is the energy divided by the number of samples. The energy and power of x[n] are:

$$\text{Energy} = 30\ \text{Power} = 30/4 = 7.5$$

Problem 17. Given a continuous-time signal x(t) = 3 * sin(2π * 2000 * t + π/3), where t is in microseconds (μs), find the phase of this signal at t = 0.5 μs.

Solution: The phase of a sinusoidal signal is the argument of the cosine or sine function. In this case, the phase is π/3.

Problem 18. Find the convolution of the continuous-time signals x(t) = u(t) and h(t) = $e^{(-t)}$ * u(t), where t is in seconds.

Solution: The convolution of two continuous-time signals is calculated by integrating the product of the signals, where one signal is flipped and shifted over the other. The convolution of x(t) and h(t) is:

$$y(t) = (1 - e^{(-t)}) * u(t)$$

Problem 19. Calculate the Discrete Fourier Transform (DFT) of the discrete-time signal x[n] = [1, 0, 1, 0, 1].

Solution: The DFT of a signal x[n] of length N is given by the formula:

$$X[k] = \Sigma \text{ from } n = 0 \text{ to } N - 1\ (x[n] * e^{(-j * 2\pi * k * n/N)})$$

Applying this formula, we get:

$$X[0] = 3X[1] = 1 - jX[2] = 0X[3] = 1 + jX[4] = 0$$

Problem 20. Find the Inverse Discrete Fourier Transform (IDFT) of the sequence X[k] = [4, 2, 0, 2].

Solution: The IDFT of a sequence X[k] of length N is given by the formula:

$$x[n] = (1/N) * \Sigma \text{ from } k = 0 \text{ to } N - 1\ (X[k] * e^{(j * 2\pi * k * n/N)})$$

Applying this formula, we get:

$$x[0] = 2\ x[1] = 3\ x[2] = 2\ x[3] = -1$$

Short Questions and Answers

Q.1. What is signal processing?

Ans. Signal processing is the manipulation and analysis of signals to extract useful information or enhance their quality.

Q.2. What are some common types of signals in signal processing?

Ans. Audio signals, image signals, video signals, sensor data, and communication signals are common types.

Q.3. Why is signal processing important?

Ans. It's essential for tasks like data compression, noise reduction, feature extraction, and communication.

Q.4. What is a digital signal?

Ans. A signal represented by discrete values, is often obtained by sampling a continuous signal.

Q.5. What is a Fourier transform used for?

Ans. It transforms a signal from the time domain to the frequency domain, revealing its frequency components.

Q.6. What is noise in signal processing?

Ans. Noise is unwanted random or undesirable information that can distort or degrade a signal.

Q.7. What is a filter in signal processing?

Ans. A filter is a system that processes a signal to modify certain frequency components.

Q.8. What is the difference between analog and digital signals?

Ans. Analog signals are continuous, while digital signals are discrete and quantized.

Q.9. What is convolution in signal processing?

Ans. Convolution combines two signals to produce a third, often used for filtering or convolutional neural networks.

Q.10. What is the Nyquist theorem?

Ans. It states that to accurately reconstruct a signal, the sampling rate must be at least twice the highest frequency in the signal.

Q.11. What is the purpose of downsampling?

Ans. Downsampling reduces the sampling rate to reduce data size or focus on specific frequency components.

Q.12. What is signal denoising?

Ans. Signal denoising is the process of removing noise from a signal while preserving useful information.

Q.13. What is an impulse response?

Ans. It is the output of a system when an impulse (Dirac delta) is applied as the input.

Q.14. What is spectral analysis?

Ans. It is the study of a signal in the frequency domain to analyze its frequency components.

Q.15. What is a digital filter's transfer function?

Ans. It is the relationship between the input and output of a filter in the frequency domain.

Q.16. What is the purpose of a low-pass filter?

Ans. A low-pass filter allows low-frequency components to pass while attenuating higher frequencies.

Q.17. What is the purpose of a high-pass filter?

Ans. A high-pass filter allows high-frequency components to pass while attenuating lower frequencies.

Q.18. What is the purpose of a band pass filter?

Ans. A band pass filter allows a specific range of frequencies to pass while attenuating frequencies outside that range.

Q.19. What is the purpose of a notch filter?

Ans. A notch filter attenuates a narrow range of frequencies while allowing others to pass.

Q.20. What is an FIR filter?

Ans. A Finite Impulse Response filter has a response that settles in a finite number of samples.

Q.21. What is an IIR filter?

Ans. An Infinite Impulse Response filter can have a response that extends indefinitely.

Q.22. What is quantization in signal processing?

Ans. Quantization is the process of mapping continuous values to discrete values, often used in digital signal representation.

Q.23. What is modulation in signal processing?

Ans. Modulation involves varying the properties of a carrier signal with a modulating signal to transmit information.

Q.24. What is demodulation?

Ans. Demodulation is the process of extracting the modulating signal from a modulated carrier signal.

Q.25. What is autocorrelation?

Ans. Autocorrelation measures the similarity of a signal to a delayed version of itself.

Q.26. What is cross-correlation?

Ans. Cross-correlation measures the similarity between two different signals as a function of time lag.

Q.27. What is the purpose of a window function?

Ans. Window functions are used in spectral analysis to reduce the spectral leakage effect.

Q.28. What is signal interpolation?

Ans. Signal interpolation is the process of estimating values between known data points.

Q.29. What is signal decimation?

Ans. Signal decimation is the process of reducing the number of samples in a signal.

Q.30. What is signal interpolation used for?

Ans. Signal interpolation can be used to increase the sampling rate, creating a smoother representation of a signal.

Q.31. What is a spectrogram?

Ans. A spectrogram is a visual representation of the spectrum of a signal over time.

Q.32. What is peak detection?

Ans. Peak detection is the process of identifying local maxima or minima in a signal.

Q.33. What is time-domain filtering?

Ans. Time-domain filtering directly operates on the input signal without transforming it to the frequency domain.

Q.34. What is a matched filter?

Ans. A matched filter maximizes the signal-to-noise ratio when detecting a known signal in the presence of noise.

Q.35. What is adaptive filtering?

Ans. Adaptive filtering adjusts the filter coefficients in real-time based on the input signal and desired output.

Q.36. What is echo cancellation?

Ans. Echo cancellation removes echoes from a signal, commonly used in telecommunication.

Q.37. What is the purpose of feature extraction in signal processing?

Ans. Feature extraction identifies relevant characteristics of a signal for further analysis or classification.

Q.38. What is the cross-power spectral density?

Ans. The cross-power spectral density shows how the power of one signal is distributed across different frequency components relative to another signal.

Q.39. What is the power spectral density (PSD)?

Ans. The power spectral density describes how the power of a signal is distributed across different frequencies.

Q.40. What is aliasing in signal processing?

Ans. Aliasing occurs when a high-frequency component in a signal appears as a low-frequency component due to inadequate sampling.

Q.41. What is the purpose of a Butterworth filter?

Ans. A Butterworth filter is designed to have a flat frequency response in the pass band.

Q.42. What is the purpose of a Chebyshev filter?

Ans. A Chebyshev filter is designed to have a steeper roll-off but may have ripples in the pass band or stop band.

Q.43. What is digital audio signal processing used for?

Ans. Digital audio signal processing is used for tasks like sound synthesis, equalization, and audio effects.

Q.44. What is the purpose of a filter bank?

Ans. A filter bank splits a signal into multiple frequency bands, often used in audio and image processing.

Q.45. What is signal thresholding?

Ans. Signal thresholding involves segmenting a signal by setting a threshold, useful for image segmentation or noise removal.

Q.46. What is time-frequency analysis?

Ans. Time-frequency analysis characterizes how the frequency content of a signal changes over time.

Q.47. What is the Fast Fourier Transform (FFT)?

Ans. The FFT is an efficient algorithm for computing the Discrete Fourier Transform (DFT) of a signal.

Q.48. What is the inverse Fourier transform?

Ans. The inverse Fourier transform converts a signal from the frequency domain back to the time domain.

Q.49. What is the purpose of zero-padding in the FFT?

Ans. Zero-padding increases the frequency resolution in the FFT output but does not add more information to the signal.

Q.50. What is speech recognition used for in signal processing?

Ans. Speech recognition converts spoken language into text, enabling human-robot interaction and transcription tasks.

Multiple Choice Questions and Answers

Q.1. What is signal processing?

(*a*) Processing of traffic signals (*b*) Processing of audio signals
(*c*) Processing of food signals (*d*) Processing of currency signals

Q.2. What type of signal is continuous and not quantized?

(*a*) Digital signal (*b*) Analog signal (*c*) Binary signal (*d*) Discrete signal

Q.3. What does the Nyquist theorem state?

(*a*) Signals must be sampled at twice the highest frequency
(*b*) Signals must be sampled at half the highest frequency
(*c*) Signals must be sampled at the same frequency
(*d*) Signals can be sampled at any frequency

Q.4. Which transform is used to convert a signal from time domain to frequency domain?

(*a*) Laplace transform (*b*) Z-transform
(*c*) Fourier transform (*d*) Convolution transform

Q.5. What is the primary use of an FIR filter?

(*a*) Infinite Impulse Response (*b*) Filtering with recursive components
(*c*) Filtering with non-recursive components (*d*) Filtering with complex components

Q.6. Which type of filter allows a specific range of frequencies to pass while attenuating others?

(*a*) Low-pass filter (*b*) High-pass filter (*c*) Bandpass filter (*d*) Notch filter

Q.7. What is the purpose of a window function in spectral analysis?

(*a*) To convert the signal to the frequency domain

(*b*) To reduce the spectral leakage effect

(*c*) To increase the noise level

(*d*) To create a linear phase response

Q.8. What does modulation involve?

(*a*) Increasing the amplitude of a signal

(*b*) Varying the properties of a carrier signal with a modulating signal

(*c*) Filtering a signal

(*d*) Downsampling a signal

Q.9. What is the purpose of matched filters?

(*a*) To attenuate high-frequency noise

(*b*) To enhance the power of a signal

(*c*) To maximize the signal-to-noise ratio when detecting a known signal

(*d*) To convert analog signals to digital

Q.10. What does the autocorrelation measure?

(*a*) The similarity of a signal to a delayed version of itself

(*b*) The frequency content of a signal

(*c*) The power of a signal

(*d*) The time delay of a signal

Q.11. What is time-frequency analysis used for?

(*a*) Converting signals from analog to digital

(*b*) Estimating values between known data points

(*c*) Characterizing how the frequency content of a signal changes over time

(*d*) Removing noise from a signal

Q.12. Which algorithm is used to compute the Discrete Fourier Transform (DFT) efficiently?

(*a*) Fourier Series

(*b*) Laplace Transform

(*c*) Z-Transform

(*d*) Fast Fourier Transform (FFT)

Q.13. What is the primary purpose of signal quantization?

(*a*) To increase the size of the signal

(*b*) To reduce the size of the signal

(*c*) To convert a continuous signal to a digital signal

(*d*) To add noise to the signal

Q.14. What is a spectrogram used for?

(*a*) Reducing noise in a signal

(*b*) Detecting peaks in a signal

(*c*) Visualizing the time and frequency content of a signal

(*d*) Converting an analog signal to a digital signal

Q.15. What is the purpose of a low-pass filter?

(*a*) To allow low-frequency components to pass while attenuating higher frequencies

(*b*) To allow high-frequency components to pass while attenuating lower frequencies

(*c*) To allow a specific range of frequencies to pass

(*d*) To attenuate a narrow range of frequencies

Q.16. What is the primary use of a Chebyshev filter?

(*a*) To have a flat frequency response in the pass band

(*b*) To have a steeper roll-off in the stop band (*c*) To remove noise from a signal

(*d*) To create a notch filter

Q.17. What type of filter splits a signal into multiple frequency bands?

(*a*) Low-pass filter (*b*) High-pass filter (*c*) Band pass filter (*d*) Filter bank

Q.18. What is the primary use of a matched filter?

(*a*) To enhance the power of a signal

(*b*) To maximize the signal-to-noise ratio when detecting a known signal

(*c*) To remove noise from a signal

(*d*) To convert a digital signal to an analog signal

Q.19. What does the cross-correlation measure?

(*a*) The similarity of a signal to a delayed version of itself

(*b*) The frequency content of a signal (*c*) The power of a signal

(*d*) The time delay of a signal

Q.20. What is the primary purpose of a window function in spectral analysis?

(*a*) To convert the signal to the frequency domain

(*b*) To reduce the spectral leakage effect (*c*) To increase the noise level

(*d*) To create a linear phase response

Q.21. What does modulation involve?

(*a*) Increasing the amplitude of a signal

(*b*) Varying the properties of a carrier signal with a modulating signal

(*c*) Filtering a signal

(*d*) Downsampling a signal

Q.22. What is the purpose of matched filters?

(*a*) To attenuate high-frequency noise (*b*) To enhance the power of a signal

(*c*) To maximize the signal-to-noise ratio when detecting a known signal

(*d*) To convert analog signals to digital

Q.23. What does the autocorrelation measure?

(*a*) The similarity of a signal to a delayed version of itself

(*b*) The frequency content of a signal (*c*) The power of a signal

(*d*) The time delay of a signal

Q.24. What is time-frequency analysis used for?

(*a*) Converting signals from analog to digital

(*b*) Estimating values between known data points

(*c*) Characterizing how the frequency content of a signal changes over time

(*d*) Removing noise from a signal

Q.25. Which algorithm is used to compute the Discrete Fourier Transform (DFT) efficiently?

(*a*) Fourier Series (*b*) Laplace Transform
(*c*) Z-Transform (*d*) Fast Fourier Transform (FFT)

Q.26. What is the primary purpose of signal quantization?

(*a*) To increase the size of the signal (*b*) To reduce the size of the signal
(*c*) To convert a continuous signal to a digital signal
(*d*) To add noise to the signal

Q.27. What is a spectrogram used for?

(*a*) Reducing noise in a signal (*b*) Detecting peaks in a signal
(*c*) Visualizing the time and frequency content of a signal
(*d*) Converting an analog signal to a digital signal

Q.28. What is the purpose of a low-pass filter?

(*a*) To allow low-frequency components to pass while attenuating higher frequencies
(*b*) To allow high-frequency components to pass while attenuating lower frequencies
(*c*) To allow a specific range of frequencies to pass
(*d*) To attenuate a narrow range of frequencies

Q.29. What is the primary use of a Chebyshev filter?

(*a*) To have a flat frequency response in the pass band
(*b*) To have a steeper roll-off in the stop band (*c*) To remove noise from a signal
(*d*) To create a notch filter

Q.30. What type of filter splits a signal into multiple frequency bands?

(*a*) Low-pass filter (*b*) High-pass filter (*c*) Band pass filter (*d*) Filter bank

Q.31. What is the primary purpose of a matched filter?

(*a*) To enhance the power of a signal
(*b*) To maximize the signal-to-noise ratio when detecting a known signal
(*c*) To remove noise from a signal
(*d*) To convert a digital signal to an analog signal

Q.32. What does the cross-correlation measure?

(*a*) The similarity of a signal to a delayed version of itself
(*b*) The frequency content of a signal
(*c*) The power of a signal
(*d*) The time delay of a signal

Q.33. What is the primary purpose of a window function in spectral analysis?

(*a*) To convert the signal to the frequency domain
(*b*) To reduce the spectral leakage effect
(*c*) To increase the noise level
(*d*) To create a linear phase response

Q.34. What does modulation involve?

(*a*) Increasing the amplitude of a signal
(*b*) Varying the properties of a carrier signal with a modulating signal
(*c*) Filtering a signal (*d*) Downsampling a signal

Q.35. What is the purpose of matched filters?

(*a*) To attenuate high-frequency noise (*b*) To enhance the power of a signal
(*c*) To maximize the signal-to-noise ratio when detecting a known signal
(*d*) To convert analog signals to digital

Q.36. What does the autocorrelation measure?

(*a*) The similarity of a signal to a delayed version of itself
(*b*) The frequency content of a signal (*c*) The power of a signal
(*d*) The time delay of a signal

Q.37. What is time-frequency analysis used for?

(*a*) Converting signals from analog to digital
(*b*) Estimating values between known data points
(*c*) Characterizing how the frequency content of a signal changes over time
(*d*) Removing noise from a signal

Q.38. Which algorithm is used to compute the Discrete Fourier Transform (DFT) efficiently?

(*a*) Fourier Series (*b*) Laplace Transform
(*c*) Z-Transform (*d*) Fast Fourier Transform (FFT)

Q.39. What is the primary purpose of signal quantization?

(*a*) To increase the size of the signal (*b*) To reduce the size of the signal
(*c*) To convert a continuous signal to a digital signal
(*d*) To add noise to the signal

Q.40. What is a spectrogram used for?

(*a*) Reducing noise in a signal (*b*) Detecting peaks in a signal
(*c*) Visualizing the time and frequency content of a signal
(*d*) Converting an analog signal to a digital signal

Q.41. What is the purpose of a low-pass filter?

(*a*) To allow low-frequency components to pass while attenuating higher frequencies
(*b*) To allow high-frequency components to pass while attenuating lower frequencies
(*c*) To allow a specific range of frequencies to pass
(*d*) To attenuate a narrow range of frequencies

Q.42. What is the primary use of a Chebyshev filter?

(*a*) To have a flat frequency response in the passband
(*b*) To have a steeper roll-off in the stopband (*c*) To remove noise from a signal
(*d*) To create a notch filter

Q.43. What type of filter splits a signal into multiple frequency bands?

(*a*) Low-pass filter (*b*) High-pass filter (*c*) Bandpass filter (*d*) Filter bank

Q.44. What is the primary purpose of a matched filter?

(*a*) To enhance the power of a signal
(*b*) To maximize the signal-to-noise ratio when detecting a known signal
(*c*) To remove noise from a signal
(*d*) To convert a digital signal to an analog signal

Q.45. What does the cross-correlation measure?

(*a*) The similarity of a signal to a delayed version of itself
(*b*) The frequency content of a signal (*c*) The power of a signal
(*d*) The time delay of a signal

Q.46. What is the primary purpose of a window function in spectral analysis?

(*a*) To convert the signal to the frequency domain
(*b*) To reduce the spectral leakage effect (*c*) To increase the noise level
(*d*) To create a linear phase response

Q.47. What does modulation involve?

(*a*) Increasing the amplitude of a signal
(*b*) Varying the properties of a carrier signal with a modulating signal
(*c*) Filtering a signal (*d*) Downsampling a signal

Q.48. What is the purpose of matched filters?

(*a*) To attenuate high-frequency noise (*b*) To enhance the power of a signal
(*c*) To maximize the signal-to-noise ratio when detecting a known signal
(*d*) To convert analog signals to digital

Q.49. What does the autocorrelation measure?

(*a*) The similarity of a signal to a delayed version of itself
(*b*) The frequency content of a signal (*c*) The power of a signal
(*d*) The time delay of a signal

Q.50. What is time-frequency analysis used for?

(*a*) Converting signals from analog to digital
(*b*) Estimating values between known data points
(*c*) Characterizing how the frequency content of a signal changes over time
(*d*) Removing noise from a signal

Key to Multiple Choice Questions

1. (*b*)	**2.** (*b*)	**3.** (*a*)	**4.** (*c*)	**5.** (*c*)	**6.** (*c*)	**7.** (*b*)	**8.** (*b*)	**9.** (*c*)
10. (*a*)	**11.** (*c*)	**12.** (*d*)	**13.** (*c*)	**14.** (*c*)	**15.** (*a*)	**16.** (*b*)	**17.** (*d*)	**18.** (*b*)
19. (*a*)	**20.** (*b*)	**21.** (*b*)	**22.** (*c*)	**23.** (*a*)	**24.** (*c*)	**25.** (*d*)	**26.** (*c*)	**27.** (*c*)
28. (*a*)	**29.** (*b*)	**30.** (*d*)	**31.** (*b*)	**32.** (*a*)	**33.** (*b*)	**34.** (*b*)	**35.** (*c*)	**36.** (*a*)
37. (*c*)	**38.** (*d*)	**39.** (*c*)	**40.** (*c*)	**41.** (*a*)	**42.** (*b*)	**43.** (*d*)	**44.** (*b*)	**45.** (*a*)
46. (*b*)	**47.** (*b*)	**48.** (*c*)	**49.** (*a*)	**50.** (*c*)				

Project Works

1. **Speech Recognition for Human-Robot Interaction:** Develop a speech recognition system that allows a robot to understand and respond to spoken commands from users. Use signal processing techniques to preprocess and analyze the audio input, enabling the robot to recognize different commands and carry out appropriate actions.
2. **Gesture Recognition for Robot Control:** Create a system that enables a robot to interpret hand gestures as control commands. Use image or depth sensors to capture the gestures, and apply signal processing algorithms to recognize and map these gestures to specific robot movements or actions.
3. **Visual Object Tracking for Robotic Manipulation:** Build a robotic arm or manipulator that can track and follow a moving object in a real-world environment. Use computer vision techniques to process the video stream, detect the object, and calculate the necessary robot movements to keep the object within the robot's grasp.
4. **Real-Time Localization and Mapping (SLAM):** Implement Simultaneous Localization and Mapping (SLAM) using sensor data from a mobile robot. Process data from cameras, LiDAR, or other sensors to create a map of the environment while simultaneously estimating the robot's position within the map.
5. **Obstacle Detection and Avoidance:** Design an autonomous robot capable of navigating through an environment while detecting and avoiding obstacles in real time. Use sensor data, such as LiDAR, ultrasonic sensors, or cameras, to process the surroundings and implement algorithms for safe navigation.
6. **Environmental Monitoring with Robotics:** Develop a robotic system that can autonomously collect and analyze environmental data, such as air quality, temperature, or humidity, in a specific area. Process the sensor data to generate real-time reports and insights about the environment.
7. **Human Activity Recognition for Robot Assistance:** Create a robot that can assist users in their daily activities by recognizing specific human actions. Use signal processing techniques on sensor data (e.g., accelerometers, gyroscopes) to identify actions like sitting, standing, walking, or performing specific tasks, and allow the robot to respond accordingly.
8. **Autonomous Drone Navigation:** Design a drone that can navigate autonomously through a predefined path or explore an unknown environment. Use vision-based navigation or sensor fusion techniques to process data from onboard cameras and sensors, ensuring safe and accurate flight.
9. **Health Monitoring and Care with Robotics:** Develop a robotic system for monitoring vital signs (e.g., heart rate, body temperature) of patients in a healthcare setting. Process the sensor data to detect anomalies and send alerts to healthcare professionals when necessary.
10. **Industrial Quality Control using Vision:** Create a vision-based quality control system for an industrial process. Use cameras to capture images or videos of products on a production line, and apply image processing and pattern recognition techniques to identify defects or ensure product conformity.

12

Optimization Process of finding the Best Possible Solution from a Set of Available Options

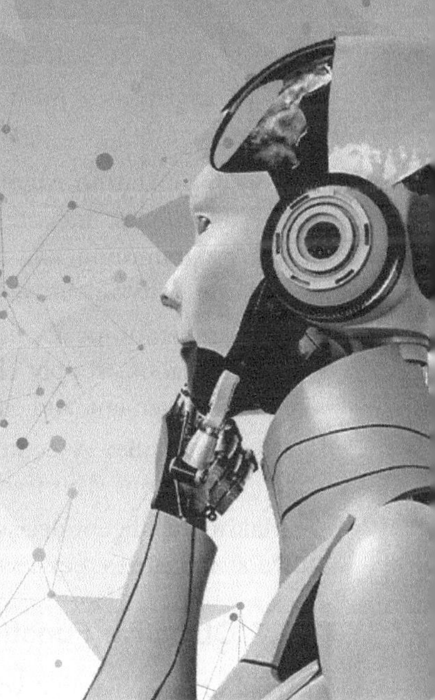

12.1 INTRODUCTION

In many disciplines, including mathematics, engineering, economics, computer science, and more, optimization is a fundamental idea. Given particular limitations and objectives, it entails the process of selecting the best option among a range of alternatives. The primary objective of optimization is to maximize or enhance a given objective function while adhering to the given restrictions. Making intelligent decisions and increasing the efficiency and effectiveness of numerous processes in diverse industries require a thorough understanding of optimization principles and methodologies.

12.2 KEY COMPONENTS

Here are some key components and concepts in optimization:

1. **Objective Function:** You want to maximize this function. It stands for the standard of success or the objective you hope to reach. A quantity can be either maximized (for example, profit, efficiency, or performance) or minimized (for example, cost, error, or time).
2. **Decision Variables:** These are the elements you can modify or control to affect how the objective function turns out. Finding these variables' values that result in the best solution is the goal of optimization.
3. **Restrictions:** Limitations or restrictions that the solution must meet are referred to as constraints. They may be inequality constraints (like capacity restrictions) or equality constraints (like fixed resources).
4. **Optimal Solution:** The collection of values for the choice variables that maximizes or minimizes the objective function while abiding by the given constraints is referred to as the optimal solution.
5. **Differentiating between Local and Global Optima:** It is vital because optimization issues may have many solutions. A local optimum is the best option in a certain area, but it might not be the overall optimal option. The absolute best option out of all those available is known as the global optimum.

6. **Optimization Algorithms:** These are logical techniques for locating the best answer. Different techniques, including gradient descent, genetic algorithms, linear programming, dynamic programming, and more, may be used, depending on the properties of the problem (such as convexity, linearity, or nonlinearity).
7. **Applications:** Optimization is utilized in many real-world situations, including resource allocation, supply chain optimization, machine learning model training, financial portfolio management, and engineering design, where selecting the optimal choices is essential.
8. **Trade-offs:** A common part of optimization is choosing between opposing objectives or limitations. This trade-off balancing is a crucial step in the optimization process.

Making intelligent decisions and increasing the efficiency and effectiveness of numerous processes in diverse industries require a thorough understanding of optimization principles and methodologies.

12.2.1 Objective Function in Optimization

The goal or objective that you want to optimize is quantified by the objective function, a key idea in optimization. It is a mathematical expression that depends on one or more decision variables, and the optimization process seeks to identify the values of these variables, subject to any given constraints, that either maximize or minimize the objective function.

The goal function in optimization is described in more depth below, along with some mathematical notation:

1. **Mathematical Notation:** Let's denote the decision variables as $x_1, x_2, \ldots, x_n$, where n is the number of decision variables. The objective function is usually represented as $f(x_1, x_2, \ldots, x_n)$, where f is a mathematical function that depends on the values of the decision variables.
2. **Goal:** Either a maximization problem or a minimization issue can be used to describe the goal. You wish to identify the values of the decision variables in a maximization problem that maximize the objective function. In mathematics, this is expressed as:

   Maximize $f(x_1, x_2, \ldots, x_n)$

   In a minimization problem, you want to find the values that minimize the objective function:

   Minimize $f(x_1, x_2, \ldots, x_n)$
3. **Example:** Let's consider a simple example. Suppose you're running a business that manufactures and sells two products, A and B, and you want to maximize the total profit. The decision variables are the quantities of each product to produce and sell, denoted as x (quantity of A) and y (quantity of B). The profit from selling one unit of A is \$5, and the profit from selling one unit of B is \$8. The objective is to maximize the total profit, which can be expressed as:

   $$f(x, y) = 5x + 8y$$

   The goal is to find the values of x and y that maximize this profit function, subject to constraints on resources, demand, or other relevant factors.
4. **Constraints:** The objective function frequently faces restrictions, which place limits on the possible values for the decision variables. Both inequalities and equalities can be used to indicate constraints. For instance, if the company in the aforementioned scenario has few resources, such as labor or raw materials accessible, you can face restrictions like:

   x &\geq 0 \quad \text{(non-negativity constraint)} \\ y &\geq 0 \quad \text{(non-negativity constraint)} \\ 2x + 3y &\leq 100 \quad \text{(labor constraint)} \\ x + y &\leq 50 \quad \

text{(raw material constraint)} \end{align*} \] These constraints restrict the feasible region where the optimal solution must lie.

The purpose is to identify the values of the choice variables that meet the constraints and either maximize or minimize the objective function. Depending on the specifics of the problem, this frequently entails applying optimization algorithms like gradient descent, linear programming solvers, evolutionary algorithms, or other techniques.

Whether the primary objective of the optimization problem is to maximize profit, minimize cost, increase efficiency, or achieve any other desired result, it is captured by the objective function. We can use mathematical and computational techniques to identify the best solution within the limits thanks to the mathematical representation.

12.2.2 Decision Variables

Basic components of optimization issues are decision variables. They indicate the amounts or values that you can modify or manage in order to get the result you want while staying within the boundaries that have been set. Decision variables play a critical role in identifying the best solution and are a significant part in formulating and solving optimization issues.

Here's a detailed explanation of decision variables, along with mathematical notation:

1. **Definition:** Decision variables are the variables or unknowns for which you wish to determine the best values. These variables stand for the amounts, configurations, or decisions you can make to optimize the objective function.
2. **Mathematical Notation:** Decision variables are frequently represented mathematically by letters, frequently x, y, or z, or other pertinent symbols. When there are numerous choice variables, subscripts like x_1, x_2, etc. are used to separate them.
3. Consider the following concrete example to better understand decision variables. Consider that you are organizing a production process and must choose how many units of two goods, A and B, to make. The quantities of product A and B can be represented by the letters x and y, respectively. The choice is yours.
   - Decision Variable 1: x (quantity of product A)
   - Decision Variable 2: y (quantity of product B)

   The values of x and y are the quantities of each product that you decide to produce, and they directly impact the objective (e.g., profit, cost, efficiency) you're trying to optimize.
4. Decision-making factors have restrictions. These limitations could be brought on by a scarcity of resources, capacity limitations, demand, budgetary constraints, or any other factors that restrict the breadth of the issue that can be resolved.

   For instance, in the manufacturing example, there may be restrictions on the work force, raw materials, or production capacity. These limitations can be modeled mathematically as inequalities affecting the choice variables.

   Example constraints:

   $x \geq 0$ (non-negativity constraint for product A)

   $y \geq 0$ (non-negativity constraint for product B)

   $2x + 3y \leq 100$ (labor constraint)

   $x + y \leq 50$ (raw material constraint)

4. **Objective Function:** As was already said, the choice variables affect the objective function. It quantifies the objective you wish to achieve, such as maximising revenue or lowering expenses. The choice variables are used in the mathematical expression of the objective function.

   Example objective function:

   $f(x, y) = 5x + 8y$ (total profit)

   In conclusion, decision variables describe the options you have while trying to solve an optimization problem. To choose the optimal result that satisfies the given constraints, they interact with the objective function and constraints. At its essence, optimization is the process of identifying the values of these decision variables that result in the best solution.

12.2.3 Constraints

Constraints are crucial elements of optimization issues that set restrictions or boundaries on the possible values for the decision variables. The collection of all legitimate solutions that satisfies the provided criteria constitutes the viable region, which is defined by these constraints. Constraints make the optimization challenge more realistic and verify that the solution complies with the precise specifications of the issue you're attempting to address. Mathematical inequalities or equalities involving the choice variables are frequently used to express constraints.

Here's a detailed explanation of constraints in optimization, along with mathematical notation:

1. **Mathematical Notation:** Constraints are frequently represented mathematically by inequality or equality conditions involving the decision variables, which are designated as $x_1, x_2, ..., x_n$. These variables can be used to create a mathematical equation that corresponds to each requirement.
2. **Constraints for Equality:** These restrictions mandate that a particular linear combination of the choice variables equals a particular value. An equality requirement can be expressed mathematically as:

   $$a_1x_1 + a_2x_2 + \ldots + a_nx_n = b$$

   Here, $a_1, a_2, \ldots, a_n$ are coefficients, and b is a constant value.
3. **Inequality Constraints:** These constraints restrict the values of the decision variables to be within certain bounds. Inequality constraints can be expressed as:

   $$g(x_1, x_2, \ldots, x_n) \le c$$

   or $$h(x_1, x_2, \ldots, x_n) \ge d$$

   Here, g and h are functions involving the decision variables, and c and d are constants.
4. **Example:** Let's continue with the manufacturing example from previous discussions, where you're producing two products, A and B, with labor and raw material constraints. We'll express these constraints mathematically:

   Labor Constraint: The total labor used in producing products A and B must not exceed 100 units.

   $$2x + 3y \le 100$$

   Raw Material Constraint: The total raw material used in producing products A and B must not exceed 50 units.

   $$x + y \le 50$$

   Non-Negativity Constraints: You cannot produce a negative quantity of products.

   $$x \ge 0 \quad \text{and} \quad y \ge 0$$

5. **Feasible Region:** The set of all decision variable values that satisfy the provided constraints is known as the feasible region. It is where all restrictions meet. In the manufacturing example, the intersection of the labor constraint, the raw material constraint, and the non-negativity constraints defines the viable zone. The most practical answer must be found within this area.

   The feasible solution space is significantly shaped by constraints, which also guarantee that the ideal solution is both practical and pertinent to the issue at hand. In order to discover the optimal answer within the range of possibilities, you must take the objective function and the constraints into account when solving an optimization problem.

12.2.4 Optimal Solution

The set of choice variable values that either maximizes or minimizes the objective function while satisfying all of the specified constraints is the optimal solution to an optimization problem. The best result that can be obtained given the parameters of the issue is the optimal solution. Finding the highest or lowest value for the objective function depends on the type of optimization issue (maximization or minimization).

The ideal response in optimization is explained in full below, along with mathematical notation:

1. **Maximization Problem:** The purpose of a maximization problem is to identify the values of the decision variables that produce the objective function's maximum value. The highest value of the objective function that can be achieved within the range of possibilities established by the constraints is the optimal solution.

   Mathematical Notation:

   Decision variables: $x_1, x_2, \ldots, x_n$

   Objective function to maximize: $f(x_1, x_2, \ldots, x_n)$

   Constraints: $g_i(x_1, x_2, \ldots, x_n) \leq c_i$ and/or $h_j(x_1, x_2, \ldots, x_n) = d_j$

   The optimal solution $(x_1^*, x_2^*, \ldots, x_n^*)$ is the set of values that maximizes the objective function while satisfying all the constraints:

   $$f(x_1^*, x_2^*, \ldots, x_n^*) \geq f(x_1, x_2, \ldots, x_n)$$

   subject to:

   $$g_i(x_1^*, x_2^*, \ldots, x_n^*) \leq c_i \quad \text{and} \quad h_j(x_1^*, x_2^*, \ldots, x_n^*) = d_j$$

2. **Minimization Problem:** The objective of a minimization problem is to identify the values of the decision variables that provide the objective function's minimal value. The lowest possible value of the objective function within the range of possibilities established by the constraints is what is meant by the best solution.

   Mathematical Notation:

   Decision variables: $x_1, x_2, \ldots, x_n$

   Objective function to minimize: $f(x_1, x_2, \ldots, x_n)$

   Constraints: $g_i(x_1, x_2, \ldots, x_n) \leq c_i$ and/or $h_j(x_1, x_2, \ldots, x_n) = d_j$

   The optimal solution $(x_1^*, x_2^*, \ldots, x_n^*)$ is the set of values that minimizes the objective function while satisfying all the constraints:

   $$f(x_1^*, x_2^*, \ldots, x_n^*) \leq f(x_1, x_2, \ldots, x_n)$$

   subject to:

   $$g_i(x_1^*, x_2^*, \ldots, x_n^*) \leq c_i \quad \text{and} \quad h_j(x_1^*, x_2^*, \ldots, x_n^*) = d_j$$

3. **Finding the Optimal Solution:** To find the optimum solution, one must first solve the optimization issue, which calls for merging the constraints and the objective function. The values of the choice variables that result in the optimal solution inside the feasible zone are chosen using a variety of mathematical and computational techniques, such as linear programming, nonlinear optimization, gradient descent, genetic algorithms, or other optimization algorithms. The best possible conclusion given the optimization problem's criteria is represented by the optimal solution, which is an essential outcome for making knowledgeable judgments in a variety of disciplines, including engineering, economics, operations research, and more.

12.2.5 Local vs. Global Optima

When working with objective functions that could have several solutions, the terms local optima and global optima are frequently used in optimization. It is essential to comprehend the distinction between these two categories of optima since doing so enables us to evaluate the effectiveness of our answers and take thoughtful action in a variety of problem-solving situations.

1. **Local Optimum:** A local optimum is a solution that, while being the greatest within a specific neighborhood or part of the feasible space, may not be the best overall. Alternatively said, a local optimum is a position where the objective function has a higher value (for maximization) or lower value (for reduction) than its neighboring points, albeit it may not be a globally optimal point.

   Mathematical Notation:

   Objective function: $f(x_1, x_2, \ldots, x_n)$

   Local Optimum: $(x_1^*, x_2^*, \ldots, x_n^*)$ is a local optimum if there exists a neighborhood around this point (let's say N) such that $f(x_1^*, x_2^*, \ldots, x_n^*)$ is the highest (for maximization) or lowest (for minimization) value within that neighborhood, but it may not be the highest (for maximization) or lowest (for minimization) value over the entire feasible space.

   Mathematically, a local optimum is defined as: $f(x_1^*, x_2^*, \ldots, x_n^*) \geq f(x_1, x_2, \ldots, x_n)$ for all $(x_1, x_2, \ldots, x_n)$ in the neighborhood N around $(x_1^*, x_2^*, \ldots, x_n^*)$.

2. **Global Optimum:** The best potential outcome for the entire feasible region is referred to as a global optimum, also known as the absolute optimum, global maximum (for maximization), or global minimum (for reduction). It is the objective function's maximum (for maximization) or minimum (for minimization) value among all feasible answers that adhere to the specified restrictions.

   Mathematical Notation:

   Global Optimum: $(x_1^*, x_2^*, \ldots, x_n^*)$ is a global optimum if it satisfies the objective function better than any other feasible solution in the entire feasible region.

   Mathematically, a global optimum is defined as: $f(x_1^*, x_2^*, \ldots, x_n^*) \geq f(x_1, x_2, \ldots, x_n)$ for all feasible $(x_1, x_2, \ldots, x_n)$ in the feasible region.

3. **Exemplification:** Imagine an objective function that resembles a mountainous or valleyous landscape. The highest peak or lowest valley are the locations where the function is lower than in the immediate vicinity but higher than at the local optimum. Local optimum points are found in the valleys. The absolute highest position (for maximization) or absolute lowest point (for reduction) on the entire landscape is the global optimum.

Finding the global optimum in optimization is a challenge since local optimum results might be deceptive and result in poor solutions if we're not attentive with our optimization strategy.

4. **Consequences:** It's critical to be mindful of the possibility of several local optima when resolving optimization issues. You might discover a local optimum that isn't the greatest overall answer, depending on the nature of the problem and the optimization algorithm utilized. To reduce the risk of becoming stuck in local optima and enhance the likelihood of discovering the global optimum, tactics can be used such global optimization algorithms, random initialization, or well planned search strategies.

12.2.6 Optimization Algorithms

In order to maximize or minimize the objective function of a given optimization problem while adhering to the provided constraints, optimization algorithms are methodical techniques. These algorithms aid in navigating the range of potential solutions in order to converge on the best one. The parameters of the problem, such as the kind of objective function, the restrictions, and the problem's dimensionality, determine the optimization procedure to be used.

Here are some common optimization algorithms used in detail, along with a brief overview and mathematical concepts:

1. **Gradient Descent:** For differentiable functions, gradient descent is a popular optimization approach. In convex issues when the objective function has a singular global minimum (or maximum), it works exceptionally well. The algorithm iteratively updates the choice variables in a way that decreases the value of the goal function by using the gradient (derivative) of the objective function.

   Mathematical Notation:

   Objective function: $f(x_1, x_2, \ldots, x_n)$

   Gradient: $\nabla f(x_1, x_2, \ldots, x_n)$

   The update rule for gradient descent is: $x_i(t+1) = x_i(t) - \alpha \cdot \nabla f(x_1(t), x_2(t), \ldots, x_n(t))$ where $x_i(t)$ is the i-th decision variable at iteration t, and α is the learning rate (step size).

2. **Newton's Technique:** Newton's technique is an additional optimisation procedure that iteratively improves the solution by using both the gradient and the Hessian matrix (second derivatives) of the objective function. Although it converges more quickly than gradient descent, high-dimensional problems may make it computationally expensive.

   Mathematical Notation:

   Hessian matrix: $\nabla^2 f(x_1, x_2, \ldots, x_n)$

   The update rule for Newton's method is: $x(t+1) = x(t) - (\nabla^2 f(x(t))) - 1 \cdot \nabla f(x(t))$ where $x(t)$ is the vector of decision variables at iteration t.

3. **Linear Programming:** Optimisation problems with linear objective functions and linear constraints are solved using linear programming (LP). It is an effective method with numerous practical applications, including resource allocation, production scheduling, and transportation issues.

   Mathematical Notation:

   Objective function: $c_1x_1 + c_2x_2 + \ldots + c_nx_n$

   Linear constraints: $ai_1x_1 + ai_2x_2 + \ldots + ai_nx_n \leq b_i$

The objective is to find the values of the decision variables $x_1, x_2, \ldots, x_n$) that optimize the linear objective function while satisfying the linear constraints.

4. **Genetic Algorithms:** The principles of natural selection and evolution serve as the foundation for genetic algorithms. They are applied to optimisation issues whose extensive or intricate solution spaces. In order to evolve the population towards better answers, genetic algorithms retain a population of potential solutions and use genetic operators (crossover, mutation, and selection).

   Mathematical Notation: Unlike gradient-based techniques, genetic algorithms do not rely on derivatives. Instead, they employ genetic operators to gradually generate a population of solutions encoded as chromosomes (genetic representations) across several generations.

5. **Particle Swarm Optimisation (PSO):** The social behaviour of fish and birds served as the inspiration for this population-based optimisation technique. In PSO, individuals (particles) move across the solution space while being influenced by both their personal best-known location and the best-known position in the population.

   Mathematical Notation: Each particle in PSO has a position vector x and a velocity vector v, and they are travelling through the solution space. The best solution discovered by the entire swarm, the particle's memory of its own best solution, and its current velocity are all combined in the PSO update rule.

   Integer programming, nonlinear constrained optimization, simulated annealing, and other specialized methods are only a few examples of optimization techniques. There are many more for various issue kinds. The kind of the problem, the features of the objective function, the presence of constraints, the dimensionality, and the computer resources available all play a role in selecting the best optimization technique.

12.2.7 Optimization Applications

The goal of optimization, which has many applications across many different domains, is to identify the best possible solution given a set of constraints and objectives. I'll explain in-depth applications of optimization in several fields below, along with pertinent mathematical ideas:

1. **Engineering Design:** Engineering design efficient and affordable systems need optimization. Engineers frequently use this technique in structural design, where they try to determine the dimensions of a structure (such a bridge or a skyscraper) that use the least amount of materials while still being able to sustain stresses. Cost reduction while meeting requirements for structural integrity is frequently the goal.

   Mathematical Notation:

   Objective function: Cost function, $C(x_1, x_2, \ldots, x_n)$

   Constraints: Structural constraints, such as stress limits, displacement limits, etc.

   Example:

   Minimize the cost of a beam subject to constraints on maximum stress, deflection, and dimensions.

2. **Finance and Portfolio Optimization:** In finance, portfolios of investments that maximize expected returns while minimizing risk are built. Investments should be spread across a variety of assets to strike a balance between possible profit and risk reduction.

Mathematical Notation:

Objective function: Expected portfolio return, $R(x_1, x_2, \ldots, x_n)$

Constraints: Budget constraints, risk constraints, asset allocation constraints.

Example:

Maximize the expected return of a portfolio while ensuring that the risk (measured by standard deviation) does not exceed a certain threshold.

3. **Supply Chain Management:** The purpose of supply chain optimization, which aims to reduce costs while satisfying supply and demand restrictions, is optimization. Decisions must be made about distribution, inventory control, transportation, and production.

   Mathematical Notation:

   Objective function: Total supply chain cost, $C(x_1, x_2, \ldots, x_n)$

   Constraints: Production capacities, demand, inventory limits, transportation capacities.

   Example:

   Minimize the total cost of producing and distributing goods while satisfying demand and capacity constraints.

4. **Machine Learning and Model Training:** Training machine learning models requires optimization, particularly when employing gradient-based techniques to reduce the loss function. The objective is to avoid over fitting when modifying model parameters to fit the data.

   Mathematical Notation:

   Objective function: Loss function, $L(\theta)$ (where θ represents model parameters)

   Constraints: Regularization constraints, parameter bounds (if any).

   Example:

   Minimize the mean squared error between predicted and actual values in a linear regression model.

5. **Operations Research and Network Design:** Network design, facility location, and resource allocation are just a few of the complicated issues that optimization is frequently employed to address in operations research.

   Mathematical Notation:

   Objective function: Cost function, $C(x_1, x_2, \ldots, x_n)$

   Constraints: Network connectivity, resource constraints, demand satisfaction.

   Example:

   Optimize the location of warehouses in a distribution network to minimize transportation costs while ensuring timely delivery to customers.

   These are only a few examples of optimization applications, and depending on the formulation of the problem, each application's mathematical notation may differ. A strong framework for making wise selections in a variety of real-world situations is provided by optimization.

12.2.8 Trade-offs

Trade-offs, which refers to the juggling act between opposing objectives or constraints, are a necessary part of optimization problems. When a system is optimized, often one aspect suffers at the expense of another. Making wise selections require being aware of and managing these trade-offs. Calculating and visualizing these trade-offs through mathematical modeling enables decision-makers to select options that fit their objectives and restrictions.

Here's a detailed explanation of trade-offs in optimization, along with mathematical concepts:

1. **Objective Function and Constraints:** In most optimization tasks, you must maximize or minimize an objective function while adhering to a set of constraints. The objective function is a reflection of the target that you're attempting to realize, such as maximizing performance, efficiency, or profit. Contrarily, constraints signify restrictions or needs that the solution must satisfy.

   Mathematical Notation:

   Objective function: $f(x_1, x_2, \ldots, x_n)$

   Constraints: $g_i(x_1, x_2, \ldots, x_n) \leq c_i, \quad h_j(x_1, x_2, \ldots, x_n) = d_j$

2. **Trade-offs in Multi-Objective Optimization:** Multi-objective optimization deals with circumstances where you want to simultaneously optimize a number of conflicting objectives. Due to intrinsic conflicts, finding a single solution that precisely optimizes all objectives may be impossible. The purpose is instead to identify a group of solutions, referred to as the Pareto front, where achieving one objective will make another worse.

   Mathematical Notation:

   Multi-objective: Optimize $f_1(x_1, x_2, \ldots, x_n)$ and $f_2(x_1, x_2, \ldots, x_n)$ simultaneously.

3. **Quantifying Trade-Offs:** Utility functions, Pareto dominance, and sensitivity analysis are a few techniques that can be used to quantify trade-offs. Sensitivity analysis looks at how changes in the values of the decision variables affect the goal function. Comparing solutions to find those that are superior in at least one objective and neutral in all others is known as Pareto dominance. Utility functions reflect the preferences of the decision-maker by aggregating numerous objectives into a single scalar value.

   Mathematical Notation:

   Sensitivity Analysis: Analyze $\partial x_i \partial f$ to understand how changing x_i impacts the objective function.

   Pareto Dominance: A solution A dominates solution B if $fk(A) \geq fk(B)$ for all objectives k and $fk(A) > fk(B)$ for at least one objective k.

   Utility Function: $U(x_1, x_2, \ldots, x_n) = w_1 \cdot f_1(x_1, x_2, \ldots, x_n) + w_2 \cdot f_2(x_1, x_2, \ldots, x_n)$, where w_1 and w_2 are weights representing the decision-maker's preferences.

   Understanding how changes in choice variables affect the objectives and constraints is a key topic in optimization, and it aids in finding the optimal solutions within the restrictions. Making practical and effective decisions when dealing with complicated real-world challenges necessitates taking these trade-offs into account.

12.3 LINEAR OPTIMIZATION

A popular mathematical method for resolving optimization issues with linear objective functions and linear constraints is linear optimization, commonly referred to as linear programming (LP). The purpose of linear optimization is to discover the optimum solution that minimizes or maximizes a linear objective while satisfying linear constraints. This technique is used in many disciplines, including operations research, economics, engineering, and resource allocation.

Here's a detailed explanation of linear optimization, along with mathematical concepts:

1. **Mathematical Formulation:** In linear optimization, we wish to minimize or maximize an objective function within a set of linear constraints. In the decision variables, the objective function and constraints are linear.

   Mathematical Notation:

   Objective function: $c_1x_1 + c_2x_2 + \ldots + c_nx_n$

   Constraints: $ai_1x_1 + ai_2x_2 + \ldots + ai_nx_n \leq b_i$ (inequalities)

   Constraints: $ai_1x_1 + ai_2x_2 + \ldots + ai_nx_n = b_i$ (equalities)

   Here, $x_1, x_2, \ldots, x_n$ are the decision variables we want to determine, $c_1, c_2, \ldots, c_n$ are coefficients in the objective function, and $ai_1, ai_2, \ldots, ai_n$ are coefficients in the constraints. The goal is to find the values of the decision variables that optimize the objective function while satisfying the linear constraints.

2. **Goals:** In linear optimization, the goals might be either minimization or maximization of the linear combination of the variables that will be used to make decisions. The decision is based on the context of the issue.

   Mathematical Notation:

   Minimization: $f(x_1, x_2, \ldots, x_n) = c_1x_1 + c_2x_2 + \ldots + c_nx_n$

   Maximization: $f(x_1, x_2, \ldots, x_n) = -c_1x_1 - c_2x_2 - \ldots - c_nx_n$ (by negating the coefficients)

3. **Constraints:** Linear constraints outline the area where the best solution is likely to be found. These restrictions guarantee that the solution complies with particular conditions or restrictions.

   Mathematical Notation:

   Linear inequality constraints: $ai_1x_1 + ai_2x_2 + \ldots + ai_nx_n \leq b_i$

   Linear equality constraints: $ai_1x_1 + ai_2x_2 + \ldots + ai_nx_n = b_i$

   The linear inequalities restrict the solution space, while the linear equalities define specific relationships among the decision variables.

4. **Resolving Linear Optimisation Issues:** Simplex and interior-point algorithms are effective ways to resolve linear optimisation issues. By iteratively travelling along the edges of the feasible zone, these algorithms enhance the answer and eventually arrive at the best one.

5. **Example:** We consider a simple linear optimization problem: Maximize $f(x, y) = 5x + 8y$ Subject to:

$$2x + 3y \leq 100$$
$$x + y \leq 50$$
$$x, y \geq 0$$

   In this example, we want to maximize the objective function $5x + 8y$ while satisfying the given linear inequality constraints.

   When there are linear correlations between the choice variables and constraints, linear optimization is an effective approach for solving practical optimization issues. Due to its effectiveness and wide range of applications, it is widely employed.

12.4 NONLINEAR OPTIMIZATION

The mathematical method of nonlinear optimization, commonly referred to as nonlinear programming (NLP), is used to resolve optimization issues where the objective function or constraints are nonlinear.

Nonlinear optimization deals with more complicated interactions that can be nonlinear in nature, in contrast to linear optimization, where the objective and constraints are linear functions of the decision variables.

A thorough description of nonlinear optimization is provided below, along with related mathematical ideas:

1. **Mathematical Formulation:** In nonlinear optimization, we seek to minimize or maximize an objective function while taking into account a number of restrictions. The decision variables' nonlinear functions may be included in both the objective function and constraints.

   Mathematical Notation:

   Objective function: $f(x_1, x_2, \ldots, x_n)$

   Constraints: $g_i(x_1, x_2, \ldots, x_n) \leq c_i$ (inequalities)

   Constraints: $h_j(x_1, x_2, \ldots, x_n) = d_j$ (equalities)

   Here, $x_1, x_2, \ldots, x_n$ are the decision variables we want to determine. The objective function and constraints can contain nonlinear terms involving these decision variables.

2. A nonlinear function of the decision variables may be minimized or maximized in nonlinear optimization, encapsulating the specific objectives of the problem.

   Mathematical Notation:

   Minimization: $f(x_1, x_2, \ldots, x_n)$

   Maximization: $-f(x_1, x_2, \ldots, x_n)$ (by negating the objective function)

3. **Constraints:** Nonlinear constraints outline the area where the best solution is likely to be found. These restrictions guarantee that the solution complies with particular nonlinear conditions.

   Mathematical Notation:

   Nonlinear inequality constraints: $g_i(x_1, x_2, \ldots, x_n) \leq c_i$

   Nonlinear equality constraints: $h_j(x_1, x_2, \ldots, x_n) = d_j$

   The nonlinear constraints restrict the solution space, while the nonlinear equalities define specific relationships among the decision variables.

4. **Solving Nonlinear Optimization Problems:** Because the nonlinear objective and constraints add complexity, solving nonlinear optimization problems is typically more difficult than solving linear optimization problems. These issues are addressed using a variety of techniques, including gradient-based techniques, genetic algorithms, interior-point techniques, and others. The properties of the problem, such as smoothness, convexity, and dimensionality, influence the approach chosen.

5. **Example:** Let's consider a simple nonlinear optimization problem: Maximize $f(x, y) = x_2 + y_2$ Subject to:

   $$x_2 + y_2 \leq 25$$
   $$x \geq 0$$
   $$y \geq 0$$

   In this example, we want to maximize the objective function $x_2 + y_2$ while satisfying the given nonlinear inequality and equality constraints.

   For the purpose of resolving optimization issues involving nonlinear relationships, nonlinear optimization is a potent tool. It has uses in many different disciplines, including physics, engineering, economics, and many other areas where the underlying linkages are more complex than simple linear interactions.

12.5 LINEAR OPTIMIZATION USEFUL FOR ROBOT PATH PLANNING

Certain kinds of robot route planning issues can benefit from linear optimization, particularly when the issue can be adequately expressed as a linear program. Although it has drawbacks, linear optimization is useful in situations where the constraints and objectives can be represented linearly. Let's explore a particular robot path planning case where linear optimization can be used:

Example: Robot Path planning in a Grid-based Environment

Consider a straightforward situation where a robot must avoid obstacles as it moves from a start cell to a goal cell in a grid-based environment. The objective is to choose the best route with the shortest overall distance. In order to formulate and resolve this problem, we'll apply linear optimization.

Mathematical Formulation:

1. **Decision Variables:**

   Let $x_{i,\,j}$ be a binary variable representing whether the robot moves from cell (i, j) to the neighboring cell in the grid.

2. **Objective Function:**

   Minimize the total travel distance, which is the sum of distances traveled between adjacent cells.

   Mathematical Notation:

   Objective function: Minimize $i, jd_{i,\,j} \cdot x_{i,\,j}$, where $d_{i,j}$ is the distance between cells (i, j) and (i + 1, j) or (i, j + 1) depending on the direction.

3. **Constraints:**

   The robot must start from the start cell and reach the goal cell.

   The robot cannot pass through obstacles (represented by blocked cells in the grid).

   Mathematical Notation:

   Start and goal constraints: $x_{start} = 1$ and $x_{goal} = 1$, where start and goal represent the respective cells.

   Obstacle avoidance constraints: $x_{i,j\,=\,0}$ for blocked cells (i, j).

4. **Motion Constraints:**

   The robot can only move to adjacent cells.

   Mathematical Notation:

   Motion constraints: $xi,j \leq xi+1,j + xi,j+1$ for all non-boundary cells (i, j).

Solving the Linear Program

While meeting the start, target, obstacle avoidance, and motion limitations, the goal is to reduce the overall journey distance. The binary values of xi,j that reflect the ideal path for the robot to follow can be found by solving this linear program using linear optimization algorithms like the simplex method or interior-point methods.

This example shows how a straightforward robot path planning problem in a grid-based environment can be solved using linear optimization. It is crucial to understand that this method works well in situations with linear constraints and discrete choice variables. Different optimization strategies can be needed for more challenging robot path planning issues with continuous state spaces, nonlinear restrictions, moving impediments, or high-dimensional spaces.

12.6 NONLINEAR OPTIMIZATION USEFUL FOR ROBOT PATH PLANNING

When the issue incorporates nonlinear objective functions, limitations, or dynamic surroundings, nonlinear optimization is a potent strategy for planning robot paths. It enables the robot to travel through complicated environments, take nonlinear dynamics into account, and deal with increasingly challenging limitations.

The use of nonlinear optimization for robot path planning is explained in depth below, along with related mathematical ideas:

Scenario: In this task, a robot must move from a starting location to a finishing position while dodging obstacles in a continuous environment. To discover the best route that minimizes the overall journey distance while avoiding obstacles, we'll apply nonlinear optimization.

Mathematical Formulation:

1. **Decision Variables:**

   $p = [x_1, y_1, x_2, y_2, \ldots, x_n, y_n]$ be the vector of robot positions, where (x_i, y_i) represents the position of the robot at discrete time step *i*.

   We'll discretize time into *n* steps, and the objective is to optimize the positions at each time step.

2. **Objective Function:**

   Minimize the total travel distance, which is the sum of Euclidean distances between consecutive positions.

   Mathematical Notation:

   Objective function: Minimize

   $$\int_{i=1}^{n-1} \sqrt{(x_{i+1} - x_i)^2 + (y_{i+1} - y_i)^2}$$

3. **Constraints:**

   The robot must start from the start position and reach the goal position.

   The robot must avoid colliding with obstacles, which can be represented by circular regions in the environment.

   Mathematical Notation:

   Start and goal constraints: $x_1 = x_{start}, y_1 = y_{start}, x_n = x_{goal}, y_n = y_{goal}$

   Obstacle avoidance constraints:

   $$(x_i - x_{obs_j})^2 + (y_i - y_{obs_j})^2 \geq R_{obs_j}^2$$

   for all *i*, *j*, where (x_{obs_j}, y_{obs_j}) is the center of obstacle *j* with radius R_{obs_j}.

4. **Smoothness Constraints:**

   We can introduce constraints to ensure smooth paths, such as limiting the rate of change of velocity or acceleration at each time step.

Solving the Nonlinear Program

Minimizing the overall trip distance is the goal, provided that the start, target, obstacle avoidance, and perhaps smoothness restrictions are met. This nonlinear program can be solved using genetic algorithms or gradient-based approaches (e.g., sequential quadratic programming, Newton's technique), which can be used to identify the best robot path that avoids obstacles and reduces overall distance.

This example shows how nonlinear optimization can be used to plan robot paths in continuous environments while taking into consideration limitations, dynamic obstacle avoidance, and nonlinearity in the objective function. The answer offers a route that guarantees the robot will get to its destination while avoiding obstacles in a way that reduces the overall distance travelled.

12.7 LINEAR TRAJECTORY OPTIMIZATION FOR ROBOTS

Finding a linear route that satisfies multiple restrictions and enables the robot to go from an initial state to a goal state is known as linear trajectory optimization for robots. Usually, the goal is to reduce the cost function, which represents a variety of things including time, energy, or control effort. In this context, we'll concentrate on the scenario in which the robot's dynamics may be roughly described as linear during a brief time period.

Mathematical Formulation

1. **Decision Variables:**

   Let $\mathbf{x}(t)$ be the state vector of the robot at time t. This vector typically includes position, velocity, and other relevant state variables.

   Let $\mathbf{u}(t)$ be the control input vector applied to the robot at time t. The control input can be forces, torques, or other commands that affect the robot's dynamics.

2. **Objective Function:**

   The objective is to minimize a cost function that reflects the desired behavior of the robot. Common objectives include minimizing travel time, energy consumption, or control effort.

   Mathematical Notation:

   Objective function: Minimize $J = \int t_0 t_f L(\mathbf{x}(t), \mathbf{u}(t))dt$

   Here, L is the instantaneous cost function, and the integral is taken over the time interval from the initial time t_0 to the final time t_f.

3. **Dynamics Constraints:** Within the short time frame, the robot's dynamics can roughly be described as linear. We may define the state at the following time step as a linear function of the current state and control input thanks to this supposition.

   Linear relationship between state and control input:

   Linear dynamics constraint: $\mathbf{x}(t + 1)=\mathbf{Ax}(t) + \mathbf{Bu}(t)$

   Here, $\mathbf{A}$ is the state transition matrix, and $\mathbf{B}$ is the control input matrix. These matrices define the linear relationship between the state and control input.

4. **State Constraints:**

   Constraints on the state variables may be imposed to ensure the robot operates within safe limits.

Mathematical Notation:

State constraints: $\mathbf{x}_{min} \leq \mathbf{x}(t) \leq \mathbf{x}_{max}$

Here, $\mathbf{x}_{min}$ and $\mathbf{x}_{max}$ are the minimum and maximum allowable values for the state variables.

5. **Control Input Constraints:**

   Constraints on the control inputs may also be imposed to ensure they fall within physical limits.

   Mathematical Notation:

   Control input constraints: $\mathbf{u}_{min} \leq \mathbf{u}(t) \leq \mathbf{u}_{max}$

   Here, $\mathbf{u}_{min}$ and $\mathbf{u}_{max}$ are the minimum and maximum allowable values for the control input.

Solving the Linear Trajectory Optimization Problem

The goal is to identify the best trajectory (state and control input) that minimizes the cost function while satisfying the linear dynamics constraint, state constraints, and control input requirements. This issue can be effectively resolved using linear programming or quadratic programming techniques like the simple method, interior-point approaches, or gradient-based methods.

The answer offers a linear path that directs the robot from its starting point to its destination while following the provided constraints and minimizing the cost function. This strategy works well when the dynamics of the robot can be roughly described as linear during the span of time under consideration. Nonlinear programming or model predictive control are examples of more sophisticated optimization approaches that should be taken into account when the system dynamics are nonlinear or the time interval is sufficiently large for linearization to be invalid.

12.8 NONLINEAR TRAJECTORY OPTIMIZATION FOR ROBOT

Finding the best trajectory (state and control input) that enables the robot to proceed from an initial state to a goal state while taking into account nonlinear dynamics, limitations, and maybe other goals is known as nonlinear trajectory optimization for robots. When linear approximations fall short in capturing the robot's behavior, this strategy can be helpful. The use of nonlinear trajectory optimization for robots is explained in depth below, along with related mathematical ideas:

Mathematical Formulation

1. **Decision Variables:**

   Let $\mathbf{x}(t)$ be the state vector of the robot at time t. This vector typically includes position, velocity, and other relevant state variables.

   Let $\mathbf{u}(t)$ be the control input vector applied to the robot at time t. The control input can be forces, torques, or other commands that affect the robot's dynamics.

2. **Objective Function:**

   The objective is to minimize a cost function that reflects the desired behavior of the robot. The cost function often involves considerations such as minimizing travel time, energy consumption, or control effort.

   Mathematical Notation:

   Objective function: Minimize

$$J = \int_{t_0}^{t_f} L(x(t), u(t))dt$$

Here, L is the instantaneous cost function, and the integral is taken over the time interval from the initial time t_0 to the final time t_f.

3. **Dynamics Constraints:**

   The robot's dynamics are described by nonlinear equations that relate the state at the next time step to the current state and control input.

   Mathematical Notation:

   Nonlinear dynamics constraint: $\mathbf{x}(t+1) = f(\mathbf{x}(t), \mathbf{u}(t))$

   Here, f represents the nonlinear dynamics function, which models how the state evolves over time based on the current state and control input.

4. **State Constraints:**

   Constraints on the state variables may be imposed to ensure the robot operates within safe limits.

   Mathematical Notation:

   State constraints: $\mathbf{x}_{min} \leq \mathbf{x}(t) \leq \mathbf{x}_{max}$

   Here, $\mathbf{x}_{min}$ and $\mathbf{x}_{max}$ are the minimum and maximum allowable values for the state variables.

5. **Control Input Constraints:**

   Constraints on the control inputs may also be imposed to ensure they fall within physical limits.

   Mathematical Notation:

   Control input constraints: $\mathbf{u}_{min} \leq \mathbf{u}(t) \leq \mathbf{u}_{max}$

   Here, $\mathbf{u}_{min}$ and $\mathbf{u}_{max}$ are the minimum and maximum allowable values for the control input.

12.8.1 Solving the Nonlinear Trajectory Optimization Problem

The goal is to identify the best trajectory (state and control input) that minimizes the cost function while satisfying the nonlinear dynamics constraint, state constraints, and control input constraints. This issue can be resolved using nonlinear optimization approaches, such as direct methods (e.g., shooting methods, direct collocation) or gradient-based methods (e.g., sequential quadratic programming, Gauss-Newton method). The answer offers a nonlinear trajectory that moves the robot along while following the stated restrictions and minimizing the cost function, guiding it from the starting point to the desired state.

This strategy enables a more accurate depiction of the robot's dynamics and is appropriate for a variety of situations where linear approximations fall short of capturing the behavior of the system. When dealing with complicated robots or settings, it is especially helpful.

Numerical Problems with Solutions

1. Robot Arm Kinematics

Problem 1. A robotic arm has three joints, and its end-effector's position in 3D space is defined by the joint angles (θ_1, θ_2, θ_3). Given the transformation matrix T that relates joint angles to end-effector position, find the joint angles that position the end-effector at a desired location.

Solution: The problem involves solving the inverse kinematics problem. Given the transformation matrix T, we need to find the joint angles θ_1, θ_2, θ_3 that satisfy the equation $T(\theta_1, \theta_2, \theta_3) = T_{desired}$, where

$T_{desired}$ is the desired end-effector transformation matrix. This is a nonlinear optimization problem, and numerical optimization methods like the Levenberg-Marquardt algorithm can be used to find the joint angles that minimize the error between *T* and *Tdesired.*

2. Robot Path Planning

Problem 2. A mobile robot needs to navigate from a starting point to a goal point while avoiding obstacles in a 2D environment. Find the optimal path that minimizes the total distance traveled.

Solution: This problem can be formulated as a path planning problem where the objective is to minimize the total distance traveled while satisfying constraints (e.g., avoiding obstacles). Numerical optimization techniques such as the A* algorithm or the Dijkstra algorithm can be used to find the optimal path through the grid while avoiding obstacles.

3. Robot Trajectory Planning

Problem 3. A robot needs to move its end-effector along a specified trajectory from an initial configuration to a final configuration. Find the joint trajectory that minimizes the time while ensuring that the end-effector follows the desired trajectory.

Solution: This problem involves solving the robot trajectory planning problem. The objective is to find the joint trajectory that minimizes the *time taken to move the end-effector along the desired trajectory* while satisfying kinematic constraints. Numerical optimization techniques like the Sequential Quadratic Programming (SQP) algorithm can be used to optimize the joint trajectory.

4. Robot Manipulator Optimization

Problem 4. A robotic manipulator needs to pick up objects of different weights and sizes. Optimize the joint torques required to perform the pick-and-place operation while minimizing energy consumption.

Solution: This problem can be formulated as an optimization problem where the objective is to minimize the energy consumption while performing the pick-and-place operation with different objects. Constraints include kinematic and dynamic constraints, torque limits, and object stability. Numerical optimization methods such as gradient-based optimization or genetic algorithms can be used to find the optimal joint torques for each pick-and-place operation.

5. Robot Calibration

Problem 5. A robot's kinematic parameters need to be calibrated based on measured data from the robot's motion. Find the optimal adjustments to the robot's kinematic parameters that minimize the error between the predicted end-effector positions and the actual measured positions.

Solution: This problem involves calibrating the robot's kinematic parameters using measured data. It can be formulated as a nonlinear optimization problem, where the objective is to minimize the error between the predicted end-effector positions (based on the current kinematic parameters) and the actual measured positions. Numerical optimization techniques like the Levenberg-Marquardt algorithm can be used to find the optimal adjustments to the kinematic parameters.

6. Robot Grasping Optimization

Problem 6. A robotic gripper needs to grasp an object of a certain shape and size while considering the contact forces and stability of the grasp. Optimize the gripper's configuration and contact forces to ensure a stable grasp.

Solution: This problem can be formulated as a grasp optimization problem, where the objective is to find the gripper's configuration and contact forces that ensure a stable grasp while considering the shape and size of the object. Constraints include force closure conditions and object stability. Numerical optimization techniques such as quadratic programming can be used to find the optimal gripper configuration and contact forces.

7. Robot Motion Planning with Constraints

Problem 7. A mobile robot needs to navigate through an environment with multiple constraints, such as avoiding certain areas or staying within specific regions. Find the optimal path that satisfies the constraints while minimizing the total distance traveled.

Solution: This problem involves motion planning with constraints. The objective is to find the optimal path for the robot while satisfying various constraints on its motion. Numerical optimization methods like the A* algorithm with constrained optimization or the RRT* algorithm with constrained sampling can be used to find the optimal path that adheres to the given constraints.

8. Robot Manipulator Workspace Optimization

Problem 8. A robotic manipulator's workspace needs to be optimized to perform specific tasks. Find the optimal joint ranges and configurations that maximize the reachable workspace while ensuring that the manipulator can perform the required tasks.

Solution: This problem involves optimizing the manipulator's workspace to maximize its reachability for a given set of tasks. The objective is to find the optimal joint ranges and configurations that maximize the manipulator's reachable workspace while considering task-specific requirements. Numerical optimization techniques, such as genetic algorithms or particle swarm optimization, can be used to find the optimal joint ranges and configurations.

9. Robot Control Parameter Tuning

Problem 9. A robot control system needs to be tuned to improve its performance, such as tracking accuracy or disturbance rejection. Find the optimal control parameters that minimize a performance metric while considering system dynamics and stability.

Solution: This problem can be formulated as a control parameter tuning problem, where the objective is to find the optimal control parameters that minimize a performance metric (e.g., tracking error) while satisfying stability and dynamic constraints. Numerical optimization methods, such as gradient-based optimization or genetic algorithms, can be used to find the optimal control parameters that improve the robot's performance.

10. Robot Sensor Placement Optimization

Problem 10. A robot needs to have sensors placed in an environment to ensure effective perception and navigation. Find the optimal sensor placement that maximizes coverage while minimizing the number of sensors used.

Solution: This problem involves sensor placement optimization, where the objective is to find the optimal locations for placing sensors in the environment to maximize coverage and minimize the number of sensors used. Constraints include sensor range, field of view, and overlaps. Numerical optimization techniques, such as genetic algorithms or simulated annealing, can be used to find the optimal sensor placement that maximizes coverage while satisfying constraints

Short Questions and Answers

Q.1. What is optimization?

Ans. Optimization is the process of finding the best solution for a given problem, typically involving maximizing or minimizing an objective function subject to certain constraints.

Q.2. What is an objective function?

Ans. An objective function is a mathematical expression that needs to be optimized, either maximized or minimized.

Q.3. What are decision variables?

Ans. Decision variables are the variables that can be adjusted in an optimization problem to find the optimal solution.

Q.4. What are constraints in optimization?

Ans. Constraints are conditions or limitations that must be satisfied in an optimization problem.

Q.5. What is the difference between local and global optima?

Ans. A local optimum is a solution that is optimal within a specific neighborhood, while a global optimum is the best solution over the entire feasible region.

Q.6. What are linear optimization problems?

Ans. Linear optimization problems involve linear objective functions and linear constraints.

Q.7. Give an example of a linear optimization problem.

Ans. Maximizing profit subject to production constraints with linear costs and demand.

Q.8. What are nonlinear optimization problems?

Ans. Nonlinear optimization problems involve nonlinear objective functions and/or constraints.

Q.9. Give an example of a nonlinear optimization problem.

Ans. Finding the shortest path between two points in a city with varying road distances.

Q.10. What is the simplex method?

Ans. The simplex method is a widely used algorithm for solving linear programming problems.

Q.11. What is the feasible region in optimization?

Ans. The feasible region is the set of points that satisfy all the constraints in an optimization problem.

Q.12. What is the Lagrange multiplier method?

Ans. The Lagrange multiplier method is used to solve constrained optimization problems by introducing Lagrange multipliers.

Q.13. What is the Karush-Kuhn-Tucker (KKT) condition?

Ans. The KKT condition is a set of necessary conditions for a solution to be optimal in nonlinear constrained optimization.

Q.14. What is convex optimization?

Ans. Convex optimization deals with optimization problems where both the objective function and the feasible region are convex.

Q.15. Give an example of a convex optimization problem.

Ans. Linear programming problems are convex since linear functions are convex, and linear constraints define a convex feasible region.

Q.16. What is dynamic programming?

Ans. Dynamic programming is a method used to solve optimization problems by breaking them into smaller subproblems and reusing solutions.

Q.17. Give an example of dynamic programming in optimization.

Ans. The knapsack problem, where you maximize the value of items in a knapsack given weight constraints.

Q.18. What is the difference between a local search and a global search algorithm?

Ans. A local search algorithm explores the solution space around an initial point, while a global search algorithm attempts to find the best solution across a broader range.

Q.19. Give an example of a local search algorithm.

Ans. Gradient descent, commonly used in machine learning.

Q.20. Give an example of a global search algorithm.

Ans. Genetic algorithms, which explore multiple solutions in parallel.

Q.21. What is the difference between linear and nonlinear regression?

Ans. Linear regression models the relationship between variables as a linear equation, while nonlinear regression uses a nonlinear equation.

Q.22. What is the objective of multi-objective optimization?

Ans. Multi-objective optimization aims to find a set of solutions that optimize multiple conflicting objectives simultaneously.

Q.23. Give an example of multi-objective optimization.

Ans. Portfolio optimization, where you seek to maximize return while minimizing risk.

Q.24. What is the Pareto front in multi-objective optimization?

Ans. The Pareto front is the set of solutions that are not dominated by any other solution in terms of all objectives.

Q.25. What is the role of heuristics in optimization?

Ans. Heuristics are problem-solving techniques that provide approximate solutions when an exact solution is impractical or unknown.

Q.26. Give an example of a heuristic optimization algorithm.

Ans. The greedy algorithm used for the traveling salesman problem.

Q.27. What is the difference between a local minimum and a saddle point?

Ans. A local minimum is a point where the objective function has lower values in its vicinity, while a saddle point has lower values along some directions but higher values along others.

Q.28. What is the difference between discrete and continuous optimization?

Ans. Discrete optimization deals with discrete decision variables, while continuous optimization involves continuous variables.

Q.29. Give an example of discrete optimization.

Ans. The traveling salesman problem, where you aim to find the shortest route visiting a set of cities.

Q.30. Give an example of continuous optimization.

Ans. Optimizing the shape of a car to minimize drag in aerodynamics.

Q.31. What is the role of constraints in optimization?

Ans. Constraints define the feasible region in an optimization problem, ensuring that the solution satisfies specific requirements.

Q.32. What is the Knapsack problem?

Ans. The Knapsack problem is a classic optimization problem where you must choose a subset of items to maximize the total value while staying within a given weight constraint.

Q.33. What is the Traveling Salesman Problem (TSP)?

Ans. The TSP is a well-known optimization problem where you seek to find the shortest possible route that visits a set of cities exactly once and returns to the starting city.

Q.34. What is the difference between deterministic and a stochastic optimization problem?

Ans. A deterministic optimization problem has known, fixed parameters, while a stochastic optimization problem involves uncertain parameters that follow a probabilistic distribution.

Q.35. Give an example of a stochastic optimization problem.

Ans. Portfolio optimization with uncertain returns on investments.

Q.36. What is simulated annealing?

Ans. Simulated annealing is a probabilistic optimization algorithm inspired by annealing in metallurgy. It explores the solution space by allowing occasional uphill moves to escape local optima.

Q.37. What is genetic programming?

Ans. Genetic programming is a machine learning technique that uses a genetic algorithm to evolve programs (solutions) to solve optimization problems.

Q.38. What is the role of sensitivity analysis in optimization?

Ans. Sensitivity analysis helps assess how changes in input parameters or constraints affect the optimal solution.

Q.39. What is the difference between a feasible solution and an optimal solution?

Ans. A feasible solution satisfies all constraints, while an optimal solution is the best solution in terms of the objective function within the feasible region.

Q.40. What is the Big-M method in linear programming?

Ans. The Big-M method is a technique used to handle constraints in linear programming that are not initially in standard form, often by introducing a large positive constant (M).

Q.41. What is integer programming?

Ans. Integer programming is a variant of linear programming where some or all of the decision variables are required to take integer values.

Q.42. What is the difference between a convex and a concave function?

Ans. A convex function curves upward, while a concave function curves downward. In optimization, convex functions have unique global minima.

Q.43. What is the difference between a convex and a non-convex feasible region?

Ans. A convex feasible region is one where any line segment connecting two points in the region lies entirely within the region. A non-convex region may have "holes" or disconnected parts.

Q.44. What is the role of the Hessian matrix in optimization?

Ans. The Hessian matrix is used to determine the curvature of the objective function at a particular point, helping identify local minima or maxima.

Q.45. What is the difference between a continuous and a discontinuous objective function?

Ans. A continuous objective function is smooth, with no sudden jumps or discontinuities, making it more amenable to traditional optimization methods.

Q.46. What is dynamic optimization?

Ans. Dynamic optimization deals with problems where decisions are made over a sequence of time steps, and the objective function and constraints evolve dynamically.

Q.47. Give an example of dynamic optimization.

Ans. Optimal control of a robotic arm's motion over time to follow a specific trajectory.

Q.48. What is black-box optimization?

Ans. Black-box optimization refers to optimization problems where the objective function is treated as a "black box," meaning its analytical form is unknown, and evaluations are based on function calls.

Q.49. What is the role of local search algorithms in optimization?

Ans. Local search algorithms explore the solution space around an initial point, aiming to find a local optimum. They are often used when the problem's solution space is large.

Q.50. What is the difference between an analytical and a numerical solution in optimization?

Ans. An analytical solution is obtained by solving equations analytically, while a numerical solution involves iterative numerical methods to approximate the optimal solution. Numerical methods are commonly used in complex optimization problems without closed-form solutions.

Multiple Choice Questions and Answers

Q.1. What is optimization?

(*a*) Maximizing the complexity of a problem (*b*) Finding a satisfactory solution (*c*) Finding the best solution

(*d*) Minimizing the number of iterations in an algorithm

Q.2. What is an objective function in optimization?

(*a*) A function that defines the constraints

(*b*) A function that measures the quality of a solution

(*c*) A function that generates random numbers

(*d*) A function that defines the feasible region

Q.3. In optimization, decision variables are:

(*a*) Fixed values

(*b*) Variables that can be adjusted

(*c*) Constraints

(*d*) Parameters of the objective function

Q.4. What is the feasible region in optimization?

(*a*) The region where the objective function is at its maximum

(*b*) The region where the objective function is at its minimum

(*c*) The region defined by the decision variables and constraints

(*d*) The region where there are no constraints

Q.5. Which type of optimization involves linear functions for both the objective and constraints?

(*a*) Linear optimization

(*b*) Nonlinear optimization

(*c*) Convex optimization

(*d*) Dynamic programming

Q.6. The simplex method is used for solving which type of optimization problems?

(*a*) Linear programming

(*b*) Nonlinear programming

(*c*) Discrete optimization

(*d*) Convex optimization

Q.7. Which optimization technique is inspired by the process of annealing in metallurgy?

(*a*) Genetic algorithm

(*b*) Gradient descent

(*c*) Simulated annealing

(*d*) Dynamic programming

Q.8. In optimization, what is the role of constraints?

(*a*) To define the objective function

(*b*) To define the feasible region

(*c*) To define the decision variables

(*d*) To define the optimal solution

Q.9. What is the Traveling Salesman Problem (TSP)?

(*a*) A problem in logistics where you find the shortest route to visit a set of cities

(*b*) A problem in manufacturing where you optimize the production process

(*c*) A problem in finance where you allocate investments

(*d*) A problem in communication networks

Q.10. What is the role of heuristics in optimization?

(*a*) To guarantee the optimal solution

(*b*) To find approximate solutions when an optimal solution is impractical
(*c*) To minimize the number of iterations in optimization algorithms
(*d*) To enforce constraints

Q.11. What is the difference between linear and nonlinear optimization?

(*a*) Linear optimization can handle only one constraint, while nonlinear optimization can handle multiple constraints.
(*b*) Linear optimization requires that the objective function and constraints be linear, while nonlinear optimization allows nonlinear functions.
(*c*) Linear optimization is faster but less accurate than nonlinear optimization.
(*d*) Linear optimization always finds the global optimum, while nonlinear optimization often finds only local optima.

Q.12. What is the objective of optimization?

(*a*) To find the most complex solution to a problem.
(*b*) To find the most commonly used solution to a problem.
(*c*) To find the optimal solution that satisfies certain criteria.
(*d*) To find a random solution to a problem.

Q.13. Which of the following is an example of a linear optimization problem?

(*a*) Maximizing profits from selling products, subject to production capacity and resource constraints.
(*b*) Minimizing the number of iterations in an algorithm.
(*c*) Finding the shortest path in a maze.
(*d*) Allocating tasks to workers to maximize job satisfaction.

Q.14. What is a constraint in optimization?

(*a*) A requirement that the objective function must be linear.
(*b*) A limitation or condition that the solution must satisfy.
(*c*) A parameter in the objective function.
(*d*) A random factor that affects the solution.

Q.15. Which of the following optimization techniques is based on the principles of natural selection and evolution?

(*a*) Gradient descent
(*b*) Simulated annealing
(*c*) Genetic algorithm
(*d*) Dynamic programming

Q.16. In optimization, what is the feasible region?

(*a*) The region where the objective function is at its maximum.
(*b*) The region where the objective function is at its minimum.
(*c*) The region defined by the decision variables and constraints.
(*d*) The region where there are no constraints.

Q.17. What is the main goal of dynamic programming?

(*a*) To solve optimization problems using a divide-and-conquer approach.

(*b*) To find the optimal solution by iteratively improving a solution until convergence.
(*c*) To optimize functions with linear constraints.
(*d*) To solve problems that involve sequential decision-making over time.

Q.18. What is the difference between local and global optima in optimization?

(*a*) Local optima are found by local search algorithms, while global optima are found by global search algorithms.
(*b*) Local optima are the best solutions within a small region, while global optima are the best solutions over the entire feasible region.
(*c*) Local optima are always better than global optima.
(*d*) Local optima can be found using gradient descent, while global optima require genetic algorithms.

Q.19. What is the purpose of the Lagrange multiplier in optimization?

(*a*) To enforce constraints by penalizing violations in the objective function.
(*b*) To represent the rate of change of the objective function.
(*c*) To determine the optimal solution in a linear programming problem.
(*d*) To handle discrete variables in an optimization problem.

Q.20. Which of the following is an example of a nonlinear optimization problem?

(*a*) Linear regression
(*b*) Portfolio optimization with linear constraints
(*c*) Traveling Salesman Problem
(*d*) Solving a system of linear equations

Q.21. Which optimization algorithm is known for its ability to handle discrete variables?

(*a*) Genetic algorithm
(*b*) Gradient descent
(*c*) Simulated annealing
(*d*) Linear programming

Q.22. What is the primary difference between convex and non-convex optimization problems?

(*a*) Convex problems have linear constraints, while non-convex problems have nonlinear constraints.
(*b*) Convex problems have a unique global minimum, while non-convex problems can have multiple local optima.
(*c*) Convex problems always have a feasible solution, while non-convex problems may not have any feasible solutions.
(*d*) Convex problems are easier to solve than non-convex problems.

Q.23. Which optimization technique aims to find an approximate solution by repeatedly making small improvements to the current solution?

(*a*) Genetic algorithm
(*b*) Simulated annealing
(*c*) Local search
(*d*) Dynamic programming

Q.24. What is the role of the Hessian matrix in optimization?

(*a*) To define the feasible region in a linear programming problem.
(*b*) To handle discrete variables in an optimization problem.

(*c*) To determine the convexity of an objective function.

(*d*) To measure the rate of change of the objective function with respect to the decision variables.

Q.25. Which of the following optimization problems involves selecting a subset of items to maximize a value while staying within a given weight constraint?

(*a*) Traveling Salesman Problem
(*b*) Linear programming
(*c*) Knapsack problem
(*d*) Nonlinear programming

Q.26. What is the primary goal of multi-objective optimization?

(*a*) To find a unique solution that optimizes multiple conflicting objectives.

(*b*) To minimize the number of iterations in an optimization algorithm.

(*c*) To find the optimal solution for a single objective.

(*d*) To find a set of solutions that optimizes multiple objectives simultaneously.

Q.27. What is the role of sensitivity analysis in optimization?

(*a*) To measure the sensitivity of the objective function to changes in the constraints.

(*b*) To find the optimal solution with the fewest number of iterations.

(*c*) To handle nonlinear constraints in optimization problems.

(*d*) To enforce the feasibility of the solution.

Q.28. Which optimization algorithm is based on the principles of natural evolution and survival of the fittest?

(*a*) Gradient descent
(*b*) Simulated annealing
(*c*) Genetic algorithm
(*d*) Dynamic programming

Q.29. What is the primary difference between continuous and discrete optimization problems?

(*a*) Continuous problems involve continuous decision variables, while discrete problems involve discrete decision variables.

(*b*) Continuous problems have linear objective functions, while discrete problems have nonlinear objective functions.

(*c*) Continuous problems always have a feasible solution, while discrete problems may not have any feasible solutions.

(*d*) Continuous problems are easier to solve than discrete problems.

Q.30. What is black-box optimization?

(*a*) An optimization technique that involves solving problems with black backgrounds.

(*b*) An optimization technique that uses a black-box model of the system, treating the objective function as an unknown function.

(*c*) An optimization technique that focuses on finding black solutions (optimal solutions).

(*d*) An optimization technique that handles only binary decision variables.

Q.31. What is the primary goal of convex optimization?

(*a*) To find a solution that optimizes multiple objectives.

(*b*) To handle discrete decision variables.

(*c*) To find the global minimum of a non-convex objective function.

(*d*) To find a solution that is both feasible and optimal in a convex optimization problem.

Q.32. Which optimization technique is inspired by the process of annealing in metallurgy and involves occasional uphill moves?

(*a*) Genetic algorithm (*b*) Gradient descent
(*c*) Simulated annealing (*d*) Dynamic programming

Q.33. What is the role of the KKT (Karush-Kuhn-Tucker) condition in nonlinear optimization?

(*a*) To handle discrete variables in an optimization problem.
(*b*) To enforce constraints in the objective function.
(*c*) To determine the convexity of the objective function.
(*d*) To provide necessary conditions for optimality in the presence of constraints.

Q.34. Which optimization problem aims to allocate resources to competing activities in a way that maximizes the overall benefit?

(*a*) Linear programming (*b*) Nonlinear programming
(*c*) Integer programming (*d*) Resource allocation problem

Q.35. Which optimization algorithm involves making probabilistic moves, simulating the annealing process in metallurgy?

(*a*) Genetic algorithm (*b*) Simulated annealing
(*c*) Gradient descent (*d*) Dynamic programming

Q.36. What is the primary goal of the Big-M method in linear programming?

(*a*) To handle large-sized linear programming problems.
(*b*) To solve nonlinear programming problems.
(*c*) To handle integer decision variables.
(*d*) To handle constraints that are not initially in standard form.

Q.37. What is the primary goal of gradient descent in optimization?

(*a*) To handle discrete decision variables.
(*b*) To find the global minimum of a non-convex objective function.
(*c*) To minimize the number of iterations in an optimization algorithm.
(*d*) To iteratively improve a solution by moving in the direction of steepest descent.

Q.38. Which optimization technique uses the concept of "crossover" and "mutation" to evolve solutions over multiple generations?

(*a*) Genetic algorithm (*b*) Gradient descent
(*c*) Simulated annealing (*d*) Dynamic programming

Q.39. What is the primary difference between deterministic and stochastic optimization problems?

(*a*) Deterministic problems have uncertain parameters, while stochastic problems have fixed parameters.
(*b*) Deterministic problems have only one feasible solution, while stochastic problems may have multiple feasible solutions.
(*c*) Deterministic problems involve continuous decision variables, while stochastic problems involve discrete decision variables.

(*d*) Deterministic problems have known, fixed parameters, while stochastic problems have uncertain parameters that follow a probabilistic distribution.

Q.40. Which optimization problem involves selecting a set of investments to maximize return while minimizing risk?

(*a*) Traveling Salesman Problem
(*b*) Knapsack problem
(*c*) Portfolio optimization
(*d*) Resource allocation problem

Q.41. What is the primary role of the Hessian matrix in optimization?

(*a*) To find the global minimum of a non-convex objective function.
(*b*) To determine the convexity of the objective function.
(*c*) To handle integer decision variables.
(*d*) To enforce constraints in the objective function.

Q.42. Which optimization technique aims to find the optimal solution by iteratively refining an initial guess?

(*a*) Genetic algorithm
(*b*) Simulated annealing
(*c*) Gradient descent
(*d*) Dynamic programming

Q.43. What is the primary goal of the Lagrange multiplier method in optimization?

(*a*) To handle discrete variables in an optimization problem.
(*b*) To find the global minimum of a non-convex objective function.
(*c*) To handle integer decision variables.
(*d*) To solve constrained optimization problems by introducing Lagrange multipliers.

Q.44. Which optimization technique uses "crossover" and "mutation" operators to create new candidate solutions in each generation?

(*a*) Genetic algorithm
(*b*) Simulated annealing
(*c*) Gradient descent
(*d*) Dynamic programming

Q.45. What is the primary goal of linear programming?

(*a*) To find the global minimum of a non-convex objective function.
(*b*) To handle discrete decision variables.
(*c*) To allocate resources efficiently to maximize or minimize an objective function.
(*d*) To solve optimization problems with nonlinear constraints.

Q.46. Which optimization technique is particularly suitable for solving problems with discrete decision variables?

(*a*) Genetic algorithm
(*b*) Gradient descent
(*c*) Simulated annealing
(*d*) Linear programming

Q.47. What is the primary goal of the Karush-Kuhn-Tucker (KKT) condition in optimization?

(*a*) To handle discrete variables in an optimization problem.
(*b*) To handle integer decision variables.
(*c*) To enforce constraints in the objective function.
(*d*) To provide necessary conditions for optimality in nonlinear constrained optimization.

Q.48. Which optimization problem involves finding the shortest path that visits a set of cities exactly once and returns to the starting city?

(*a*) Traveling Salesman Problem (*b*) Knapsack problem
(*c*) Linear programming (*d*) Resource allocation problem

Q.49. What is the primary goal of simulated annealing in optimization?

(*a*) To handle discrete decision variables.
(*b*) To find the global minimum of a non-convex objective function.
(*c*) To iteratively refine an initial guess by moving in the direction of steepest descent.
(*d*) To explore the solution space by allowing occasional uphill moves, aiming to escape local optima.

Q.50. Which optimization technique is particularly suitable for problems with a large solution space and multiple local optima?

(*a*) Linear programming (*b*) Gradient descent
(*c*) Simulated annealing (*d*) Dynamic programming

Key to Multiple Choice Questions

1. (*c*)	**2.** (*b*)	**3.** (*b*)	**4.** (*c*)	**5.** (*a*)	**6.** (*a*)	**7.** (*c*)	**8.** (*b*)	**9.** (*a*)
10. (*b*)	**11.** (*b*)	**12.** (*c*)	**13.** (*a*)	**14.** (*b*)	**15.** (*c*)	**16.** (*c*)	**17.** (*d*)	**18.** (*b*)
19. (*a*)	**20.** (*c*)	**21.** (*a*)	**22.** (*b*)	**23.** (*c*)	**24.** (*c*)	**25.** (*c*)	**26.** (*d*)	**27.** (*a*)
28. (*c*)	**29.** (*a*)	**30.** (*b*)	**31.** (*d*)	**32.** (*c*)	**33.** (*d*)	**34.** (*a*)	**35.** (*b*)	**36.** (*d*)
37. (*d*)	**38.** (*a*)	**39.** (*d*)	**40.** (*c*)	**41.** (*b*)	**42.** (*c*)	**43.** (*d*)	**44.** (*a*)	**45.** (*c*)
46. (*a*)	**47.** (*d*)	**48.** (*a*)	**49.** (*d*)	**50.** (*c*)				

Project Works

1. **Robot Path Planning and Optimization:** Develop a system that allows a mobile robot to plan its path through a dynamic environment while optimizing for factors like travel time, energy consumption, or obstacle avoidance. Use optimization techniques such as A*, RRT*, or potential field methods to find the optimal path.
2. **Automated Warehouse Optimization:** Design any optimization algorithm for the efficient management of an automated warehouse. Optimize the assignment of tasks to autonomous robots, considering factors like task priority, robot availability, and travel distances, to maximize warehouse throughput.
3. **Robot Arm Calibration:** Create a system that calibrates the kinematic parameters of a robotic arm using optimization techniques. The goal is to minimize the discrepancy between the expected and actual end-effector positions, improving the robot's accuracy.
4. **Robotic Grasping Optimization**: Develop a robotic grasping system that optimizes the gripper's configuration and contact forces for stable object manipulation. Use machine learning or simulation-based optimization to enhance the grasp success rate for various object shapes and sizes.

5. **Robot Trajectory Optimization:** Build a system that optimizes robot trajectories for tasks such as pick-and-place operations or painting. Consider factors like smoothness, time efficiency, and avoiding joint limits to improve the robot's motion.
6. **Automated Drone Routing:** Create an optimization algorithm for routing drones to perform tasks in a large area, such as surveillance or delivery. Optimize the drone paths to minimize travel time, energy consumption, or maximize area coverage.
7. **Optimal Resource Allocation in Smart Factories:** Design optimization framework that allocates resources (robots, machines, manpower) in a smart factory to maximize production efficiency while meeting demand and considering resource constraints.
8. **Robotic Swarming and Formation Control:** Develop an optimization-based approach for controlling a swarm of robots to achieve desired formations or behaviors while optimizing communication efficiency, obstacle avoidance, or energy consumption.
9. **Energy-Efficient Robot Motion:** Create a system that optimizes the motion of a robot to minimize energy consumption while performing a given task. Consider the robot's dynamics and control parameters to achieve energy-efficient motion.
10. **Sensor Placement Optimization:** Design an algorithm that optimizes the placement of sensors in an industrial or environmental monitoring scenario. Optimize sensor locations to maximize coverage, minimize redundancy, and reduce blind spots.
11. **Resource-Optimized Robotic Surgery:** Develop an optimization approach for robotic-assisted surgery to optimize the robot's movements, tool trajectories, and tissue interactions to minimize surgery time, reduce trauma, and improve surgical outcomes.
12. **Optimal Control for Robotic Manipulators:** Create an optimal control strategy for a robotic manipulator performing complex tasks with varying constraints. Use trajectory optimization methods to find the control inputs that achieve the desired motion while considering dynamics and constraints.

13

Computer Vision as a Field of Artificial Intelligence

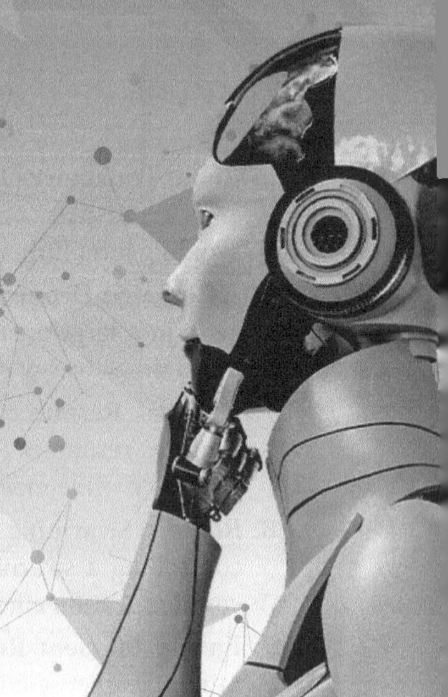

13.1 INTRODUCTION

Computer vision is a branch of computer science and artificial intelligence (AI) that focuses on giving computers the ability to interpret and comprehend the visual world. It attempts to mimic the extraordinary capacity of the human visual system to take in and analyze visual data from the environment. Handling differences in illumination, viewpoint, scale, occlusions, and noise, as well as creating models that can generalize well to new, unseen data, are important issues in computer vision. In order to overcome these difficulties, scientists and engineers in this sector are always working to improve algorithms, gather substantial amounts of data, and develop novel solutions.

13.2 PRIMARY GOAL OF COMPUTER VISION

The development of algorithms and models that can analyse and interpret photos, videos, and other visual data in order to extract meaningful information and make decisions based on that information is the main objective of computer vision. This field's applications span a wide number of industries and disciplines, including but not limited to:

1. Classifying and identifying objects, sceneries, or patterns in photographs by image recognition. Numerous applications, including facial recognition, object identification, and image categorization, are built on this principle.
2. **Video Analysis:** Examining and comprehending video footage, including tracking moving objects or people, identifying movements or activities, and spotting irregularities.
3. **Medical Imaging:** Assisting in the interpretation of diagnostic pictures, such as X-rays, MRIs, and CT scans.
4. **Autonomous Vehicles:** Giving self-driving automobiles the ability to sense their environment and make decisions in real-time to travel safely.
5. **Augmented Reality (AR) and Virtual Reality (VR):** These technologies improve user experiences by superimposing digital data on the physical world or by building realistic virtual worlds.

6. **Robotics:** Giving machines the ability to use visual data to comprehend and interact with their surroundings.

Combining methods from many disciplines, such as image processing, machine learning, deep learning, and neural networks, is what computer vision entails. Convolutional neural networks (CNNs), in particular, have made major strides in deep learning recently, enhancing the accuracy and robustness of computer vision systems and opening up new possibilities in a variety of applications.

13.3 IMAGE PROCESSING AS APPLIED TO ROBOTIC AND AUTOMATION

Robots and automated systems may interact with their surroundings by exploiting visual information thanks to image processing, which is a vital component of robotics and automation. Several important uses of image processing in robotics and automation are listed below:

Object Detection and Recognition: Using image processing methods like feature extraction and pattern recognition, a robot's environment can be searched for and recognized items. This is necessary when picking and arranging objects, classifying goods on a conveyor belt, or identifying particular pieces in a manufacturing process. The identification and localization of certain items inside an image is a component of object detection and recognition in image processing. A wide range of applications, including robots, surveillance, autonomous cars, and more depend on this fundamental computer vision job. Convolutional neural networks (CNNs), in particular, have considerably increased the accuracy of object detection and recognition in recent years.

Let's explore the process in detail, including some key mathematical concepts:

1. **Object Localization:** Localizing identified objects typically entails drawing bounding boxes around them in addition to identifying them. The position of the object within the image is determined by its bounding box. Convolutional neural networks (CNNs) are frequently employed for object detection and recognition.
2. **CNNs and Feature Extraction:** The network can recognize pertinent patterns since they automatically pick up features from the input photos. Convolutional, pooling, and fully linked layers are among the many layers that make up CNNs.
3. **Sliding Window Approach:** One straightforward method for detecting objects entails moving a fixed-size window across the image at various scales and positions, applying a classifier (typically a CNN) to the contents of the window, and then figuring out whether an object is present. However, this strategy may be computationally expensive.
4. **Region Proposal Networks (RPN):** Contemporary object detection methods, such as Faster R-CNN, employ RPN to produce prospective object regions. The presence of an object is then determined by classifying these locations. With this method, processing fewer windows more quickly increases efficiency.
5. **Non-maximum Suppression (NMS):** NMS is frequently used to get rid of duplicate or overlapped bounding boxes after the initial detection. For each object, NMS makes sure that only the most certain detections are maintained, producing cleaner results.
6. **Mathematical Notation:**

   Input Image: Usually written as $I(x, y)$, this image is represented as a 2D array of pixel values.

   Bounding Box: A bounding box's top-left (x_1, y_1) and bottom-right (x_2, y_2) corners serve as its boundaries.

CNN Output: A CNN's object detection output frequently comprises bounding box coordinates (B) and a class label (C) for each probable object.

Intersection over Union (IoU): IoU is a crucial metric used to evaluate the accuracy of bounding box predictions. It measures the overlap between two bounding boxes.

The formula for IoU is:

$$\text{IoU} = \text{Area of Overlap}/\text{Area of Union}$$

7. **Loss Functions:** Appropriate loss functions are required for training object detection models. You can use smooth L_1 loss for bounding box regression and cross-entropy loss for classification.
8. **Backpropagation:** Gradients are produced with respect to the loss function and used to update the model's parameters using optimisation methods like stochastic gradient descent (SGD) during the training of deep learning models, including object identification models.

It's vital to remember that there are different object detection architectures, each with its own distinct method of fusing object localization and classification, such as Faster R-CNN, YOLO (You Only Look Once), and SSD (Single Shot Multi Box Detector). These structures have helped to progress object detection and recognition research.

Navigation and Localization: Robots can independently travel by using computer vision, which interprets data from cameras or other sensors to comprehend their environment. For mobile robots, drones, and autonomous *vehicles to avoid obstacles, map their surroundings*, and localize themselves inside them, this is essential. For robots, autonomous vehicles, and other systems to understand their surroundings and navigate in their environment, navigation and localization in image processing are essential. These activities entail using visual data, frequently obtained by cameras or sensors, to locate the robot (localization) and map out its path (navigation). These procedures can involve a variety of algorithms and are somewhat sophisticated. Below is a summary of the essential ideas, along with some pertinent mathematical notations:

1. **Localization:** The first step in localization is to establish the robot's location inside a predetermined map or environment. Techniques like feature matching, sensor fusion, and probabilistic methods like the Monte Carlo Localization (Particle Filter) algorithm can be used to accomplish this. Here's an overview:

   Mathematical Notation:

   Robot Position: Denoted as (x, y, θ), where (x, y) is the 2D position, and θ is the orientation or heading of the robot.

   Map: A representation of the environment, often as a grid or feature map.

   Feature Matching: To determine the robot's location, compare features (landmarks) found in sensor data (such as camera images) with features on the map.

   Sensor fusion: Improve localisation accuracy by combining data from many sensors, such as cameras, lidar, and inertial sensors.

   Monte Carlo Localization (Particle Filter): This probabilistic method uses a set of particles to represent the robot's potential positions. Based on command inputs and sensor readings, particles spread. Based on the likelihood of the sensor data given the map, each particle's weight is updated.
2. **Navigation:** Planning the robot's journey from its current location to a desired destination while avoiding obstacles is referred to as navigation. Maps and obstacle identification are only two

examples of the useful information that image processing may provide about the environment. Here is a summary:

Mathematical Notation:

Path: A sequence of (x, y) points that define the robot's planned trajectory.

Obstacle Map: A binary map indicating obstacle locations.

Obstacle Detection: From camera data or sensor measurements, use image processing to identify barriers in real-time.

Path Planning: To identify a collision-free route from the robot's current position to the target position on the map, one can utilise algorithms such as A* (A-star), Dijkstra's method, or RRT (Rapidly Exploring Random Trees).

Motion Control: After a path is chosen, the robot's control system must carry out the chosen course, making adjustments for things like steering, acceleration, and velocity.

Combining image processing, sensor fusion, probabilistic techniques, and path planning algorithms is required for these jobs. Although I may give a general picture here, the precise implementation details and mathematical equations can change depending on the selected algorithms and the complexity of the surrounding environment.

Quality Control and Inspection: To check items for flaws and ensure high-quality production, image processing is frequently employed in the manufacturing industry. Robots with cameras may inspect things visually, spot flaws, and reject or categorize products according to predetermined standards. Image processing is a critical application in manufacturing and other sectors for quality control and inspection. Automated image analysis is used to identify flaws, guarantee product quality, and uphold uniform production standards. Here is a thorough summary that includes several essential mathematical ideas:

1. **Defect Detection:** Finding flaws or irregularities in products is the main objective of quality control and inspection. Surface flaws, colour differences, missing parts, and other issues could fall under this category. Algorithms for image processing are used to analyse images and spot these flaws. **This is how it goes:**

   Image Acquisition: Using cameras or other imaging tools, acquire clear photographs of the products to be examined.

   Preprocessing: Improve the quality of the photograph by reducing noise, modifying the lighting, and using filters to draw attention to important details.

   Defect Detection Algorithms: Use algorithms to find flaws in images based on their edges, textures, variations in color, and shapes.

   Thresholding: To segment the image and isolate the defect patches from the background, employ thresholding techniques.

   Related Component Analysis: Locate pixels that are related together and probably represent flaws. The morphological procedures of mathematics can be used for this.

   Mathematical Notation:

   Image: Represented as a 2D array of pixel values, often denoted as I(x, y).

   Thresholding: A binary operation that separates pixels into two classes based on a threshold value.

   Defect Detection Metrics: You can assess the effectiveness of the defect detection algorithm using metrics like accuracy, precision, recall, and F1-score.

2. **Quality evaluation:** Along with defect detection, quality evaluation entails assessing other quality factors, like dimensions, forms, and color consistency.
   Here is how to accomplish it:
   Feature Extraction: Take note of the size, shape, and colour distribution of items in the image and extract essential information from it.
   Statistical Analysis: To determine quality, compare the retrieved features to standard values or statistical models. The mean and standard deviation, for instance, can be used to determine whether particular parameters fall within a reasonable range.
   Machine Learning: Develop models for machine learning to categorise products as "acceptable" or "defective" based on the features that were retrieved.
   Mathematical Notation:
   Features: Represented as a vector, denoted as $F = [f_1, f_2, ..., f_n]$, where f_i is a specific feature.
   Quality Control Metrics: To quantify the discrepancy between extracted features and reference values, metrics such as Mean Squared Error (MSE), Mean Absolute Error (MAE), or other domain-specific metrics may be utilized.
3. **Making a Decision:** The items can be accepted, rejected, or reworked depending on the findings of fault detection and quality assessment. Based on pre-set criteria, this decision-making process can be automated.
   Thresholds – Define thresholds for various quality metrics, such as defect size and shape irregularity. A product is regarded as faulty if its features are more than these limits.
   Rules: Create rules based on knowledge from experts. A product might be rejected, for instance, if it contains too many flaws.

Systems for quality control and inspection frequently combine image processing methods with statistical analysis and, occasionally, machine learning to achieve high accuracy and efficiency in identifying flaws and ensuring the quality of the produced goods. The particular mathematical formulas and algorithms employed will depend on the type of the products being examined and the desired level of quality.

Robot Vision: Cameras and image processing are used by robot vision systems to direct robots in tasks that call for accurate positioning and manipulation. This is essential in fields like manufacturing where robots must manage delicate or difficult jobs based on visual signals. Robot vision, often referred to as machine vision, is a field that uses cameras and image processing methods to give robots the ability to perceive and comprehend their.surroundings. This is necessary for activities like manipulating objects, navigation, and interacting with the environment.

Let's explore the specifics of robot vision, including fundamental mathematical ideas and elements:

1. **Robotic Vision:** It begins with the acquisition of pictures from cameras or other imaging sensors. The visual information about the robot's surroundings provided by these photos can be analyzed to yield useful data.
   Preprocessing: To improve the quality of the data and improve the efficiency of subsequent processing, preprocessing techniques are frequently used before analyzing the photos. Typical preprocessing procedures include:
   Image noise reduction methods include median filtering or Gaussian noise reduction.
   Image Enhancement: Changing the brightness, contrast, and applying filters to draw attention to important details.
   Image Calibration: Fixing issues such sensor misalignment and lens distortion.

2. **Feature Extraction:** Feature extraction entails locating crucial visual cues or patterns in the image that are pertinent to the task at hand for the robot. These features may have edges, corners, textures, and other characteristics.

   Edge Detection: Canny edge detector and other algorithms can be used to discover important edges in an image.

   Corner Detection: Techniques to identify corner points in the image, such as Harris corner detection.

   Texture Analysis: Methods for examining texture patterns, such as Gabor filters.

3. **Object Detection and Recognition:** Essential elements of robot vision are object detection and recognition. These challenges entail locating particular items or interesting areas in the image.

   Template Matching: Finding matches by comparing the image to pre-established templates.

   Haar Cascades: An object detection technique that makes use of machine learning to pinpoint particular features in an image.

   Deep Learning: Object identification using convolutional neural networks (CNNs) with Faster R-CNN, YOLO, and SSD architectures.

4. **Pose Estimate:** Pose estimate establishes how items in the image should be positioned in relation to the robot's coordinate system (translation and orientation).

   Homography: A transformation that converts points in one image into equivalent points in another, frequently used to determine the pose of planar objects.

   PnP (Perspective-n-Point): This technique calculates the camera posture using a set of 3D points and the corresponding 2D projections in the image.

5. **Mathematical Notation:**

   Image: Represented as a 2D array of pixel values, often denoted as I(x, y).

   Feature: A specific characteristic of the image, denoted as F.

6. **Integration with Robotics:** Robot vision information is used to make judgments and direct the operations of the robot. For instance, a robot with vision capabilities may recognize items, figure out where they are, and devise a path to interact with them.

Robot vision combines methods from image processing, computer vision, and robotics to provide robots the ability to comprehend and communicate with their surroundings using visual data. The particular techniques and mathematical formulas employed will vary depending on the application and the difficulty of the tasks the robot must carry out.

Gesture Recognition: Robots can interact with people more naturally if image processing is utilized to recognize and understand human gestures. This has uses in user interfaces and human-robot collaboration. The technique of recognizing and understanding human gestures, such as hand movements, as a means of human-computer interaction is known as gesture recognition in image processing. With the use of this technology, gadgets can recognize and react to gestures, simplifying user interfaces. Here is a thorough explanation of gesture recognition, complete with mathematical ideas and important elements:

1. **Image Acquisition:** The first step in gesture recognition is to take pictures or video frames of the gestures. Usually, imaging technologies like cameras, depth sensors, or others are used for this.

   Preprocessing: In order to get the image data ready for gesture analysis, preprocessing is crucial. Typical preprocessing procedures include:

Background Subtraction: The process of removing the static background from the moving foreground (the person making the gesture).

Skin Detection: Locating areas of the image that correlate to the tones of human skin, which is frequently important for hand motion identification.

Image Noise: Image noise reduction methods include median filtering or Gaussian noise reduction.

2. **Feature Extraction:** Identifying pertinent details from the image that depicts the gesture is known as feature extraction. These characteristics could include hand shapes, motion trajectories, and other things.

   Hand Contour: Spotting the hand's contour in an image, which is frequently shown as a series of points (x, y).

   Motion Vectors: This technique determines the hand's direction and speed between two frames, which can reveal details about the dynamics of a gesture.

3. **Gesture Classification:** The classification of motions based on the extracted information is the core of gesture recognition. Usually, machine learning algorithms are used for this.

   Support Vector Machines (SVM): A well-liked technique for multi- or binary-class classification applications, such as gesture recognition.

   Random Forests: When dealing with a large number of classes or complex gesture patterns, this ensemble learning technique can be successful for gesture identification.

   Deep Learning: For difficult gesture identification tasks, convolutional neural networks (CNNs) can be utilised to directly learn hierarchical features from the picture input.

4. **Mathematical Notation:**

   Image: Represented as a 2D array of pixel values, often denoted as I(x, y).

   Hand Contour: A set of points denoted as $C = \{(x_1, y_1), (x_2, y_2), .\}$ that defines the contour of the hand in the image.

   Motion Vector: A vector denoted as $V = (\Delta x, \Delta y)$, representing the change in hand position between consecutive frames.

5. **Integration with Human-Computer Interaction:** In the context of human-computer interaction, a gesture can be mapped to certain commands or actions once it has been recognized. A recognizable "swipe left" gesture, for instance, could be read as an order to go back to the previous screen.

In order to enable intuitive and natural interactions between people and computers, the intriguing topic of gesture recognition combines image processing, feature extraction, machine learning, and human-computer interaction. Depending on the intricacy of the motions being recognized and the desired level of accuracy, different techniques and mathematical formulas will be employed.

Pick-and-Place Applications: Robots with vision systems have the ability to recognize things on a crowded surface, devise a picking plan, and carry out the pick-and-place operation precisely. This is crucial for production lines, logistics, and warehouse automation. Automation and robotics frequently use pick-and-place techniques, particularly in the manufacturing and shipping industries. These applications include picking objects up from one area and placing them in another using robotic devices, frequently requiring precise positioning and manipulation. In order to help the robot recognize and handle objects effectively, image processing is essential.

Let's explore pick-and-place applications in detail, including relevant mathematical concepts:

1. **Object Localization and Detection:** Image processing is utilized to locate and identify things in the workspace of the robot. This is necessary for the robot to recognize the object it needs to pick.

   Bounding Box: A bounding box is a popular representation for localizing things. Its top-left corner (x_1, y_1) and bottom-right corner (x_2, y_2) form the bounding box.

   Notation for Mathematics:

   Object Position: Shown as a point (x, y) designating the centroid or reference point of the object.

2. **Feature Extraction:** Relevant features are extracted after the items have been found to help with object identification and pose estimation. These traits may include the objects' edges, corners, or other distinguishing qualities.

   Edge Features: To discriminate between object borders, edges can be detected using techniques like the canny edge detector.

   Shape Parameters: To categorize and distinguish between various things, parameters like the area, perimeter, or aspect ratio of the object are calculated.

3. **Pose Estimation:** Proper pick-and-place procedures require knowledge about the object's pose, or its position and orientation. It is necessary to compute the transformation between the coordinate systems of the object and the robot.

   Homogeneous Transformations: Using homogeneous matrices to represent an object's translation and rotation. This allows the robot to determine the rights grasp point and orientation.

   Mathematical Notation:

   The Homogeneous Transformation Matrix (sometimes spelt T) is a 4 by 4 matrix that includes translation and rotational information.

4. **Grasp Planning:** A proper grasp point and orientation for the object to be selected must be planned by the robot. The attitude of the object and the geometry of the robot's end-effectors are used to calculate this.

   Gripper Opening: Using the object's size to determine the gripper's opening width.

   Approach Vector: Choosing the angle from which the robot should approach the target in order to secure a firm grip.

5. **Control and Execution:** The robot does the pick-and-place action in accordance with these calculations after the object has been recognized, its pose has been calculated, and a grab point has been planned.

   Inverse Kinematics: If the robot is equipped with a robotic arm with many joints, inverse kinematics is utilized to determine the joint angles required to achieve the desired grip point and orientation.

Pick-and-place applications necessitate a fusion of control theory, robotics, and computer vision. For precise and dependable pick-and-place operations, the previously described mathematical ideas, such as homogeneous transformations and inverse kinematics, are crucial. The kinematics of the robot, the objects it is handling, and the desired level of precision will all affect the specific implementation specifics.

Medical Robotics: Medical robotics uses image processing to help with minimally invasive operations, analyze medical pictures (such MRI or CT scans), and guide surgery. A multidisciplinary discipline called medical robotics integrates robotics, image processing, and medical expertise to create tools and systems for performing, diagnosing, and treating medical procedures. In areas including image-guided surgery, medical imaging analysis, and computer-assisted treatments, image processing is crucial to medical robotics.

Here's a detailed overview, including mathematical concepts and key components:

1. **Image-Guided Surgery:** Using real-time ultrasound, MRI, or CT scans, image-guided surgery uses medical pictures to direct surgical processes. The accuracy of surgical operations is increased and significant guidance is provided by image processing, which overlays these images on the surgeon's vision.

   Image Registration: Mathematical methods for real-time alignment of preoperative images with the operating field.

   Transformation Models: Affine and non-rigid transformations that are used to convert points between two different coordinate systems, such as surgical space and an image.

   Intensity-Based Registration: Finding the optimal alignment between images by optimising a similarity metric (such as mutual information).

2. **Medical Imaging Analysis:** Medical picture analysis and information extraction for diagnosis and therapy planning are done using image processing.

   Segmentation: The division of a picture into discernible areas, frequently used to isolate organs or lesions.

   Level Set Techniques: Development of a curve to divide an image based on edge or intensity data.

3. **Active Contours:** deformable image segmentation models those are especially helpful for spherical object forms. Identifying particular features from medical images for quantification and analysis is known as "feature extraction."

   Texture Analysis: calculating statistical measures that characterise the texture of the structures in the image, such as mean, variance, and entropy.

   Shape Descriptors: Extraction of shape-based features such as area, circumference, and perimeter.

   Computer-Assisted Interventions: Medical robotics helps in minimally invasive operations by giving the surgeon real-time feedback and direction.

   Navigation: The use of real-time tracking and preoperative photographs to direct the surgical tools.

   Kalman Filters: Mathematical filters used in surgery to track the location and orientation of implants or tools.

   Robotic surgery is a form of minimally invasive surgery that combines robotics with image guiding.

   Kinematics: Inverse kinematics is used to control the robot's movement in accordance with surgical objectives and patient feedback.

4. **Mathematical Notation:**

   Image: Represented as a 2D or 3D array of intensity values, often denoted as I(x, y, z).

   Transformation Matrix: Used to represent geometric transformations (translation, rotation, scaling, etc.) between image spaces.

Contour: A set of points denoted as C = {(x_1, y_1), (x_2, y_2), .} representing the boundary of a structure in the image.

Feature Vector: A vector of numerical values representing quantifiable features extracted from the image, denoted as F.

Medical imaging modalities, surgical techniques, and mathematical underpinnings must all be thoroughly understood in order to perform medical robotics and image processing. The efficient development and use of these technologies in healthcare depends on the cooperation of medical experts, computer scientists, and engineers.

Agricultural Robotics: Robots with image processing capabilities can recognise and harvest ripe fruits with precision, keep an eye on the health of crops, and carry out other jobs in agriculture that call for visual analysis. Robotics and automation are applied to different parts of agriculture in the constantly expanding subject of agricultural robotics. Agricultural robots relies heavily on image processing to perform tasks like crop monitoring, precision farming, weed detection, and yield calculation.

The application of image processing in agricultural robotics is explained in depth below, along with pertinent mathematical ideas and important elements:

1. **Crop Monitoring and Health Assessment:** By examining photos of fields taken by drones, robotic vehicles, or stationary cameras, image processing aids in the monitoring of the health and growth of crops.

   Normalized Difference Vegetation Metric (NDVI): This popular metric measures the vigor and health of vegetation by examining how well it reflects near-infrared (NIR) and visible (VIS) light.

   NDVI: Determined by dividing the reflectance values in the near-infrared and visible bands by two, as follows: NDVI = (NIR – VIS)/(NIR + VIS).

   Plant Counting: Identifying and counting specific plants in an image to calculate the population of those plants.

   Plant Health Indicators: Identifying stressed or ill plants by analyzing color variations in the photos.

2. **Weed Detection and Management:** To differentiate weeds from crops and carry out targeted weed management, image processing is used.

   Image Segmentation: Dividing an image into sections that represent weeds and crops.

   Color-Based Segmentation: This method analyses the image's colour data to distinguish between crops and weeds.

   Training machine learning models (such as convolutional neural networks) to categorize picture regions as crops or weeds.

3. **Yield Estimation:** Using photos of the field and information on the size of individual plants, image processing can be used to estimate crop yield.

   Object Detection and Sizing: Locating certain crops within the photos and determining their sizes.

   Geometric Models: Using geometric models (such as assuming a specific form or volume model) to estimate the dimensions of plants.

   Statistical Analysis: Developing predictive models by relating image-based metrics with known yield data.

4. **Path Planning and Navigation:** For autonomous operation, image-based path planning and navigation technologies are needed for robotic agricultural vehicles.

   Obstacle Detection: Using visual data from sensors, locate obstructions in the robot's path.

   Path Planning Algorithms: These algorithms determine the robot's best course based on observed impediments and predetermined objectives.

5. **Mathematical Notation:**

   Image: Represented as a 2D or 3D array of pixel values, often denoted as I(x, y) or I(x, y, z).

   Reflectance: Represented as values between 0 and 1, denoting the fraction of light reflected by a surface.

Remote sensing, plant physiology, and mathematical concepts must be thoroughly understood in order to use agricultural robotics and image processing. These technological advancements could increase crop yields, optimize resource use, and boost the productivity and sustainability of agriculture. Realizing the full potential of agricultural robots requires cooperation between professionals in agriculture, robotics engineering, and computer vision.

Complex algorithms, machine learning, and deep learning approaches are frequently used in the integration of image processing with robotic systems. Robots can now learn from data, adjust to changing circumstances, and continuously improve thanks to these technologies. Robotics and image processing as a combination are developing increasingly powerful and adaptable automated systems in a variety of industries.

13.4 TECHNIQUES FOR IMAGE FILTERING

A key step in image processing called image filtering modifies an image's pixel values by applying a specific mathematical operation to particular neighborhoods of pixels. Filtering can improve an image's quality, extract certain elements from it, lessen noise, and carry out other forms of image editing. Here is a detailed explanation of popular image filtering methods together with the necessary mathematics:

13.4.1 Linear Filtering

Convolution, a mathematical technique that entails swiping a tiny kernel across the image and computing a weighted sum of pixel values within the kernel for each output pixel, forms the foundation of linear filters.

Mathematical Notation:

Input Image: Represented as a 2D array of pixel values, often denoted as I(x, y).

Kernel: A small matrix representing the filter, denoted as K(i, j).

Smoothing (Blurring) Filters:

Box Filter: Each output pixel in a square neighborhood (kernel) is the average of all the pixels inside that neighborhood.

Gaussian Filter: This blurring effect is smoother and more realistic since the weights in the kernel are chosen by a Gaussian distribution.

Notation for Mathematics:

Box Filter: For a 3×3 filter, the kernel has 1/9 weights, for example.

Gaussian Filter: Based on the Gaussian distribution, the kernel weights are computed.

Edge Detection Filters:

The Sobel filter highlights edges in the image by computing the gradient in the x and y dimensions.

Laplacian Filter: improves edges in areas where there are frequent changes in intensity.

Notation for Mathematics:

Sobel Filter: The x and y gradients are computed using two different kernels.

Laplacian Filter: The kernel draws attention to intensity variations.

13.4.2 Non-Linear Filtering

Instead of using a straightforward weighted sum, non-linear filters alter the values of pixels based on a specific statistical characteristic of the neighborhood. These filters are frequently employed for tasks like edge-preserving smoothing and noise reduction.

Median Filter: Replaces each pixel in a window with its median value.

Notation for Mathematics:

Median Filter: The median value is used to replace the pixel value at the kernel's centre.

Filtering in Frequency Domain:

The Fourier transform is used for converting a picture to its frequency representation, which is followed by the application of a filter before being converted back to the spatial domain.

Low-Pass Filter: This filter smoothes out the image by reducing high-frequency noise and details.

High-Pass Filter: Highlights edges and details in high-frequency components.

Notation for Mathematics:

Frequency Representation: Image in the frequency domain, often denoted as F(u, v).

Filter: A filter function in the frequency domain, denoted as H(u, v).

For image processing tasks including picture enhancement, noise reduction, edge detection, and more, it is crucial to comprehend these image filtering algorithms, their mathematical representations, and their implementations. The precise task at hand and the properties of the image data will determine which filter is the best fit.

13.5 FEATURE EXTRACTION

A vital step in image processing is feature extraction, which includes choosing pertinent details or patterns from an image in order to express it more succinctly while preserving key details. Various applications, including classification, object identification, and image retrieval, use these attributes as input.

Here's a detailed overview of feature extraction in image processing, including relevant mathematical concepts:

13.5.1 Spatial Domain Features

Spatial domain features are computed directly from the pixel values in the image.

Intensity-Based Features:

Mean: The average pixel intensity of the image.

Mathematical Notation: Mean = $(\Sigma I(x, y))/N$, where I(x, y) is the intensity of the pixel at position (x, y), and N is the total number of pixels.

Standard Deviation: A measure of the spread of pixel intensities.

Mathematical Notation: StdDev = sqrt(Σ(I(x, y) – Mean)^2/N), where I(x, y) is the intensity of the pixel at position (x, y), and N is the total number of pixels.

Histogram Features:

Histogram: A representation of the frequency distribution of pixel intensities.

Mathematical Notation: Histogram is often presented as a histogram vector, where each element represents the number of pixels with a specific intensity value.

13.5.2 Texture-Based Features

Texture features capture information about spatial patterns and structures in the image.

Gray-Level Co-occurrence Matrix (GLCM):

Contrast: Measures the local variations in pixel intensities.

Energy (Uniformity): Measures the homogeneity of the image.

Entropy: Represents the randomness or complexity of the texture.

Mathematical Notation: These texture features are computed from the GLCM, a matrix that describes how often pairs of pixel intensities occur together at a specified offset.

13.5.3 Frequency Domain Features

Frequency domain features are computed after transforming the image from the spatial domain to the frequency domain using techniques like the Fourier transform.

- **Frequency Spectrum:** Represents the amplitude of different spatial frequency components in the image.
- **Mathematical Notation:** Frequency Spectrum = abs(F(u, v)), where F(u, v) is the complex-valued frequency representation of the image at frequency coordinates (u, v).

13.5.4 Edge-Based Features

Edge features capture information about the presence of edges and contours in the image.

Canny Edge Detection:

Number of Edges: Count the number of detected edges in the image.

Average Edge Length: Calculate the average length of detected edges.

Mathematical Notation: These edge-based features can be calculated from the output of the canny edge detection algorithm, which identifies edges in the image.

These are only a handful of the numerous feature extraction methods that are applied to image processing. The particular task at hand and the properties of the image data influence the features that are selected. These features act as the basis for further analysis, such as classification or machine learning techniques, making it possible to handle picture data effectively and efficiently.

13.6 OBJECT DETECTION

One of the most important steps in image processing is object detection, which includes locating and recognizing particular items within an image. Due to the diverse variety of applications for this task, including computer vision, autonomous driving, spying, and more, it has attracted considerable attention. Convolutional neural networks (CNNs), a type of deep learning model, are frequently used in contemporary object detection approaches.

Let's explore object detection in detail, including key concepts and mathematical representations:

1. **Definition of the Issue:** Object detection attempts to locate various items inside an image and provide the bounding box coordinates for each one. The objective is to categorise the identified objects and locate them spatially.
2. **Sliding Window Approach:** A straightforward method for detecting objects entails moving a fixed-size window across the image at various scales and positions, then using a classifier (typically a CNN) to assess whether any objects are present. However, when dealing with various object scales, this strategy might be computationally expensive.
3. **Region Proposal Networks (RPN):** To identify probable object regions, contemporary object detection systems like Faster R-CNN, YOLO (You Only Look Once), and SSD (Single Shot MultiBox Detector) use region proposal networks. The presence of an object is then determined by classifying these locations. With this method, processing fewer windows more quickly increases efficiency.
4. **Intersection over Union (IoU):** IoU is an essential indicator for assessing the precision of bounding box predictions. It calculates the amount of overlap between the expected and actual bounding boxes. The IoU formula is:

   $$\text{IoU} = \text{Area of Overlap/Area of Union}$$
5. **Non-maximum Suppression (NMS):** NMS is frequently used to eliminate redundant or overlapping bounding boxes after initial object detection. For each object, NMS makes sure that only the most certain detections are maintained, producing cleaner results.
6. **Mathematical Notation:**

   Input Image: Represented as a 2D array of pixel values, often denoted as $I(x, y)$.

   Bounding Box: A bounding box is defined by its top-left corner (x_1, y_1) and bottom-right corner (x_2, y_2).

   CNN Output: A CNN's object detection output frequently comprises bounding box coordinates (B) and a class label (C) for each probable object.
7. **Loss Functions:** Appropriate loss functions are required for training object detection models. You can use smooth L_1 loss for bounding box regression and cross-entropy loss for classification.
8. **Back Propagation:** Gradients are produced with respect to the loss function and used to update the model's parameters using optimization methods like stochastic gradient descent (SGD) during the training of deep learning models, including object identification models.

The accuracy and effectiveness of object identification tasks have considerably increased thanks to modern object detection models like YOLO and Faster R-CNN. To manage the difficulties of detecting objects of various sizes, orientations, and spatial connections inside an image, these models employ deep learning and specialized approaches.

13.7 CAMERA CALIBRATION NECESSARY FOR ACCURATE 3D RECONSTRUCTION

For accurate 3D picture reconstruction, camera calibration is indeed required. Calibration makes sure that both the camera's extrinsic parameters (position and orientation in the 3D world) and intrinsic parameters (such as focus length, primary point, and lens distortion) are precisely known. The resulting

3D reconstruction could have scale ambiguities, errors, and distortions without adequate camera calibration.

Here's why camera calibration is crucial for accurate 3D reconstruction:

1. **Removes Lens Distortion:** Lens distortion introduced by cameras can impair the precision of measurements in the image. This distortion is corrected via calibration, allowing for a more precise estimation of 3D positions from image coordinates.
2. **Accurate Scale:** Camera calibration aids in determining the scale at which the 2D image and the 3D environment are to be compared. The reconstructed 3D picture might not have an exact scale without adequate calibration, making it challenging to determine distances or sizes.
3. **Accurate Projection:** Calibration offers precise data on how 3D points project onto the 2D image plane. This accuracy is necessary to convert the associated 2D picture coordinates of points to their corresponding 3D positions.
4. **Enhances Depth Estimation:** When using stereo or multi-view image configurations to create 3D points, inaccurate camera parameters might cause mistakes in depth estimation. More dependable depth estimation is made possible by accurate calibration.
5. **Consistency:** Accurate calibration settings for all the cameras involved are crucial for successfully matching the reconstructions when recreating a 3D scene from several images (stereo or multi-view reconstruction).

In order to calibrate a camera, photos of a calibration pattern with known 3D geometry, such as a checkerboard, are taken from various angles. The camera parameters that best fit the observed distortions and transformations are then estimated. The pinhole camera model (for perspective cameras) and more advanced models for addressing lens aberrations and complex camera geometries are common camera calibration models. The reprojection error—the discrepancy between the observed 2D image points and their corresponding projections from the known 3D points—is often minimized by mathematical optimization approaches used throughout the calibration process. In conclusion, accurate camera calibration is a crucial requirement for attaining dependable and accurate 3D reconstruction from images, particularly when handling computer vision jobs that demand exact comprehension of the 3D world.

13.7.1 Mathematics behind the Camera Calibration

The pinhole camera model and the ideas of intrinsic and extrinsic parameters can be used to quantitatively describe camera calibration, which is indeed required for effective 3D reconstruction from images.

Let's get into the specifics of the mathematics:

Model of pinhole camera The pinhole camera model is a condensed illustration of how light beams move through an aperture in a camera to create an image. The link between 3D world points and their 2D projections on the image plane of the camera is explained mathematically by this model.

In the pinhole camera model, the key components are:

1. **Intrinsic Parameters:**

   Focal Length (f): The distance between the camera's center and the image plane, often represented in pixel units.

   Principal Point (cx, cy): The coordinates of the optical axis intersection with the image plane (usually the center of the image).

   Lens Distortion (k_1, k_2, k_3, p_1, p_2): Coefficients that account for lens distortion, such as radial and tangential distortions.

2. **Extrinsic Parameters:**

   Rotation Matrix (R): Describes the camera's orientation in 3D space.

   Translation Vector (T): Specifies the position of the camera in the world coordinate system.

Mathematical Equations: The mathematical relationship between a 3D point in the world coordinate system (X, Y, Z) and its projection onto the image plane (x, y) can be expressed using the pinhole camera model:

$$x = (X/Z) * f + cx + dx \; y = (Y/Z) * f + cy + dy$$

Here:

(x, y) are the coordinates of the 2D image point.

(X, Y, Z) are the coordinates of the corresponding 3D world point.

f is the focal length.

(cx, cy) is the principal point.

(dx, dy) are the lens distortion corrections.

Why Calibration is Necessary:

1. **Correcting Distortion:** Images may have nonlinear effects due to lens distortion. To enable distortion correction and precise picture measurements, calibration aids in the determination of the distortion parameters.
2. **Accurate Scale:** Calibration facilitates the precise establishment of the link between real-world distances and picture distances.
3. **Precise Projection:** For accurate 3D reconstruction, intrinsic characteristics from calibration are employed to precisely project 3D points onto the image plane.
4. **Constant Transformations:** Extrinsic parameters (rotation and translation) allow for the alignment of numerous cameras or views, resulting in the constant spatial relationships in multi-view reconstruction.

In order to calibrate a camera, you would typically take pictures of a well-known calibration pattern from various angles. Then, using a pinhole camera model, you would minimize the difference between the observed image points and their corresponding projections by using mathematical optimization techniques (like the Levenberg-Marquardt algorithm). In conclusion, camera calibration is essential for accurate 3D reconstruction from images because it ensures that the geometric characteristics and distortions of the camera system are properly taken into account. This is especially true when using the pinhole camera model with intrinsic and extrinsic parameters.

Numerical Problems with Solutions

Problem 1: Given an image with dimensions 640 pixels wide and 480 pixels tall, calculate the total number of pixels in the image.

Solution: Total pixels = Width * Height = 640 * 480 = 307,200 pixels.

Problem 2: A camera with a focal length of 50 mm is capturing an object 2 meters away. Calculate the size of the object in the image if the object is 1 meter tall.

Solution: The magnification (M) can be calculated using the formula: M = Focal Length/Distance to Object M = 50 mm/2000 mm (2 meters converted to millimeters) = 0.025

The size of the object in the image is given by: Size in Image = Actual Size * Magnification Size in Image = 1 meter * 0.025 = 0.025 meters (25 mm)

Problem 3: You have an image with a resolution of 1024 × 768 pixels and a color depth of 24 bits per pixel (8 bits per channel). Calculate the total size of the image in bytes.

Solution: Each pixel has 24 bits, which is 3 bytes (8 bits per byte) since there are 3 color channels (RGB). Total bytes = Width * Height * Bytes per pixel Total bytes = 1024 * 768 * 3 = 2,359,296 bytes.

Problem 4: An image has a signal-to-noise ratio (SNR) of 30 dB. Calculate the ratio of the original signal power to the noise power.

Solution: SNR (in decibels) = 10 * log10(Psignal/Pnoise) 30 dB = 10 * log10(Psignal/Pnoise) 3 dB = log10(Psignal/Pnoise) Psignal/Pnoise = $10^{(3)}$ = 1000

Problem 5: You have a grayscale image with intensity values ranging from 0 to 255. Apply a threshold of 128 to the image to convert it into a binary image (0 for intensities < 128, and 1 for intensities >= 128). Calculate the total number of white (1) pixels in the resulting binary image.

Solution: Since the threshold is set at 128, all intensities greater than or equal to 128 will be converted to white (1) pixels. The total number of pixels in the image is the same as the original image. Total white pixels = Width * Height = 1024 * 768 = 786,432 pixels.

Problem 6: An image is represented using 8 bits per pixel. Calculate the total number of different grayscale levels (intensity values) that can be represented in the image.

Solution: For 8 bits per pixel, the number of different intensity values is 2^8 = 256.

Problem 7: You have an image with a resolution of 1920 × 1080 pixels. If you want to downsample the image by a factor of 2 (reduce both width and height by half), calculate the dimensions of the resulting image.

Solution: New width = Original width/Downsample factor New height = Original height/Downsample factor New width = 1920/2 = 960 pixels New height = 1080/2 = 540 pixels

Problem 8: A convolutional neural network (CNN) has an input image with dimensions 256 × 256 pixels. If the first convolutional layer uses filters of size 3 × 3 and the padding is set to keep the spatial dimensions the same, calculate the dimensions of the feature map (output) after this layer.

Solution: With padding, the spatial dimensions remain the same if the stride is 1. Feature map width = (Input width – Filter width + 2 * Padding)/Stride + 1 Feature map height = (Input height – Filter height + 2 * Padding)/Stride + 1

Feature map width = (256 – 3 + 2 * 1)/1 + 1 = 256 Feature map height = (256 – 3 + 2 * 1)/1 + 1 = 256

The feature map after the first convolutional layer is 256 × 256 pixels.

Problem 9: An image has a resolution of 800 × 600 pixels. If you apply a 5 × 5 Gaussian blur filter to the image, calculate the dimensions of the resulting image (assuming no padding and a stride of 1).

Solution: Applying a 5 × 5 Gaussian blur filter does not change the image dimensions, assuming no padding. The resulting image will have the same dimensions as the original image: 800 × 600 pixels.

Problem 10: You are performing image segmentation on a 640 × 480 grayscale image. If you use a segmentation algorithm that assigns each pixel to one of 5 different classes (labels), calculate the total number of possible segmentation outcomes (label assignments).

Solution: Each pixel can be assigned to one of 5 different classes, so the total number of possible label assignments is $5^{(640 * 480)}$.

$$\text{Total outcomes} = 5^{(640 * 480)}$$

$$\text{Total outcomes} \approx 1.6 * 10^{(1380)} \text{ (a very large number)}$$

Short Questions and Answers

Q.1. What is computer vision?

Ans. Computer vision is a field of artificial intelligence that focuses on enabling computers to interpret and understand visual information from the world, similar to how humans perceive and analyze images and videos.

Q.2. What are the main goals of computer vision?

Ans. The main goals are image recognition, object detection, image segmentation, 3D reconstruction, motion analysis, and scene understanding.

Q.3. What is image segmentation?

Ans. Image segmentation is the process of dividing an image into meaningful regions, often used to separate objects or identify boundaries within an image.

Q.4. What is object recognition?

Ans. Object recognition is the task of identifying and classifying objects or specific patterns within an image or a video stream.

Q.5. What are convolutional neural networks (CNNs)?

Ans. CNNs are a class of deep neural networks particularly effective for image-related tasks due to their ability to automatically learn hierarchical features from the data.

Q.6. What is optical character recognition (OCR)?

Ans. OCR is a technology that converts printed or handwritten text within images into machine-readable text.

Q.7. What is facial recognition?

Ans. Facial recognition is a technology that identifies and verifies individuals based on their facial features.

Q.8. What is stereo vision?

Ans. Stereo vision involves using two or more cameras to create a 3D view of a scene, similar to how human depth perception works.

Q.9. What is the depth map in computer vision?

Ans. A depth map is an image where each pixel represents the distance from the camera to the corresponding point in the scene.

Q.10. What is image registration?

Ans. Image registration is the process of aligning multiple images or views of the same scene to a common coordinate system.

Q.11. What is feature extraction?

Ans. Feature extraction is the process of identifying and representing specific visual features (e.g., edges, textures, keypoints) from an image.

Q.12. What is the purpose of edge detection?

Ans. Edge detection highlights the boundaries or transitions between different regions in an image.

Q.13. What is template matching?

Ans. Template matching is a technique used to find occurrences of a predefined template (pattern) within a larger image.

Q.14. What is motion tracking in computer vision?

Ans. Motion tracking is the process of following the movement of objects or points of interest in consecutive frames of a video.

Q.15. What is the difference between image recognition and image classification?

Ans. Image recognition involves identifying and categorizing objects or scenes within an image, while image classification specifically assigns a single label to an entire image.

Q.16. What is the purpose of data augmentation in computer vision?

Ans. Data augmentation involves applying transformations (e.g., rotation, scaling, cropping) to the training dataset to increase its diversity and improve the generalization of machine learning models.

Q.17. What is mean shift clustering in image processing?

Ans. Mean shift is a clustering algorithm used to identify regions of high data density within an image, often used for image segmentation.

Q.18. What is histogram equalization?

Ans. Histogram equalization is a technique used to enhance the contrast of an image by redistributing the intensity levels to cover the full range.

Q.19. What is the Hough transform used for?

Ans. The Hough transform is used for detecting shapes (lines, circles, etc.) within an image, even if they're not well-defined by pixel values.

Q.20. What is non-maximum suppression?

Ans. Non-maximum suppression is a technique used in edge detection to thin out the detected edges by keeping only the local maxima (strongest edges) and suppressing weaker responses.

Q.21. What is the role of convolution in CNNs?

Ans. Convolution is a key operation in CNNs that involves sliding a filter (kernel) over an input image to compute feature maps, which capture local patterns and relationships.

Q.22. What is transfer learning in computer vision?

Ans. Transfer learning involves using pre-trained neural network models (often trained on large datasets) as a starting point for a new task, allowing for faster and more effective training on smaller datasets.

Q.23. What is the difference between object detection and object recognition?

Ans. Object detection not only identifies objects in an image but also provides their location by drawing bounding boxes around them, while object recognition only assigns class labels to the entire image.

Q.24. What is the role of anchor boxes in object detection?

Ans. Anchor boxes are pre-defined boxes of different shapes and sizes that are used in object detection models (e.g., YOLO) to predict the location and shape of objects in the image.

Q.25. What is the IoU (Intersection over Union) metric used for?

Ans. IoU is used to evaluate the accuracy of object detection by measuring the overlap between predicted and ground truth bounding boxes.

Q.26. What is the purpose of data preprocessing in computer vision?

Ans. Data preprocessing involves tasks like resizing, normalizing, and augmenting images to make them suitable for input to machine learning models.

Q.27. What is the concept of overfitting in computer vision?

Ans. Overfitting occurs when a model learns to perform well on the training data but fails to generalize to new, unseen data. It often indicates that the model has learned the noise in the training data.

Q.28. What is image super-resolution?

Ans. Image super-resolution is the process of increasing the resolution of an image, often using deep learning techniques to generate high-resolution details.

Q.29. What is image denoising?

Ans. Image denoising is the process of removing noise or unwanted artifacts from an image, improving its clarity.

Q.30. What are Gabor filters used for in image processing?

Ans. Gabor filters are used for texture analysis, detecting features with specific orientations and frequencies in an image.

Q.31. What is the concept of scale-invariant feature transform (SIFT)?

Ans. SIFT is a feature detection and description method that is invariant to changes in scale, rotation, and illumination, making it robust for object recognition.

Q.32. What is the purpose of principal component analysis (PCA) in image processing?

Ans. PCA is used for dimensionality reduction, finding the most important features or components in a dataset.

Q.33. What is the difference between global and local image features?

Ans. Global features represent characteristics of the entire image, while local features focus on specific regions or keypoints.

Q.34. What is background subtraction in computer vision?

Ans. Background subtraction is used to extract moving objects from a video stream by identifying pixels that differ from a static background model.

Q.35. What is the concept of mean-shift tracking?

Ans. Mean-shift tracking is a technique used to follow the movement of an object in consecutive frames by finding the mode (peak) of the color or feature distribution around the object's current position.

Q.36. What is the purpose of watershed segmentation in image processing?

Ans. Watershed segmentation is a technique used for image segmentation, often applied to separate touching or overlapping objects.

Q.37. What is the use of morphological operations in image processing?

Ans. Morphological operations, such as dilation and erosion, are used for image preprocessing, noise removal, and feature extraction.

Q.38. What is background modeling in video processing?

Ans. Background modeling involves creating a representation of the static background in a video, enabling the detection of foreground objects and motion.

Q.39. What is the concept of homography in computer vision?

Ans. A homography is a transformation that describes the planar projective mapping between two images of the same scene taken from different viewpoints, often used for image stitching and registration.

Q.40. What is the purpose of histogram-based image segmentation?

Ans. Histogram-based segmentation divides an image into regions based on intensity levels, aiming to separate objects with distinct intensity distributions.

Q.41. What is the concept of active contour models (snakes) in image processing?

Ans. Active contour models are used for image segmentation by defining an initial contour that iteratively adjusts to the edges or boundaries of objects.

Q.42. What is the role of Gaussian filters in image processing?

Ans. Gaussian filters are used for blurring (smoothing) images, reducing noise, and preparing images for further analysis.

Q.43. What is mean filtering in image processing?

Ans. Mean filtering is a form of spatial filtering that replaces the intensity of each pixel with the average intensity of its neighborhood, reducing noise.

Q.44. What is the purpose of texture analysis in computer vision?

Ans. Texture analysis involves quantifying the visual patterns (textures) in an image, which can be useful for image classification and segmentation.

Q.45. What is background removal in image processing?

Ans. Background removal is the process of isolating the foreground object from the background, often used for creating transparent or isolated object images.

Q.46. What is multi-view geometry in computer vision?

Ans. Multi-view geometry deals with the relationships between multiple views (images) of a scene, enabling tasks such as 3D reconstruction and camera calibration.

Q.47. What is the purpose of edge linking and boundary detection in image processing?

Ans. Edge linking connects edge points to form object boundaries, which is useful for object recognition and segmentation.

Q.48. What is the concept of motion segmentation in computer vision?

Ans. Motion segmentation involves identifying distinct motion patterns in a video, separating objects with different motion trajectories.

Q.49. What is camera pose estimation in computer vision?

Ans. Camera pose estimation determines the position and orientation of a camera relative to a 3D scene, often used for augmented reality and camera tracking.

Q.50. What are the challenges in computer vision?

Ans. Some challenges include handling variations in illumination, occlusions, complex backgrounds, large datasets, computational efficiency, and robustness to real-world conditions.

Multiple Choice Questions and Answers

Q.1. What is computer vision?

(*a*) A branch of psychology
(*b*) A field of artificial intelligence
(*c*) A type of computer graphics
(*d*) A hardware component

Q.2. Which of the following is NOT a goal of computer vision?

(*a*) Object recognition
(*b*) Scene understanding
(*c*) Image compression
(*d*) Image segmentation

Q.3. What is image segmentation?

(*a*) Identifying objects in an image
(*b*) Dividing an image into regions
(*c*) Removing noise from an image
(*d*) Converting a grayscale image to color

Q.4. What does OCR stand for in computer vision?

(*a*) Optical Character Recognition
(*b*) Object Counting and Recognition
(*c*) Object Color Recognition
(*d*) Overlapping Circle Removal

Q.5. What is the main purpose of convolutional neural networks (CNNs) in computer vision?

(*a*) Speech recognition
(*b*) Image compression
(*c*) Object detection and recognition
(*d*) Natural language processing

Q.6. What does SIFT stand for in thc contcxt of computer vision?

(*a*) Single Image Feature Transformation
(*b*) Scale-Invariant Feature Transform
(*c*) Spatial Image Filtering Technique
(*d*) Segmentation and Image Filtering Toolkit

Q.7. What is the purpose of image super-resolution in computer vision?

(*a*) Reducing image resolution
(*b*) Increasing image size
(*c*) Enhancing image quality and details
(*d*) Applying image segmentation

Q.8. What is the primary function of non-maximum suppression in object detection?

(*a*) Removing noisy pixels from an image
(*b*) Suppressing low-intensity regions
(*c*) Reducing the number of false positives
(*d*) Enhancing image contrast

Q.9. Which technique is used to transform an image from the spatial domain to the frequency domain?

(*a*) Histogram equalization
(*b*) Convolution
(*c*) Fourier transform
(*d*) Edge detection

Q.10. What is the primary purpose of data augmentation in computer vision?

(*a*) Increasing the size of the dataset (*b*) Reducing overfitting
(*c*) Improving model training speed (*d*) Enhancing model interpretability

Q.11. What is the primary goal of image registration in computer vision?

(*a*) Image compression (*b*) Image resizing
(*c*) Image alignment (*d*) Image segmentation

Q.12. What is the key idea behind the Hough transform in image processing?

(*a*) Detecting edges (*b*) Image blurring
(*c*) Identifying patterns (*d*) Removing noise

Q.13. What is the purpose of template matching in computer vision?

(*a*) Object recognition (*b*) Object tracking
(*c*) Feature extraction (*d*) Image compression

Q.14. Which of the following is a common preprocessing step in computer vision before feeding data to a machine learning model?

(*a*) Data augmentation (*b*) Object detection
(*c*) Template matching (*d*) Histogram equalization

Q.15. What is the main goal of edge detection in computer vision?

(*a*) Identifying objects in an image (*b*) Removing noise from an image
(*c*) Enhancing image contrast (*d*) Highlighting boundaries and transitions

Q.16. What is the purpose of the watershed algorithm in image segmentation?

(*a*) Reducing image noise (*b*) Enhancing image quality
(*c*) Separating touching or overlapping objects (*d*) Applying morphological operations

Q.17. Which technique is used to separate foreground objects from a static background in a video?

(*a*) Object detection (*b*) Image segmentation
(*c*) Template matching (*d*) Background subtraction

Q.18. What is the primary purpose of principal component analysis (PCA) in image processing?

(*a*) Image enhancement (*b*) Dimensionality reduction
(*c*) Noise reduction (*d*) Object recognition

Q.19. What is the role of anchor boxes in object detection models like YOLO (You Only Look Once)?

(*a*) Removing background noise (*b*) Initializing model weights
(*c*) Determining the location and shape of objects
(*d*) Applying non-maximum suppression

Q.20. What is the primary purpose of mean-shift clustering in image processing?

(*a*) Image compression (*b*) Image enhancement
(*c*) Noise reduction (*d*) Identifying regions of high data density

Q.21. What does IoU (Intersection over Union) measure in object detection?

(*a*) Model accuracy (*b*) Model complexity
(*c*) Overfitting (*d*) Bounding box overlap

Q.22. Which of the following is a common challenge in computer vision?

(*a*) Underfitting (*b*) Overfitting
(*c*) Data augmentation (*d*) Model training

Q.23. What is the primary goal of image denoising in computer vision?

(*a*) Increasing image resolution (*b*) Enhancing image details
(*c*) Removing noise from an image (*d*) Segmenting objects in an image

Q.24. Which method is used to track the movement of an object in consecutive frames of a video?

(*a*) Object recognition (*b*) Image registration
(*c*) Mean-shift tracking (*d*) Edge detection

Q.25. What is the concept of camera pose estimation in computer vision?

(*a*) Determining the position and orientation of a camera
(*b*) Calculating image resolution (*c*) Identifying objects in a scene
(*d*) Removing lens distortion

Q.26. Which technique is used for image stitching to create panoramic images?

(*a*) Template matching (*b*) Stereo vision
(*c*) Homography (*d*) Histogram equalization

Q.27. What is the primary purpose of image thresholding in image processing?

(*a*) Noise reduction (*b*) Image enhancement
(*c*) Object recognition (*d*) Segmentation

Q.28. Which of the following describes the term "feature extraction" in computer vision?

(*a*) The process of generating high-resolution images
(*b*) The process of reducing the number of pixels in an image
(*c*) The process of identifying and representing specific visual characteristics
(*d*) The process of adding artificial noise to an image

Q.29. What is the main challenge in image recognition and object detection in varying lighting conditions?

(*a*) Overfitting (*b*) Lens distortion
(*c*) Illumination invariance (*d*) Background subtraction

Q.30. Which of the following is NOT a morphological operation in image processing?

(*a*) Dilation (*b*) Erosion (*c*) Opening (*d*) Blurring

Q.31. What does the acronym R-CNN stand for in object detection?

(*a*) Region Classification Neural Network
(*b*) Recurrent Convolutional Neural Network

(*c*) Region-Based Convolutional Neural Network
(*d*) Randomized Convolutional Neural Network

Q.32. Which type of neural network architecture is commonly used for sequence-to-sequence tasks in computer vision, such as image captioning?
(*a*) Convolutional Neural Network (CNN) (*b*) Recurrent Neural Network (RNN)
(*c*) Fully Connected Neural Network (FCNN) (*d*) Generative Adversarial Network (GAN)

Q.33. What is the main purpose of a loss function in training a computer vision model?
(*a*) Increasing the model's complexity
(*b*) Reducing the number of parameters in the model
(*c*) Evaluating the model's performance and guiding the learning process
(*d*) Adding regularization to the model

Q.34. What is the primary benefit of using transfer learning in computer vision?
(*a*) Reducing the need for data augmentation (*b*) Avoiding the use of convolutional layers
(*c*) Training a model from scratch on a small dataset
(*d*) Leveraging pre-trained models to improve performance on new tasks

Q.35. Which type of layer is commonly used to downsample the spatial dimensions of feature maps in convolutional neural networks?
(*a*) Convolutional layer (*b*) Fully connected layer
(*c*) Pooling layer (*d*) Batch normalization layer

Q.36. Which activation function is commonly used in convolutional neural networks to introduce non-linearity?
(*a*) Linear activation (*b*) Sigmoid activation
(*c*) ReLU (Rectified Linear Unit) activation (*d*) Softmax activation

Q.37. What is the role of the back propagation algorithm in training deep neural networks for computer vision tasks?
(*a*) Initializing the model's weights (*b*) Regularizing the model
(*c*) Calculating the gradient and updating the model's weights
(*d*) Reducing the learning rate

Q.38. What is the primary challenge in fine-tuning a pre-trained neural network for a new computer vision task?
(*a*) Underfitting (*b*) Overfitting
(*c*) Determining the architecture of the network
(*d*) Lack of labeled data for the new task

Q.39. What is the purpose of a validation set in training a computer vision model?
(*a*) Evaluating the model's performance on the training data
(*b*) Selecting the best hyperparameters for the model
(*c*) Checking the model's performance on unseen data during training
(*d*) Initializing the model's weights

Q.40. What is the primary goal of a learning rate schedule in training deep neural networks for computer vision tasks?

(*a*) Increasing the learning rate over time
(*b*) Reducing the learning rate over time
(*c*) Keeping the learning rate constant
(*d*) Determining the optimal batch size

Q.41. What is the concept of early stopping in training a computer vision model?

(*a*) Terminating the training process before reaching the maximum number of epochs
(*b*) Stopping the model's learning entirely
(*c*) Introducing additional noise to the input data
(*d*) Reducing the model's complexity

Q.42. Which technique is commonly used to prevent overfitting in a convolutional neural network?

(*a*) Increasing the model's complexity
(*b*) Reducing the size of the validation set
(*c*) Adding dropout layers
(*d*) Increasing the learning rate

Q.43. What is the purpose of the Adam optimizer in training deep neural networks?

(*a*) Regularizing the model
(*b*) Initializing the model's weights
(*c*) Reducing the learning rate
(*d*) Performing gradient-based weight updates

Q.44. What is the concept of gradient clipping in training deep neural networks?

(*a*) Reducing the size of the validation set
(*b*) Constraining the magnitude of the gradients during backpropagation
(*c*) Regularizing the model
(*d*) Increasing the learning rate

Q.45. What does the term "transfer learning" mean in the context of deep learning?

(*a*) Transferring model weights from one task to another
(*b*) Transferring labeled data from one task to another
(*c*) Transferring model architectures from one task to another
(*d*) Transferring the entire training process from one task to another

Q.46. Which of the following is NOT a common computer vision application?

(*a*) Object detection
(*b*) Image recognition
(*c*) Speech recognition
(*d*) Image segmentation

Q.47. What is the primary goal of image captioning in computer vision?

(*a*) Identifying objects in an image
(*b*) Enhancing image quality
(*c*) Converting images to text
(*d*) Generating descriptive captions for images

Q.48. What is the primary purpose of data annotation in computer vision?

(*a*) Reducing the size of the dataset
(*b*) Increasing the model's complexity
(*c*) Labeling images with ground truth information
(*d*) Adding noise to the input data

Q.49. Which computer vision application involves identifying the spatial location and orientation of a camera in a 3D scene?

(*a*) Object detection (*b*) Image registration
(*c*) Image segmentation (*d*) Object recognition

Q.50. What is the role of non-maximum suppression in object detection?

(*a*) Reducing image noise (*b*) Suppressing low-intensity regions
(*c*) Suppressing duplicate detections (*d*) Enhancing image contrast

Key to Multiple Choice Questions

1. (*b*)	**2.** (*c*)	**3.** (*b*)	**4.** (*a*)	**5.** (*c*)	**6.** (*b*)	**7.** (*c*)	**8.** (*c*)	**9.** (*c*)
10. (*b*)	**11.** (*c*)	**12.** (*c*)	**13.** (*a*)	**14.** (*a*)	**15.** (*d*)	**16.** (*c*)	**17.** (*d*)	**18.** (*b*)
19. (*c*)	**20.** (*d*)	**21.** (*d*)	**22.** (*b*)	**23.** (*c*)	**24.** (*c*)	**25.** (*a*)	**26.** (*c*)	**27.** (*d*)
28. (*c*)	**29.** (*c*)	**30.** (*d*)	**31.** (*c*)	**32.** (*b*)	**33.** (*c*)	**34.** (*d*)	**35.** (*c*)	**36.** (*c*)
37. (*c*)	**38.** (*d*)	**39.** (*c*)	**40.** (*b*)	**41.** (*a*)	**42.** (*c*)	**43.** (*d*)	**44.** (*b*)	**45.** (*a*)
46. (*c*)	**47.** (*d*)	**48.** (*c*)	**49.** (*b*)	**50.** (*c*)				

Project Works

1. **Object Detection and Robotic Grasping:** Design a system where a robotic arm equipped with a camera can detect and grasp objects in a cluttered environment. Use techniques such as deep learning-based object detection to identify objects and plan robotic grasping strategies.
2. **Visual Simultaneous Localization and Mapping (SLAM):** Create a visual SLAM system that allows a robot to navigate and build a map of its environment using only visual information from a camera. Implement a mapping algorithm and use the map for navigation and exploration.
3. **Autonomous Drone Navigation:** Develop a drone that can autonomously navigate through indoor or outdoor environments using computer vision techniques. Implement obstacle avoidance, path planning, and object tracking to enable safe and efficient navigation.
4. **Automated Inspection and Quality Control:** Build a system that uses computer vision to inspect and ensure the quality of manufactured products on an assembly line. Implement then the algorithms for defect detection, measurement and sorting of items.
5. **Human-Robot Interaction:** Create a robot that can interact with humans using computer vision. Develop a system that can recognize gestures, facial expressions, and natural language, enabling the robot to understand and respond to human commands.
6. **Visual Pick-and-Place for Robotic Assembly:** Design a robotic system that can automatically pick and place objects in a specified configuration using visual feedback. Implement object detection, pose estimation, and motion planning for precise assembly tasks.
7. **Vision-Based Navigation for Autonomous Vehicles:** Build an autonomous ground vehicle that can navigate through a complex environment using computer vision. Implement lane detection, traffic sign recognition, and obstacle avoidance to ensure safe and efficient navigation.
8. **Surveillance and Security Robotics:** Create a surveillance robot equipped with cameras that can autonomously patrol an area, detect intruders, and report suspicious activities. Implement object detection and tracking for effective surveillance.

9. **Agricultural Robotics:** Develop a robotic system for agricultural tasks, such as identifying and classifying crops, monitoring plant health, or even automated harvesting. Use computer vision to analyze the field and make informed decisions.
10. **Robotic Vision for Industrial Automation:** Design a vision-based automation system for industrial processes, such as sorting, packaging, or material handling. Implement image analysis techniques to optimize the workflow and improve efficiency.

14

Quantum Computing as Applied to Robotics and Automation

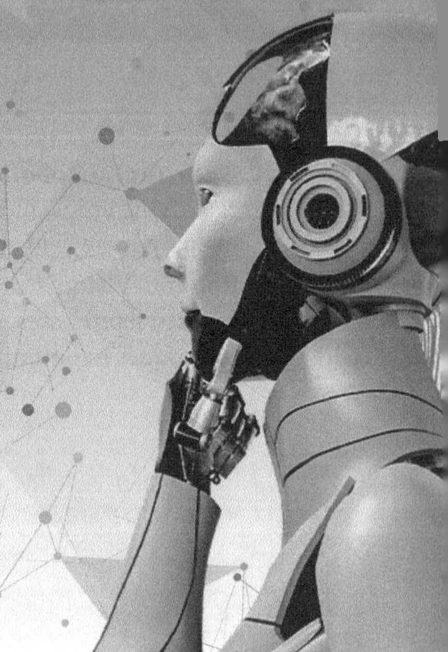

14.1 INTRODUCTION

Numerous industries, including robotics and automation, stand to benefit from the development of quantum computing. Although it is still in its early phases of development, academics are looking into how quantum computing might be applied in various fields. In this thorough presentation, we'll go through the advantages of applying quantum computing to robotics and automation as well as some potential difficulties. Despite the enormous potential of quantum computing for robotics and automation, its practical application and mass acceptance will necessitate resolving technical issues and making more progress in quantum hardware and software. As quantum technology develops further, it has the potential to completely transform the industry by enabling robotic and automation systems that are quicker, more effective, and more intelligent.

14.2 OPTIMIZATION PROBLEMS

Complex optimization issues are well suited for quantum computing. Many jobs in robotics and automation require process optimization, including path planning for factory robots or task scheduling for autonomous vehicles. In comparison to conventional computers, quantum algorithms such as the Quantum Approximate Optimization Algorithm (QAOA) and the Quantum Annealing method may be able to offer quicker and more effective solutions for various optimization issues.

A unique computational paradigm used in quantum computing allows it to simultaneously explore various answers while being adept at solving complex optimization challenges. Let's break down how quantum computing addresses optimization challenges to better comprehend this idea:

Quantum Bits (Qubits): Bits (0s and 1s) are the basic units of information used in traditional computers, which can only have one state at a time. In contrast, qubits are used in quantum computers. The superposition concept enables qubits to simultaneously represent 0 and 1. Quantum computers may investigate various solutions to a problem at the same time thanks to their intrinsic duality.

Quantum Parallelism: Superposition is a technique used by quantum computers to process numerous potential solutions at once. A quantum algorithm can investigate several potential solutions simultaneously when solving an optimization problem, drastically lowering the number of steps required to discover the best option. This extreme parallelism is a significant benefit when handling challenging optimization issues.

Entanglement: Qubits can also be entangled, which means that even when they are physically separated, the state of one qubit affects the state of another. Quantum computers can efficiently explore correlated solutions thanks to this phenomena. Entanglement can be used in optimization problems to find connections between variables or constraints, resulting in better search tactics.

Quantum Gates and Quantum Algorithms: Qubits are manipulated by quantum computers using quantum gates. The special characteristics of qubits are specifically tapped into by algorithms created for quantum computers, such as Grover's algorithm and the Quantum Approximate Optimization Algorithm (QAOA). For instance, QAOA iteratively modifies qubit states to locate the best answer to an optimization issue.

Quantum Annealing: A specialized class of quantum computer called a quantum annealer was created specifically to solve optimization issues. They work by simulating the annealing process, which entails gradually cooling a system to its lowest energy state. Qubits are used by quantum annealers to explore energy landscapes and discover low-energy states that correspond to the best answers to optimization issues.

Quantum Speedup: The brute-force approach, which involves checking each potential answer one by one, is frequently used by traditional computers to solve optimization problems. Through superposition and entanglement, quantum computers are able to investigate numerous options at once. Due to the exponential speedup that results from this, quantum computing is revolutionizing a variety of industries, including materials research, logistics, finance, and more.

Limitations and Challenges: Despite the fact that quantum computing has many benefits for optimization issues, it is important to recognize its existing limitations. With few qubits and significant error rates, quantum hardware is still in its infancy. In addition, translating real-world issues onto quantum circuits can be difficult and requires domain-specific knowledge.

In conclusion, quantum computing's use of quantum features like superposition and entanglement allows it to handle challenging optimization problems. These characteristics allow quantum computers to investigate numerous answers at once, potentially accelerating the search for the best one exponentially. As quantum technology develops, it holds significant promise for revolutionizing industries that heavily rely on optimization, such as resource allocation, drug development, and supply chain management.

14.3 MACHINE LEARNING AND AI USING QUANTUM COMPUTING

Robotics machine learning techniques can be sped up using quantum computing. Large datasets may be processed more quickly by quantum machine learning models, which is advantageous for applications like object recognition, natural language processing, and decision-making in autonomous systems. Quantum-enhanced algorithms, such as quantum SVMs and quantum neural networks, have the potential to increase the learning and inference functions of AI systems.

By utilizing its special abilities to complete some machine learning tasks quicker than conventional computers, quantum computing offers the potential to speed up machine learning algorithms used in robotics. Let's investigate how quantum computing might be used for robotics machine learning and examine its benefits:

Quantum Speedup in Linear Algebra: Complex linear algebraic operations like matrix multiplications and eigen decompositions are employed in many machine learning methods, including those used in robotics. When applying quantum algorithms like the Quantum Singular Value Transformation (QSVD) or quantum matrix inversion techniques, quantum computers can carry out these tasks tenfold quicker than conventional computers. Principal component analysis (PCA) and support vector machines (SVMs), which largely rely on linear algebra, can especially benefit from this speedup.

Quantum Machine Learning Models: Quantum computing's parallelism and interference capabilities are used to create quantum machine learning models like quantum neural networks (QNNs) and quantum support vector machines (QSVMs). These models are particularly suited for robotics applications where quick decision-making and adaptability are essential since they can process and learn from data more effectively.

Quantum Neural Networks (QNNs): QNNs can simultaneously investigate a variety of features and associations since they analyze and encode data using quantum circuits. They may enhance robotics tasks including pattern recognition, object detection, and sensor fusion.

Quantum Support Vector Machines (QSVMs): QSVMs make effective use of quantum computing to locate the best hyper planes for classification jobs. By managing complicated, high-dimensional data, they can improve the classification precision of robotic systems.

Quantum Data Processing: Large datasets used in machine learning may be processed and analyzed effectively by quantum computers. This expedites the integration of sensor data for robotics, speeds up decision-making, and enhances real-time navigation and object recognition. Data preprocessing, feature selection, and grouping activities can be accelerated by quantum algorithms.

Quantum Sampling and Optimization: Choosing samples from sizable datasets or optimizing intricate objective functions are frequent tasks in machine learning. Quantum amplitude amplification and quantum annealing are two methods by which quantum computers can surpass their classical counterparts. These techniques help expedite processes like hyper parameter tuning, which is essential for developing robotics-related machine learning models.

Quantum Quantum-enhanced Search: Grover's method and other quantum algorithms can improve search tasks, which are crucial for path planning and task scheduling in robotics. Grover's approach offers a quadratic speedup over traditional search algorithms by searching an unsorted database of N items in O(N) time.

14.3.1 Advantages of Quantum Computing in Robotics and Machine Learning

(*a*) **Speed:** The training and inference phases of machine learning models used in robotics can be considerably accelerated by quantum computing, allowing for quicker responses and decisions.

(*b*) **Higher Model Accuracy and Better Robotic System Performance:** Quantum machine learning models can more accurately capture complex patterns and relationships in data.

(*c*) **Effective Data Processing:** Due to quantum computers' ability to interpret and analyze sensor data effectively, robotic systems are now more responsive and environment-adaptive.

(*d*) Path planning, task scheduling, and resource allocation in robots are all improved by the use of quantum algorithms, which can solve optimization and search problems more quickly.

14.3.2 Challenges and Considerations

(*a*) **Quantum Hardware:** Due to its restricted qubit counts and high error rates, quantum hardware is still in its infancy and may have an impact on the viability of quantum machine learning in robotics.

(*b*) **Algorithm Mapping:** It can be difficult and requires specialized knowledge to adapt classical machine learning algorithms to quantum circuits and map real-world robotic challenges onto quantum hardware.

In conclusion, quantum computing has the potential to speed up robotics' machine learning algorithms, enabling autonomous systems to make decisions more quickly and accurately. Researchers are currently investigating how to use quantum advantages for robotics to address practical issues and improve robotic capabilities, even while quantum hardware and algorithms are developing.

14.4 SIMULATIONS AND MODELING

The design and testing of robotic systems can benefit from quantum computers' superior ability to imitate physical systems. For improved design and optimization, engineers can use quantum simulators to mimic the behavior of molecules in chemical manufacturing processes or simulate the dynamics of complicated robotic systems.

The behavior of complicated quantum systems, such as molecules in chemical reactions or the dynamics of robotic systems, can be mimicked using specific quantum computers or algorithms known as quantum simulators. They are a potent instrument with enormous potential in many branches of science and engineering, such as robotics and chemistry. Here is a detailed description of how quantum simulators function and how they can be used to improve design and optimization:

Quantum Simulation Basics: Quantum simulators model the quantum dynamics and states of a target system using quantum gates and quantum bits (qubits). Quantum simulators are better suited for simulating and comprehending quantum phenomena than conventional computers, which find it difficult to accurately represent and control quantum states.

Modeling Molecular Behavior in Chemical Processes: In chemistry, quantum simulators are especially useful. They are capable of accurately simulating chemical reactions, electrical structure, and energy states associated with molecular behavior at the quantum level. For a number of reasons, this skill is essential in chemical manufacturing processes:

Materials Discovery: Quantum simulators aid in the discovery of new substances and materials with desirable properties for use as catalysts, superconductors, or medicinal molecules, among other things.

Reaction Paths: Quantum simulations can be used by engineers to investigate various reaction paths, improve reaction conditions, and forecast reaction yields, resulting in more effective and affordable chemical processes.

Catalyst Design: By comprehending the electronic structure and reactivity of catalyst materials, quantum simulations help in the design of effective catalysts, increasing the efficiency of chemical reactions.

Robotic system simulation: Quantum simulators are a useful tool for robotics engineers to model and enhance the behaviour of complex robotic systems. They can be useful in the following ways:

Dynamic Systems: Quantum simulators are capable of simulating the dynamics of robotic systems, which includes interactions between parts, sensors, and control logic. Engineers can use this to mimic and examine how robots behave in various settings and scenarios.

Path Planning: Quantum simulations can help with robot motion control and path planning. By simulating the robot's interactions with its environment, engineers can improve trajectories and prevent crashes.

Sensor Fusion: Quantum simulators are capable of combining data from a variety of sensors, including cameras, LIDAR, and IMUs (Inertial Measurement Units). Engineers are now able to create more precise perception systems for robots by simulating sensor data fusion.

Optimization: Engineers can determine the best configurations, control schemes, or settings for increased performance by simulating the full robotic system, including hardware and software components.

14.4.1 Benefits of Quantum Simulators

(*a*) **Precision:** To improve design and optimization, quantum simulators can offer highly accurate quantum-level insights into the behavior of molecules or robotic systems.

(*b*) **Speed:** For quantum systems, simulators can outperform conventional computers, enabling engineers to quickly investigate a wider variety of potential outcomes.

(*c*) **Cost Reduction:** By eliminating the need for pricey laboratory tests or physical prototypes, quantum simulations may be able to cut back on the time and money spent on research and development.

(*d*) **Quantum Advantage:** Quantum simulators use quantum features to more correctly simulate quantum systems than traditional computers, giving researchers a distinct advantage when examining quantum phenomena.

14.4.2 Challenges and Considerations

(*a*) **Hardware Restrictions:** In order to use practical quantum simulators, one needs access to quantum hardware, which is still being developed and might not be widely available in all applications just yet.

(*b*) **Algorithm Development:** Complex chemical or robotic systems may require algorithm development and specialized knowledge in order to be modified for use with quantum simulators.

In conclusion, quantum simulators are a potent instrument that may assist engineers and scientists in accurately simulating the dynamics of complicated robotic systems as well as the behavior of molecules in chemical processes. Through improved design, optimization, and comprehension of quantum and dynamic systems, these simulations have the potential to advance numerous fields of science and engineering.

14.5 CRYPTOGRAPHY AND SECURITY

Security becomes a major worry as automation systems become more interconnected and dependent on data flow. Cryptography also has consequences for quantum computing. Since quantum computers have the ability to crack existing encryption techniques, it is necessary to develop quantum-resistant encryption algorithms to protect communication between robots and automation systems.

In a world where quantum computers could potentially compromise current encryption techniques, quantum-resistant encryption algorithms are essential for guaranteeing the security of communication between robots and automation systems. Understanding how conventional encryption is threatened by quantum computers and how quantum-resistant algorithms overcome this difficulty is crucial to understanding the need for quantum-resistant encryption.

Quantum Computers and Cryptographic Vulnerabilities: There is a chance that quantum computers will compromise the safety of conventional encryption techniques. Two quantum algorithms in particular are the source of this threat:

Shor's Algorithm: Because it is effective at factoring huge numbers, Shor's algorithm poses a serious threat to popular encryption techniques like RSA and ECC (Elliptic Curve Cryptography). The difficulty of factoring big numbers into their prime factors, a challenge that quantum computers can tackle exponentially more quickly than classical computers, is the basis for these encryption schemes.

Grover's Algorithm: Four times faster than conventional algorithms, Grover's algorithm can search through unsorted databases and solve the pre-image problem for hash functions. While hash-based security, a crucial part of many encryption systems, is not directly threatened, it may become more vulnerable.

The Need for Quantum-Resistant Encryption: The time required to factor huge numbers or to solve other mathematical issues that support classical encryption will drastically decrease as quantum computing technology develops. This implies that communications that are encrypted, including those in robotics and automation systems, may be susceptible to eavesdropping or decryption by attackers with access to quantum technology.

Algorithms for Quantum-Resistant Encryption: Post-quantum or quantum-safe encryption algorithms are examples of quantum-resistant encryption algorithms that are made to withstand attacks from quantum computers. These algorithms make advantage of mathematical puzzles that are thought to be challenging for quantum computers to efficiently solve. Several well-known instances of quantum-resistant encryption methods include:

Lattice-based Cryptography: Several quantum-resistant encryption techniques are based on lattice problems. Finding the shortest vector in a lattice is a computationally difficult task that affects both classical and quantum computers.

Multivariate Polynomial Cryptography: This method employs sets of multivariate polynomial equations that are thought to be challenging for quantum computers to solve.

Code-based Cryptography: Code-based encryption depends on how challenging it is to solve specific issues with error-correcting codes. These issues are thought to be immune to quantum attacks.

Hash-based Cryptography: Because they rely on features of hash functions that are difficult for quantum computers to reverse, hash-based signatures and encryption techniques are resistant to quantum attacks.

Transition to Quantum-Resistant Encryption: It takes a lot of work to switch from conventional encryption to quantum-resistant encryption. It entails implementing new encryption algorithms, amending standards, and updating cryptographic protocols. In order to ensure the security of current systems and get ready for the quantum computing era, the transition must be properly handled.

14.5.1 Benefits of Quantum-Resistant Encryption

(*a*) **Future-Proofing:** Even in a world when quantum computers are widely used, quantum-resistant encryption maintains the long-term security of data and communication.

(*b*) **Data Confidentiality:** It guards against potential quantum attacks on sensitive data, including intellectual property and delicate robotic and automation control signals.

(*c*) **Compliance:** Industries and organizations will need to adhere to quantum-resistant encryption standards as they are developed in order to meet security requirements.

14.5.2 Challenges

(*a*) **Algorithm Selection:** Choosing the best encryption algorithm that is resistant to quantum entanglement and guaranteeing its security is a difficult task.

(*b*) **Complexity of Implementation:** Making the switch to quantum-resistant encryption necessitates major software and hardware modifications, which can be expensive and time-consuming.

In conclusion, classical encryption techniques, such as those employed in robotics and automation systems, are seriously threatened by quantum computers. To safeguard communication and data secrecy in the face of quantum threats, quantum-resistant encryption techniques are being developed. The security of vital systems and sensitive data must be maintained by preparing for the quantum computing age.

14.6 SENSOR DATA PROCESSING

For robotics and automation systems that depend on sensor inputs for decision-making, quantum computers can process and analyze sensor data more effectively. Robots and automated systems can operate more accurately thanks to quantum algorithms that improve sensor fusion and data filtering procedures.

14.6.1 Benefits of Quantum Computing in Robotics and Automation

(*a*) **Speed:** Quantum computers have the potential to handle difficult problems far more quickly than traditional computers, allowing autonomous robots and automation systems to make decisions in real time.

(*b*) Efficient resource allocation, task scheduling, and path planning are possible using quantum algorithms, which can lower energy usage and operational expenses.

(*c*) **Accuracy:** Quantum simulations can produce extremely accurate models of physical systems, improving the design and performance of robotic systems.

(*d*) **Machine Learning Acceleration:** Quantum machine learning can improve AI systems' capacities, making them more flexible and intelligent in a variety of automation activities.

14.6.2 Challenges and Limitations

(*a*) Building and maintaining quantum hardware can be costly and difficult because quantum computers are still in the early phases of development.

(*b*) **Error Correction:** Because quantum computers are prone to faults, it is difficult to implement error correction in quantum computing systems for intricate tasks like robotics and automation.

(*c*) **Integration:** It can be challenging to integrate quantum computing with current robotic and automation systems, necessitating substantial software and hardware alterations.

(*d*) **Quantum Expertise:** The use of quantum technology in robotics and automation may be hampered by the lack of experts in the field of quantum computing.

14.7 POTENTIAL QUANTUM ALGORITHMS TO ENHANCE SENSOR FUSION AND DATA FILTERING PROCESSES

In robotics and automated systems, quantum algorithms have the potential to improve sensor fusion and data filtering procedures, ultimately resulting in greater accuracy and effectiveness. Let's examine the main ideas and advantages to better comprehend how quantum algorithms accomplish this:

Quantum parallelism: Quantum algorithms make use of the special superposition property of quantum bits (qubits). Qubits have the capacity to exist in several states at once, in contrast to conventional bits, which can only represent 0 or 1 at a time. As a result, quantum computers are able to process and modify several possibilities simultaneously.

Robotics' use of sensor fusion: It involves merging data from various sensors to provide a more complete and accurate picture of the environment. This is done in robots by combining data from several sensors, including cameras, LIDAR, IMUs, and others. For operations like localization, mapping, object recognition, and navigation, sensor fusion is essential.

Quantum-Enhanced Sensor Fusion: Quantum algorithms can enhance sensor fusion processes in the following ways:

Parallel Processing: Quantum computers are capable of processing and analysing data from numerous sensors concurrently, taking into account different sensor readings and their combinations. Robots can react to changes in their environment more quickly thanks to this parallelism, which can considerably speed up the sensor fusion process.

Optimization: Problems with optimization, which are frequently faced in sensor fusion, are well-suited for quantum algorithms. A common example of an optimization problem is choosing the most effective technique to aggregate data from many sensors in order to lower noise or increase accuracy. Compared to conventional approaches, quantum algorithms are more effective at identifying the best fusion strategies.

Quantum algorithms can manage complex sensor data interactions, including nonlinear dependencies and linked dependencies. This is especially helpful in scenarios where complex data patterns are difficult for classical approaches to capture.

Machine Learning Integration: By integrating quantum machine learning models with sensor fusion techniques, robots may adapt to and learn from the data coming from their sensors. Algorithms for quantum machine learning can find hidden patterns in data and enhance sensor-based decision-making.

Data Filtering and Noise Reduction: It's common for sensor data to contain noise, mistakes, or outliers. Processes for data filtering and noise reduction can benefit from quantum algorithms:

Quantum Filtering Algorithms: By extracting usable information from noisy sensor data more effectively, quantum versions of conventional filtering algorithms can assist robots. They can spot and eliminate outliers or incorrect readings, producing sensor data that is more accurate.

Error Correction: To reduce mistakes caused during the measurement and data collecting operations, quantum error correction techniques can be used. This makes sure that the sensor fusion data is as trustworthy as feasible.

Improved Localization and Mapping: Robot localization and mapping activities require accurate sensor fusion. Better localization accuracy can be achieved even in demanding and dynamic contexts with the help of quantum-enhanced sensor fusion. Robots can create more accurate maps and navigate with more accuracy because to it.

14.7.1 Benefits

(*a*) **Accuracy:** Quantum-enhanced sensor fusion generates data that is more accurate and trustworthy, which improves robots' and automated systems' ability to make decisions.

(*b*) **Speed:** Due to quantum parallelism's ability to process data more quickly, robots are better able to react quickly to changing circumstances.

(*c*) **Complex Data Handling:** Quantum algorithms are better able to manage complex, high-dimensional data linkages and dependencies than conventional approaches.

(*d*) Robots are more adaptive and able to learn from their sensor data thanks to the integration of quantum machine learning models into sensor fusion techniques.

14.7.2 Challenges and Considerations

(*a*) **Quantum Hardware:** There are currently no commercially available quantum computers with enough qubits and low error rates.

(*b*) **Algorithm Development:** To develop quantum algorithms for particular sensor fusion tasks, both quantum computing and robotics knowledge are needed.

In conclusion, quantum algorithms have great promise for improving sensor fusion and data filtering techniques in robotics and automated systems. Quantum-enhanced sensor fusion can increase the precision and effectiveness of robot perception and decision-making, making them more adept and flexible in a variety of settings and applications. This is accomplished by utilizing quantum parallelism and optimization capabilities.

14.8 TECHNIQUES, INSTRUMENTATION AND MATHEMATICS ON QUANTUM COMPUTING

In order to do some tasks more effectively than traditional computers, quantum computing blends the principles of quantum physics with computer science. It is a challenging and fast developing topic. In the sections below, we give an overview of some basic methods, tools, and mathematical ideas related to quantum computing. Please be aware that this is a very difficult field, requiring a thorough background of both mathematics and quantum physics.

14.8.1 Quantum Bits (Qubits)

Fundamental Unit: The qubit, which is the quantum equivalent of a classical bit (0 or 1), serves as the fundamental unit of information in quantum computing. Qubits, on the other hand, are capable of simultaneously representing 0 and 1.

Superposition: A qubit can be in a linear combination of its basis states, denoted as $|0\rangle$ and $|1\rangle$. The state of a qubit can be represented as $\alpha|0\rangle + \beta|1\rangle$, where α and β are complex numbers, and $|\alpha|^2 + |\beta|^2 = 1$.

Entanglement: Even when they are physically isolated, qubits can be entangled, which means that the state of one depends on the state of another. In quantum computing, entanglement is a fundamental characteristic.

14.8.2 Quantum Gates

Quantum gates are quantum operations used to control qubits. The Pauli-X, Pauli-Y, Pauli-Z, Hadamard, CNOT, and Toffoli gates are a few examples of common gates.

Unitary Matrices: Unitary matrices are used to represent quantum gates and guarantee the reversibility and conservation of quantum probability of quantum operations.

14.8.3 Quantum Circuits

Quantum Circuits: Circuits made of qubits and quantum gates are frequently used to represent quantum algorithms. These circuits display the order in which gates are applied to qubits in order to carry out a particular computation.

Quantum Circuit Notation: Qubit states are represented as $|0\rangle$, $|1\rangle$, etc..., and gates are applied to qubits in the circuit following a specific order.

14.8.4 Quantum Algorithms

Quantum Algorithms: These are algorithms made to effectively carry out particular computations by taking advantage of quantum physics' characteristics. Examples include Grover's algorithm (for unstructured search), the Quantum Fourier Transform (used in many quantum algorithms), and Shor's method (for factoring big numbers).

14.8.5 Quantum Hardware

Quantum Processors: Several technologies, such as topological qubits, trapped ions, and superconducting qubits, are used to build quantum computers. Each technology has unique advantages and disadvantages.

Quantum Gate Operations: In order to operate on qubits, quantum processors utilize physical implementations of gates, such as microwave pulses, lasers, or ion traps.

14.8.6 Quantum Error Correction

Error Correction: Noise and environmental factors can cause errors on quantum computers. To safeguard quantum information against errors and increase the accuracy of quantum computations, quantum error correction codes, such the surface code, are used.

14.8.7 Quantum Mathematics

Linear Algebra: A key component of quantum computing is linear algebra. comprehension quantum gates and states requires a comprehension of key ideas such as inner products, matrix multiplication, and eigen decomposition.

Complex Numbers: Qubit states and quantum operations are frequently represented by complex numbers. The foundation of quantum mechanics is the idea of probability amplitudes (also known as complex numbers).

Dirac Notation: Also called bra-ket notation, Dirac notation is a succinct mathematical language that is used to express quantum states, operators, and measurements.

Quantum Amplitudes and Probabilities: Calculating the anticipated quantum amplitudes and probabilities at various phases of the computation is a common step in the analysis of quantum algorithms.

14.8.8 Quantum Fourier Transforms

A fundamental mathematical process used in many quantum algorithms, notably Shor's algorithm, is the quantum Fourier transforms (QFT). Modular arithmetic and manipulation of complex numbers are involved. Understanding the mathematical underpinnings of entanglement and Bell states is necessary to comprehend the principles of quantum computing.

Being proficient in linear algebra, complex analysis, and quantum physics is necessary for a thorough grasp of quantum computing, which is a highly mathematical field. The study of sophisticated topics like quantum information theory, quantum cryptography, and quantum error correction is necessary in order to properly understand quantum computing. It is an interdisciplinary topic that blends theory and practical application, and it makes use of specialized hardware and software tools to program, simulate, and run quantum algorithms on quantum processors.

Numerical Problems with Solutions

Problem 1. If a quantum computer has 10 qubits in a state of superposition, how many classical bits would be needed to represent the same information?

Solution: To represent the same information classically, you would need 10 bits.

Problem 2. What is the maximum number of states a quantum system with 3 qubits can represent due to superposition?

Solution: The maximum number of states is $2^3 = 8$.

Problem 3. If a quantum computer has an entangled pair of qubits and you measure one, what is the state of the other qubit?

Solution: The state of the other qubit will be instantaneously correlated with the measured qubit, but its exact state cannot be determined until measured.

Problem 4. If a quantum computer can factor large numbers exponentially faster than a classical computer and can factor a 2048-bit number in 1 hour, how long would it take a classical computer to do the same task?

Solution: It is practically infeasible for a classical computer to factor a 2048-bit number in a reasonable time frame. It could take many years or even centuries.

Problem 5. In Grover's algorithm, if you have a database of 1 million items, how many iterations would it take to find the target item using Grover's algorithm?

Solution: Grover's algorithm requires approximately sqrt(N) iterations to find the target item, where N is the number of items. In this case, it would take approximately sqrt(1,000,000) ≈ 1,000 iterations.

Problem 6. If a quantum computer uses Shor's algorithm to factor a 2048-bit number, and it takes 10 minutes to complete the task, how long would it take a classical computer to factor the same number using the best-known classical algorithm?

Solution: The best-known classical algorithms for factoring large numbers of this size would take an impractical amount of time, likely more than millions of years.

Problem 7. A quantum computer has a quantum volume of 4096. What does this quantum volume represent?

Solution: Quantum volume is a measure of a quantum computer's capability, and it quantifies the size and complexity of quantum circuits that the computer can handle. The computer can simulate quantum circuits of that size and complexity if it has a quantum volume of 4096.

Problem 8. If a quantum computer has 20 qubits, how many possible states can it represent in a quantum register?

Solution: A quantum register with 20 qubits can represent $2^{20} = 1{,}048{,}576$ possible states.

Problem 9. What is the minimum number of qubits required to factor a 256-bit number using Shor's algorithm efficiently?

Solution: Shor's algorithm efficiently factors a number N using approximately *2log2(N) qubits. In this case,* 2log2(2^{256}) = 512 qubits would be required.

Problem 10. If a quantum computer has 5 logical qubits and uses error correction to achieve fault tolerance, how many physical qubits (including error-correcting qubits) might be needed, assuming a certain error-correction code with overhead?

Solution: The number of physical qubits needed for fault tolerance depends on the specific error correcting code and its overhead. For example, a commonly used surface code might require several physical qubits per logical qubit, resulting in a significantly larger number of physical qubits.

Problem 11. A quantum computer can perform 10^{12} quantum operations per second. If it needs to run for 1 hour to complete a quantum simulation, how many quantum operations can it perform during that time?

Solution: The quantum computer can perform 10^{12} operations per second, so in 1 hour (3600 seconds), it can perform $10^{12} \times 3600 = 3.6 \times 10^{15}$ operations.

Problem 12. If a quantum computer has an entangled pair of qubits, and one qubit is measured, what is the state of the other qubit if they were initially in a Bell state?

Solution: If the qubits were initially in a Bell state and one qubit is measured, the other qubit's state will be instantaneously known and correlated with the measurement outcome, even if separated by large distances.

Problem 13. If a quantum computer has 8 qubits and they are all entangled with each other, how many possible entangled states can be represented?

Solution: The number of possible entangled states for 8 qubits is exponential, so there are $2^8 = 256$ possible entangled states.

Problem 14. A quantum computer has a coherence time of 10 milliseconds. What is its coherence time in seconds?

Solution: 10 milliseconds is equivalent to 0.01 seconds.

Problem 15. If a quantum computer can factor a 1024-bit number in 1 day using Shor's algorithm, how long would it take to factor a 2048-bit number using the same computer and algorithm?

Solution: Factoring a 2048-bit number using Shor's algorithm would take significantly longer than factoring a 1024-bit number, likely many times longer.

Problem 16. A quantum computer can simulate a quantum circuit with a depth of 100. What does the depth of a quantum circuit represent?

Solution: The depth of a quantum circuit represents the number of consecutive quantum gates that cannot be parallelized. In this case, a depth of 100 means that the quantum circuit consists of 100 consecutive gates.

Problem 17. If a quantum computer can perform 10^9 gate operations per second and executes a quantum circuit with 100 gates, how long does it take to complete the circuit in seconds?

Solution: It takes 100 gates/10^9 gates per second = $1/10^7$ seconds, which is equivalent to 0.1 microseconds.

Problem 18. A quantum computer has a quantum volume of 8192. What does this quantum volume indicate about its capabilities compared to other quantum computers with lower quantum volumes?

Solution: A quantum computer with a higher quantum volume can handle larger and more complex quantum circuits than those with lower quantum volumes. In this case, it can handle larger and more complex computations than quantum computers with lower quantum volumes.

Problem 19. If a quantum computer has 32 qubits and uses Shor's algorithm to factor a 2048-bit number, how many qubits does it need for the quantum part of the algorithm?

Solution: Shor's algorithm typically requires a number of qubits proportional to the size of the number to be factored. In this case, it would need 32 qubits for the quantum part of the algorithm.

Problem 20. A quantum computer has a quantum volume of 16384. If it can simulate quantum circuits with 128 qubits and a depth of 64, what is the quantum volume of each simulated circuit?

Solution: The quantum volume of each simulated circuit is given by the formula: Quantum Volume = (Number of Qubits)^2 × Depth. – In this case, it's (128^2) × 64 = 2,097,152.

Short Questions and Answers

Q.1. What is quantum computing?

Ans. Quantum computing is a type of computing that leverages the principles of quantum mechanics to perform certain computations much faster than classical computers.

Q.2. What is a qubit?

Ans. A qubit is the basic unit of quantum information, analogous to a classical bit, but it can exist in multiple states simultaneously due to superposition.

Q.3. What is superposition in quantum computing?

Ans. Superposition is a quantum phenomenon that allows qubits to exist in a combination of 0 and 1 states simultaneously.

Q.4. What is entanglement in quantum computing?

Ans. Entanglement is a quantum property where the states of two or more qubits become correlated, even when separated by large distances.

Q.5. What is a quantum gate?

Ans. A quantum gate is a fundamental quantum operation that manipulates qubits to perform specific quantum computations.

Q.6. What is a quantum circuit?

Ans. A quantum circuit is a graphical representation of quantum computations, showing the sequence of gates applied to qubits.

Q.7. What is the No-Cloning Theorem in quantum computing?

Ans. The No-Cloning Theorem states that it's impossible to create an exact copy of an arbitrary unknown quantum state.

Q.8. What is quantum parallelism?

Ans. Quantum parallelism refers to the ability of quantum computers to explore multiple solutions to a problem simultaneously.

Q.9. What is a quantum algorithm?

Ans. A quantum algorithm is an algorithm specifically designed to be run on a quantum computer to solve certain problems more efficiently than classical algorithms.

Q.10. What is the Quantum Fourier Transform (QFT)?

Ans. The QFT is a fundamental quantum algorithm used in various quantum computations, including Shor's algorithm.

Q.11. What is Shor's algorithm?

Ans. Shor's algorithm is a quantum algorithm that can factor large numbers exponentially faster than classical algorithms, posing a threat to classical encryption methods.

Q.12. What is Grover's algorithm?

Ans. Grover's algorithm is a quantum algorithm that can perform an unstructured search over a database of N items in $O(\sqrt{N})$ time, providing a quadratic speedup over classical search algorithms.

Q.13. What is the significance of topological qubits in quantum computing?

Ans. Topological qubits are a type of qubit that is more robust against errors and noise, making them promising candidates for building fault-tolerant quantum computers.

Q.14. What is quantum error correction?

Ans. Quantum error correction is a set of techniques and codes used to protect quantum information from errors caused by environmental factors and noise.

Q.15. What is the concept of quantum supremacy?

Ans. Quantum supremacy refers to the point at which a quantum computer can perform a specific task faster than the best classical computer, demonstrating a clear advantage in a particular computation.

Q.16. What is quantum annealing?

Ans. Quantum annealing is a quantum computing approach used for optimization problems by simulating the annealing process of physical systems to find low-energy states.

Q.17. What is quantum teleportation?

Ans. Quantum teleportation is a quantum communication process that allows the transfer of quantum information from one location to another, relying on entanglement and classical communication.

Q.18. What is the significance of the Church-Turing thesis in the context of quantum computing?

Ans. The Church-Turing thesis suggests that any computation that can be performed by a classical Turing machine can also be performed by a quantum computer, implying that quantum computing can solve problems beyond classical computing.

Q.19. What is quantum cryptography?

Ans. Quantum cryptography is a field that uses quantum properties to secure communication, offering protection against eavesdropping and ensuring the integrity of transmitted data.

Q.20. What is the role of quantum key distribution (QKD) in quantum cryptography?

Ans. QKD is a method of securely sharing encryption keys between parties using quantum properties, making it resistant to attacks by quantum computers.

Q.21. What is the significance of the Deutsch-Jozsa algorithm in quantum computing?

Ans. The Deutsch-Jozsa algorithm is one of the earliest quantum algorithms, demonstrating exponential speedup over classical algorithms for a specific problem.

Q.22. What is quantum entropic uncertainty?

Ans. Quantum entropic uncertainty is a fundamental quantum phenomenon stating that certain pairs of quantum observables cannot be precisely measured simultaneously.

Q.23. What is the concept of quantum walk in quantum computing?

Ans. Quantum walk is a quantum analog of classical random walk, which plays a crucial role in algorithms related to search and optimization.

Q.24. What is the quantum Hadamard transform?

Ans. The quantum Hadamard transform is a quantum gate that creates superposition, often used in quantum algorithms.

Q.25. What is quantum supremacy?

Ans. Quantum supremacy refers to the point at which a quantum computer can perform a specific task faster than the best classical computer, demonstrating a clear advantage in a particular computation.

Q.26. What is quantum machine learning?

Ans. Quantum machine learning is the application of quantum computing to enhance machine learning algorithms, potentially solving certain problems more efficiently.

Q.27. What is quantum annealing used for?

Ans. Quantum annealing is used to find solutions to optimization problems by simulating physical annealing processes.

Q.28. What is the HHL algorithm used for in quantum computing?

Ans. The HHL algorithm is a quantum algorithm designed for solving linear systems of equations, with potential applications in quantum chemistry and machine learning.

Q.29. What is quantum teleportation?

Ans. Quantum teleportation is a quantum communication technique that allows the transmission of quantum information from one location to another, using entanglement and classical communication.

Q.30. What is quantum tunneling in quantum computing?

Ans. Quantum tunneling is a quantum mechanical phenomenon that enables particles to pass through energy barriers that would be insurmountable in classical physics. It is leveraged in certain quantum algorithms.

Q.31. What is the role of quantum algorithms in quantum machine learning?

Ans. Quantum algorithms are used to accelerate machine learning tasks by exploiting quantum properties such as superposition and entanglement.

Q.32. What are the challenges of building practical quantum computers?

Ans. Challenges include error correction, qubit stability, decoherence, and scaling up the number of qubits while minimizing errors.

Q.33. What is the difference between quantum annealing and gate-based quantum computing?

Ans. Quantum annealing focuses on optimization problems, while gate-based quantum computing uses gates to perform general-purpose quantum computations.

Q.34. What is the significance of the Grover-Boykin oracle in quantum algorithms?

Ans. The Grover-Boykin oracle is used to mark the solution states in Grover's algorithm for unstructured search.

Q.35. What is the role of decoherence in quantum computing?

Ans. Decoherence is the loss of quantum information due to interactions with the environment, which can negatively impact the stability of quantum computations.

Q.36. What is the concept of quantum supremacy?

Ans. Quantum supremacy is a theoretical point at which a quantum computer can perform a specific task faster than the best classical computer.

Q.37. What is quantum gates used for in quantum circuits?

Ans. Quantum gates manipulate the states of qubits in a quantum circuit to perform specific quantum operations.

Q.38. What is the Deutsch problem in quantum computing?

Ans. The Deutsch problem is a simple quantum problem used to illustrate the quantum advantage over classical computing.

Q.39. What is the role of quantum error correction in quantum computing?

Ans. Quantum error correction is used to protect quantum information from errors and noise, improving the reliability of quantum computations.

Q.40. What are some potential applications of quantum computing?

Ans. Applications include cryptography, materials science, optimization problems, drug discovery, and machine learning.

Q.41. What is the difference between quantum entanglement and classical correlation?

Ans. Quantum entanglement is a stronger form of correlation that exists between qubits, where the state of one qubit is dependent on the state of another, even when separated.

Q.42. What is the significance of quantum teleportation in quantum information theory?

Ans. Quantum teleportation enables the transfer of quantum information from one location to another, relying on entanglement and classical communication.

Q.43. What is quantum supremacy and its implications for cryptography?

Ans. Quantum supremacy is the point at which a quantum computer can perform a specific task faster than the best classical computer, potentially threatening classical encryption methods.

Q.44. What is the role of the Church-Turing thesis in quantum computing?

Ans. The Church-Turing thesis suggests that any computation that can be performed by a classical Turing machine can also be performed by a quantum computer, implying that quantum computing can solve problems beyond classical computing.

Q.45. What is the quantum advantage in computing?

Ans. The quantum advantage refers to the superiority of quantum computers in solving certain problems more efficiently than classical computers.

Q.46. What are the key quantum algorithms used in quantum computing?

Ans. Key algorithms include Shor's algorithm, Grover's algorithm, and quantum variants of classical algorithms.

Q.47. What is quantum key distribution (QKD) in quantum cryptography?

Ans. QKD is a method of securely sharing encryption keys between parties using quantum properties, making it resistant to attacks by quantum computers.

Q.48. What is the role of the Quantum Approximate Optimization Algorithm (QAOA) in quantum computing?

Ans. QAOA is used to approximate solutions to optimization problems, which have applications in fields like logistics and resource allocation.

Q.49. What is the concept of quantum walk and its applications in quantum algorithms?

Ans. Quantum walk is a quantum analog of classical random walk and plays a crucial role in algorithms related to search and optimization.

Q.50. What is the significance of quantum algorithms in the field of quantum machine learning?

Ans. Quantum algorithms are used to accelerate machine learning tasks by exploiting quantum properties such as superposition and entanglement.

Multiple Choice Questions and Answers

Q.1. What is a qubit in quantum computing?

(*a*) A classical bit (*b*) A quantum bit
(*c*) A quantum gate (*d*) A quantum algorithm

Q.2. What property allows qubits to exist in multiple states simultaneously?

(*a*) Superposition (*b*) Entanglement (*c*) Parallelism (*d*) Uncertainty

Q.3. Which quantum property refers to the correlation between the states of two or more qubits?

(*a*) Superposition (*b*) Parallelism (*c*) Entanglement (*d*) Uncertainty

Q.4. What is a quantum gate in quantum computing?

(*a*) A hardware component in a quantum computer
(*b*) A quantum algorithm (*c*) A unit of quantum information
(*d*) A quantum operation that manipulates qubits

Q.5. Which of the following represents the basic unit of quantum information?

(*a*) Classical bit (*b*) Classical byte (*c*) Qubit (*d*) Quantum byte

Q.6. What is the No-Cloning Theorem in quantum computing?

(*a*) The ability to create multiple copies of a quantum state
(*b*) The impossibility of creating an exact copy of an arbitrary unknown quantum state
(*c*) The duplication of quantum gates
(*d*) The efficient cloning of quantum circuits

Q.7. What is the primary advantage of quantum parallelism in quantum computing?

(*a*) Speeding up classical computations
(*b*) Reducing the complexity of quantum algorithms
(*c*) Exploring multiple solutions simultaneously
(*d*) Decreasing the number of qubits required

Q.8. What is the primary focus of quantum annealing in quantum computing?

(*a*) Unstructured search
(*b*) Solving linear equations
(*c*) Optimization problems
(*d*) Factoring large numbers

Q.9. Which quantum algorithm is known for factoring large numbers efficiently, posing a threat to classical encryption methods?

(*a*) Deutsch-Jozsa algorithm
(*b*) Grover's algorithm
(*c*) Shor's algorithm
(*d*) Quantum Fourier Transform (QFT)

Q.10. What is the role of quantum error correction in quantum computing?

(*a*) Enhancing the speed of quantum computations
(*b*) Protecting quantum information from errors and noise
(*c*) Correcting classical computation errors
(*d*) Improving quantum gates

Q.11. What is the concept of quantum teleportation in quantum computing?

(*a*) Instantaneous travel between quantum states
(*b*) Secure communication using quantum properties
(*c*) Simultaneous computation of multiple quantum algorithms
(*d*) Quantum data compression

Q.12. What is the Church-Turing thesis in the context of quantum computing?

(*a*) A theorem proving the superiority of quantum computers
(*b*) A hypothesis stating that quantum computers are equivalent to classical computers
(*c*) A theory on quantum teleportation
(*d*) A theorem on quantum error correction

Q.13. What is quantum cryptography used for in quantum computing?

(*a*) Simulating quantum algorithms
(*b*) Securing communication using quantum properties
(*c*) Enhancing quantum error correction
(*d*) Improving quantum gate operations

Q.14. What quantum algorithm is known for its quadratic speedup in unstructured search tasks?

(*a*) Deutsch-Jozsa algorithm
(*b*) Grover's algorithm
(*c*) Shor's algorithm
(*d*) Quantum Fourier Transform (QFT)

Q.15. What are topological qubits known for in quantum computing?

(*a*) Robustness against errors and noise –
(*b*) Speeding up quantum computations
(*c*) Solving complex mathematical problems
(*d*) Simulating quantum teleportation

Q.16. What does the term "quantum supremacy" refer to in quantum computing?

(*a*) Achieving quantum teleportation
(*b*) Superiority of quantum computers over classical computers
(*c*) Proving quantum error correction theorems
(*d*) Developing quantum cryptography techniques

Q.17. What is quantum annealing primarily used for in quantum computing?

(*a*) Simulating quantum algorithms (*b*) Secure quantum communication

(*c*) Solving optimization problems (*d*) Factoring large numbers

Q.18. What is the primary function of quantum error correction in quantum computing?

(*a*) Accelerating quantum computations

(*b*) Minimizing the number of qubits required

(*c*) Protecting quantum information from errors and noise

(*d*) Increasing the speed of quantum gates

Q.19. What does the term "quantum key distribution (QKD)" refer to in quantum cryptography?

(*a*) Distribution of quantum gates (*b*) Distribution of quantum algorithms

(*c*) Secure distribution of encryption keys using quantum properties

(*d*) Distribution of quantum hardware components

Q.20. Which quantum algorithm is known for solving linear systems of equations efficiently?

(*a*) Deutsch-Jozsa algorithm (*b*) Grover's algorithm

(*c*) Shor's algorithm (*d*) HHL algorithm

Q.21. What is the primary role of the Deutsch-Jozsa algorithm in quantum computing?

(*a*) Factoring large numbers (*b*) Solving linear equations

(*c*) Demonstrating quantum parallelism (*d*) Optimizing functions

Q.22. What is quantum entropic uncertainty in quantum computing?

(*a*) A measure of quantum gate efficiency

(*b*) The impossibility of measuring quantum states

(*c*) A fundamental quantum phenomenon related to measurement uncertainty

(*d*) A property of quantum teleportation

Q.23. What role does quantum walk play in quantum algorithms?

(*a*) Simulating quantum teleportation (*b*) Speeding up quantum computations

(*c*) Enhancing quantum error correction

(*d*) Playing a crucial role in search and optimization algorithms

Q.24. What does the Quantum Fourier Transform (QFT) do in quantum computing?

(*a*) Encrypts quantum data (*b*) Decrypts quantum data

(*c*) Performs quantum teleportation

(*d*) Plays a fundamental role in many quantum algorithms

Q.25. What is the role of quantum gates in quantum circuits?

(*a*) Manipulating qubits to perform specific quantum operations

(*b*) Representing classical bits in quantum computations

(*c*) Measuring quantum states

(*d*) Creating quantum entanglement

Q.26. What is quantum supremacy in quantum computing?

(*a*) The ability to teleport between quantum states

(*b*) The point at which a quantum computer can perform a specific task faster than the best classical computer

(*c*) The superiority of quantum gates over classical gates

(*d*) The efficiency of quantum algorithms

Q.27. What is quantum machine learning in quantum computing?

(*a*) Using quantum algorithms to enhance classical machine learning

(*b*) Using classical algorithms on a quantum computer

(*c*) Simulating quantum teleportation

(*d*) Encrypting machine learning models

Q.28. Which of the following is a key challenge in building practical quantum computers?

(*a*) Quantum entanglement

(*b*) Quantum parallelism

(*c*) Quantum teleportation

(*d*) Quantum error correction

Q.29. What is the primary difference between quantum annealing and gate-based quantum computing?

(*a*) Quantum annealing uses gates, while gate-based quantum computing does not.

(*b*) Quantum annealing focuses on optimization problems, while gate-based quantum computing is general-purpose.

(*c*) Quantum annealing is faster than gate-based quantum computing.

(*d*) Quantum annealing is used for cryptography, while gate-based quantum computing is not.

Q.30. What is the significance of the Grover-Boykin oracle in quantum algorithms?

(*a*) It is a quantum error correction code.

(*b*) It is used to mark the solution states in Grover's algorithm.

(*c*) It encrypts quantum data.

(*d*) It represents a quantum gate.

Q.31. What is the primary role of decoherence in quantum computing?

(*a*) Speeding up quantum computations

(*b*) Enhancing quantum gate operations

(*c*) Protecting quantum information from errors

(*d*) Causing the loss of quantum information due to interactions with the environment

Q.32. What is quantum tunneling in quantum computing?

(*a*) The ability to tunnel through time

(*b*) The process of teleportation using quantum states

(*c*) A quantum mechanical phenomenon that allows particles to pass through energy barriers

(*d*) A quantum error correction technique

Q.33. What does the term "quantum advantage" refer to in computing?

(*a*) The superiority of classical computers over quantum computers

(*b*) The ability of quantum computers to perform all tasks faster than classical computers
(*c*) The advantage gained by using quantum computers to solve specific problems more efficiently
(*d*) The inherent uncertainty in quantum computations

Q.34. What are the key quantum algorithms used in quantum computing?
(*a*) Sorting algorithms (*b*) Optimization algorithms
(*c*) Quantum Fourier Transform algorithms (*d*) Quantum key distribution algorithms

Q.35. What is the role of quantum key distribution (QKD) in quantum cryptography?
(*a*) Distributing quantum gates (*b*) Distributing quantum algorithms
(*c*) Secure distribution of encryption keys using quantum properties
(*d*) Factoring large numbers

Q.36. Which quantum algorithm is known for its ability to efficiently solve linear systems of equations?
(*a*) Deutsch-Jozsa algorithm (*b*) Grover's algorithm
(*c*) Shor's algorithm (*d*) HHL algorithm

Q.37. What is the primary function of the Deutsch-Jozsa algorithm in quantum computing? –
(*a*) Factoring large numbers (*b*) Solving linear equations
(*c*) Demonstrating quantum parallelism (*d*) Optimizing functions

Q.38. What does the term "quantum entropic uncertainty" refer to in quantum computing? –
(*a*) A measure of quantum gate efficiency
(*b*) The impossibility of measuring quantum states
(*c*) A fundamental quantum phenomenon related to measurement uncertainty
(*d*) A property of quantum teleportation

Q.39. What role does quantum walk play in quantum algorithms?
(*a*) Simulating quantum teleportation (*b*) Speeding up quantum computations
(*c*) Enhancing quantum error correction
(*d*) Playing a crucial role in search and optimization algorithms

Q.40. What does the Quantum Fourier Transform (QFT) do in quantum computing?
(*a*) Encrypts quantum data (*b*) Decrypts quantum data
(*c*) Performs quantum teleportation
(*d*) Plays a fundamental role in many quantum algorithms

Q.41. What is the role of quantum gates in quantum circuits?
(*a*) Manipulating qubits to perform specific quantum operations
(*b*) Representing classical bits in quantum computations
(*c*) Measuring quantum states (*d*) Creating quantum entanglement

Q.42. What is quantum supremacy in quantum computing?
(*a*) The ability to teleport between quantum states

(*b*) The point at which a quantum computer can perform a specific task faster than the best classical computer
(*c*) The superiority of quantum gates over classical gates
(*d*) The efficiency of quantum algorithms

Q.43. What is quantum machine learning in quantum computing?
(*a*) Using quantum algorithms to enhance classical machine learning
(*b*) Using classical algorithms on a quantum computer
(*c*) Simulating quantum teleportation
(*d*) Encrypting machine learning models

Q.44. Which of the following is a key challenge in building practical quantum computers?
(*a*) Quantum entanglement
(*b*) Quantum parallelism
(*c*) Quantum teleportation
(*d*) Quantum error correction

Q.45. What is the primary difference between quantum annealing and gate-based quantum computing?
(*a*) Quantum annealing uses gates, while gate-based quantum computing does not.
(*b*) Quantum annealing focuses on optimization problems, while gate-based quantum computing is general-purpose.
(*c*) Quantum annealing is faster than gate-based quantum computing.
(*d*) Quantum annealing is used for cryptography, while gate-based quantum computing is not.

Q.46. What is the significance of the Grover-Boykin oracle in quantum algorithms?
(*a*) It is a quantum error correction code.
(*b*) It is used to mark the solution states in Grover's algorithm.
(*c*) It encrypts quantum data.
(*d*) It represents a quantum gate.

Q.47. What is the primary role of decoherence in quantum computing?
(*a*) Speeding up quantum computations
(*b*) Enhancing quantum gate operations
(*c*) Protecting quantum information from errors
(*d*) Causing the loss of quantum information due to interactions with the environment

Q.48. What is quantum tunneling in quantum computing?
(*a*) The ability to tunnel through time
(*b*) The process of teleportation using quantum states
(*c*) A quantum mechanical phenomenon that allows particles to pass through energy barriers
(*d*) A quantum error correction technique

Q.49. What does the term "quantum advantage" refer to in computing?
(*a*) The superiority of classical computers over quantum computers
(*b*) The ability of quantum computers to perform all tasks faster than classical computers
(*c*) The advantage gained by using quantum computers to solve specific problems more efficiently
(*d*) The inherent uncertainty in quantum computations

Q.50. What are the key quantum algorithms used in quantum computing?

(*a*) Sorting algorithms (*b*) Optimization algorithms

(*c*) Quantum Fourier Transform algorithms (*d*) Quantum key distribution algorithms.

Key to Multiple Choice Questions

1. (*b*)	**2.** (*a*)	**3.** (*c*)	**4.** (*d*)	**5.** (*c*)	**6.** (*b*)	**7.** (*c*)	**8.** (*c*)	**9.** (*c*)
10. (*b*)	**11.** (*b*)	**12.** (*b*)	**13.** (*b*)	**14.** (*b*)	**15.** (*a*)	**16.** (*b*)	**17.** (*c*)	**18.** (*c*)
19. (*c*)	**20.** (*d*)	**21.** (*c*)	**22.** (*c*)	**23.** (*d*)	**24.** (*d*)	**25.** (*a*)	**26.** (*b*)	**27.** (*a*)
28. (*d*)	**29.** (*b*)	**30.** (*b*)	**31.** (*d*)	**32.** (*c*)	**33.** (*c*)	**34.** (*b*)	**35.** (*c*)	**36.** (*d*)
37. (*c*)	**38.** (*c*)	**39.** (*d*)	**40.** (*d*)	**41.** (*a*)	**42.** (*b*)	**43.** (*a*)	**44.** (*d*)	**45.** (*b*)
46. (*b*)	**47.** (*d*)	**48.** (*c*)	**49.** (*c*)	**50.** (*b*)				

Project Works

1. **Quantum Algorithm Implementation**

   Project: Implement a well-known quantum algorithm such as Grover's search algorithm or the Quantum Fourier Transform (QFT) and provide a step-by-step explanation of how it works.
   Hints: You can find code examples and detailed explanations of quantum algorithms in quantum programming frameworks like Qiskit, Cirq, or Quipper.

2. **Quantum Error Correction Simulator**

   Project: Create a simulator for quantum error correction codes, such as the surface code or the Shor code, and demonstrate error correction in a quantum circuit.

   Hints: You can implement this project using a quantum programming framework and simulate error correction processes.

3. **Quantum Cryptography Simulation**

   Project: Develop a quantum cryptography simulation that showcases the principles of quantum key distribution (QKD) like BB84. Simulate key generation and secure communication between quantum parties.

   Hints: You can use Python libraries like Qiskit to simulate QKD protocols and demonstrate secure key exchange.

4. **Quantum Machine Learning Model**

   Project: Create a quantum machine learning model using quantum algorithms to solve a specific problem, such as classification or regression.

   Hints: You can use quantum machine learning libraries like Pennylane or Qiskit's Aqua for this project.

5. **Quantum Game Development**

   Project: Develop a quantum game that introduces quantum concepts in an interactive and engaging manner.
   Hints: Unity3D and Qiskit provide a framework for developing quantum games that can be run on quantum simulators.

6. **Quantum Random Number Generator (QRNG)**

   Project: Build a quantum random number generator using quantum circuits to generate truly random numbers.

   Hints: Use quantum gates and measurements in a quantum programming framework to create a QRNG.

7. **Quantum Chemistry Simulation**

   Project: Simulate the behavior of molecules and chemical reactions using quantum computing. Calculate molecular energies or reaction pathways.

   Hints: Quantum chemistry libraries like Psi4, OpenFermion, and Qiskit Chemistry can be used for such simulations.

8. **Quantum Optimization**

   Project: Implement and test quantum optimization algorithms like the Quantum Approximate Optimization Algorithm (QAOA) on real-world optimization problems, such as portfolio optimization or logistics.

   Hints: Use quantum optimization libraries and problem-solving platforms to apply QAOA to various optimization problems.

9. **Quantum Cloud Computing**

   Project: Create a cloud-based quantum computing platform that allows users to run quantum programs on remote quantum hardware.

   Hints: Build a web-based interface using APIs provided by quantum cloud computing providers like IBM Quantum or Google Quantum.

10. **Quantum Hardware Development**

    Project: Design and build a simple quantum device, such as a superconducting qubit or a trapped ion qubit, and demonstrate basic quantum operations.

    Hints: Access quantum development kits and hardware platforms provided by companies like IBM, Rigetti, or IonQ for prototyping and experimentation.